Ehrhard Behrends

Introduction to Markov Chains

vieweg

Ehrhard Behrends

Introduction to Markov Chains

with Special Emphasis on Rapid Mixing

Prof. Dr. Ehrhard Behrends
Freie Universität Berlin
Institut für Mathematik I
Arnimallee 2 – 6
14195 Berlin
Germany

behrends@math.fu-berlin.de

Die Deutsche Bibliothek – CIP-Cataloguing-in-Publication-Data
A catalogue record for this publication is available from
Die Deutsche Bibliothek

© Springer Fachmedien Wiesbaden 2000
Originally published by Friedr. Vieweg & Sohn Verlagsgesellschaft mbH, Braunschweig/Wiesbaden in 2000.

http://www.vieweg.de

Cover design: Ulrike Weigel, www.CorporateDesignGroup.de

Printed on acid-free paper

ISBN 978-3-528-06986-5   ISBN 978-3-322-90157-6 (eBook)
DOI 10.1007/978-3-322-90157-6

Preface

The aim of this book is to answer the following questions:

- **What is a Markov chain?** We start with a naive description of a Markov chain as a memoryless random walk, turn to rigorous definitions and develop in the first part the essential results for homogeneous chains on finite state spaces. The connections with linear algebra will be particularly emphasized, matrix manipulations, eigenvectors and eigenvalues will play an important role.

  One of the main results will be the fact that some chains forget all information about the starting position and the length of the walk after "sufficiently many" steps. Chains where this happens within reasonable time are called *rapidly mixing*.

- **What methods are available to prove that a chain is rapidly mixing?** Several techniques have been proposed to deal with this problem: eigenvalue estimation, conductance, couplings, strong uniform times; in the case of Markov chains on groups also representation theory comes into play. These methods are presented in part II.

- **Why should it be interesting to know these things?** Those readers who are interested in applications of Markov chain techniques will find some examples in part III. Markov chains are mainly used to produce samples from huge spaces in accordance with a prescribed probability distribution. To illustrate why this could be important we discuss the connections between random generation and counting, the problem of sampling from Gibbs fields, the Metropolis sampler and simulated annealing.

The book is written for readers who have never met Markov chains before, but have some familiarity with elementary probability theory and linear algebra. Therefore the material has been selected in such a way that it covers the relevant ideas, it is complemented by many – mostly easy – exercises.

As far as the presentation is concerned, extensive efforts have been made to motivate the new concepts in order to facilitate the understanding of the rigorous definitions. Also, our investigations are completely self-contained. E.g., in chapters 15 and 16, we develop the foundations of harmonic analysis on finite groups to be able to reduce rapid mixing to certain properties of characters or representations.

The history of Markov chains began one hundred years ago, the leading pioneering figures of the "classical" period in the first half of the twentieth century were Markov, Doeblin and Kolmogorov. For a long time, however, the theory of Markov chains was mainly interesting as a theory for its own sake. Really important applications to other fields of mathematics or to other sciences had to wait until – some decades ago – computer power became widely available. Nowadays Markov chains are present in all applied parts of mathematics, in physics, biology and also in the social sciences.

The idea to write this book was born several years ago when I had some seminars together with Emo Welzl and his group where we tried to understand certain techniques concerned with rapid mixing. Later I continued to discuss these problems with specialists from various fields, these efforts led to a course on "Rapidly mixing Markov chains" given in the winter term 1997/98 at Free University of Berlin.

It is a pleasure to acknowledge the help of several colleagues from which I have benefited during the preparation and the realization of this book: Stefan Felsner, Peter Mathé, Bernd Schmidt, Christian Storbeck, Emo Welzl, and Dirk Werner. I am especially grateful to Dirk Werner for giving advice at various stages, for reading the whole manuscript and for his patience in explaining all subtleties which are necessary to transform a manuscript of a book into a LaTeX-file.

Ehrhard Behrends, Berlin 1999.

Contents

Part I

Finite Markov chains (the background)

Part I is devoted to a self-contained development of the relevant aspects of finite Markov chains. *Chapter 1* provides the fundamental definition: what is a Markov chain? Examples are studied in *chapter 2*, and in *chapter 3* it is pointed out how some notions from linear algebra – like matrices and eigenvectors – come into play. In *chapter 4* we begin with a systematic study by introducing certain definitions which will be indispensable when investigating Markov chains: states which communicate, the period of a state, recurrent und transient states. The latter are discussed in some detail in *chapter 5*. Then it is time for a digression, in *chapter 6* we will prove an analytical lemma which is a necessary prerequisite to describe the limit behaviour of recurrent states in *chapter 7*.

A summary of the various techniques to analyse a chain can be found in *chapter 8*. This chapter also contains some supplementary material.

Nobody learns mathematics just by reading a book; it is crucial to have an experience of one's own with the theory under consideration. Therefore it is recommended to solve as many as possible of the *exercises* which can be found at the end of each chapter.

1   Markov chains: how to start?

In order to understand the simple idea which underlies Markov chains we remind the reader of the well-known *random walks* from elementary probability. We have in mind a walk on a finite interval $\{1, 2, \ldots, N\}$ of integers. The walk starts at 2, say, and every step is to the left or to the right with equal probability 0.5; if the walk is in position 1 or N, some extra rule has to be applied (e.g., one may prescribe that the next position is the starting position).

There are numerous other possibilities for random walks, here are two samples:

1. Start at zero. Then "walk" on $\{0, \ldots, 999\}$ according to the following rule: whenever you are in position i, move to $2i + i'$ mod 1000, where $i' \in \{1, 2, 3, 4, 5, 6\}$ is obtained by throwing a fair die. For example, if the die shows successively the numbers $2, 1, 1, 6, \ldots$, your positions – including the starting position – will be $0, 2, 5, 11, 28, \ldots$

2. As a preparation throw a fair coin 4 times and count the number of heads. *This* will be your starting point on $\{0, \ldots, 999\}$, then continue with an ordinary random walk on $\{0, \ldots, 999\}$ mod1000, i.e., the moves will be from i to $i+1$ mod 1000 or to $i-1$ mod 1000 with equal probability 0.5.

The *common feature* is the following. First, there is a prescribed set of possible positions (in our case an interval of integers). Second, there is a deterministic or random procedure to determine where to start. And finally, with every position there is associated a random generator which has to be applied before moving next.
This observation motivates the following definition:

Definition 1.1 A *finite Markov chain* consists of

- a non-void finite set S, the *state space*; the elements of S are called *states*, they are the possible positions of our "random walk"; usually we will identify S with a set $\{1, \ldots, N\}$;

- a *probability vector*, i.e., numbers $(p_i)_{i \in S}$ with $p_i \geq 0$ for all i and $\sum p_i = 1$; these numbers determine the random generator for the starting position, with probability p_i the walk starts at position i;

- a *stochastic matrix* $P = (p_{ij})_{i,j \in S}$: all p_{ij} are nonnegative and $\sum_j p_{ij} = 1$ for every i; the matrix P is nothing but a convenient abbreviation of a description of the random generators associated with the states: a walk which is now at i will be at j after the next step with probability p_{ij}.

Remarks:

1) Those who get acquainted with this definition for the first time might ask why we have extracted from our examples precisely the preceding properties. Why not an arbitrary state space? Why did we restrict the rules for the walk in precisely this way? Why, e.g, don't we allow path-dependent random generators, that is rules of the form "Use a fair die until you have been at the origin for three times; then switch to a loaded die"?

The reason is simply a pragmatic one as in other branches of mathematics, too. On the one hand the chosen properties are sufficiently rich to allow the development of an interesting theory. And at the same time they are so general that numerous applications can be studied.

2) Many of our results will have a natural generalization to the case of *countable Markov chains* (which are defined as above with the only modification that the state space S might be countable). In view of the applications we have in mind we will restrict ourselves to the finite case; remarks concerning the general situation can be found in the notes and remarks at the end of part I (chapter 8).

A number of typical examples will be presented in the next section. Here we will only restate our second walk on $\{0,\ldots,999\}$: the probabilities $p_0,\ldots,p_{999}$ to start on $\{0,\ldots,999\}$ are $1/16, 4/16, 6/16, 4/16, 1/16, 0,\ldots,0$, and $P = (p_{ij})_{i,j=0,\ldots,999}$ is defined by

$$
p_{ij} = \begin{cases} 0.5 & : \quad i - j \bmod 1000 = \pm 1 \\ 0 & : \quad \text{otherwise.} \end{cases}
$$

In all parts of mathematics it is of crucial importance to associate an appropriate visualization with an abstract concept. The reader is invited always to imagine some kind of walk when dealing with Markov chains. Of course, the abstract definition will be the basis of the investigations to come, but the meaning of the concepts we are going to introduce can hardly be understood if a Markov chain is nothing but *a set plus a probability vector plus a stochastic matrix*. Every reader should be able to manage the translation into both directions: given the rules, what are S, the p_i and the p_{ij}? And conversely, given these data, what will a "typical" walk look like?

Even more important is the following point. We deal with probabilities, we want to transform ideas into mathematical definitions and to give rigorous proofs for the results to be stated. Thus it will be indispensable to use the machinery of probability theory, and therefore the question is:

What has the "walk" which we want to associate with S, (p_i), (p_{ij}) to be considered in the framework of probability spaces, random variables etc.?

The rest of this chapter is devoted to the discussion of this question. We will present and explain the relevant notions, and it will be shown how Kolmogorov's theorem comes into play. Some readers might be satisfied to know that there *is* a rigorous foundation; they are invited to continue now with chapter 2 and to check the connections with probability theory later.

Probability spaces and random variables

Here we only want to fix notation. We assume the reader to be familiar with the definiton of a probability space $(\Omega, \mathcal{A}, \mathbb{P})$ and the other basic concepts of probability theory like random variables, independence, conditional probability and so on. We prefer to denote probability measures by "$\mathbb{P}$" instead of the more common "P", since this letter will be used here for the stochastic matrices under consideration.

Whereas in many applications real valued random variables occur, here they usually will have their values in a finite set S (our state space). Note that a mapping $X : \Omega \to S$ is a random variable if and only if all preimages $X^{-1}(i)$, $i \in S$, are in $\mathcal{A}$.

Stochastic processes

As already stated it is important to know both: the underlying ideas and the mathematical formalization. The idea with stochastic processes – more precisely with S-valued stochastic processes in discrete time – is the following. We are given a finite set S, and we observe a "walk" at "times" $0, 1, 2, \ldots$. Suppose that we have made this observation very, very often. Then we have a more or less precise estimate of the probabilities associated with such walks. We roughly know the probability that a walk will start at a certain $i \in S$, or the probability that for a walk selected at random the fifth position is i and the 111'th position is j, or even more complicated *joint probabilities*. Thus we can also evaluate *conditional probabilities* by considering quotients: $\mathbb{P}(A \mid B) = \mathbb{P}(A \cap B)/\mathbb{P}(B)$. To have a *mathematical model* to deal with this situation we need something where "the probability that the walk will start at a certain $i \in S$" and all the other probabilities have a mathematical meaning. It is natural to model the position after the k'th move by a random variable X_k, and in this way we naturally arrive at the following

Definition 1.2 Let S be finite set. An *S-valued stochastic process* is a probability space $(\Omega, \mathcal{A}, \mathbb{P})$ together with random variables $X_k : \Omega \to S$, $k \in \mathbb{N}_0 := \{0, 1, \ldots\}$.

As one knows from elementary probability the fact that $\mathcal{A}$ is a σ-algebra implies that, for a given stochastic process, *every* event E in connection with the X_k lies in $\mathcal{A}$ and thus its probability $\mathbb{P}(E)$ has a well-defined meaning. We can, e.g., speak of $\mathbb{P}(X_4 = i)$ or $\mathbb{P}(X_{19} = i, X_{122} \neq j)$, where we have used the common short–hand notation for $\mathbb{P}(\{\omega \mid X_4(\omega) = i\})$ and $\mathbb{P}(\{\omega \mid X_{19}(\omega) = i, \ X_{122}(\omega) \neq j\})$. Also note that similarly one abbreviates conditional probabilities. For example, $\mathbb{P}(X_5 = i \mid X_2 = i', \ X_4 = i'')$ stands for $\mathbb{P}(A \mid B)$ with $A = \{\omega \mid X_5(\omega) = i\}, B = \{\omega \mid X_2(\omega) = i', \ X_4(\omega) = i''\}$.

This is precisely what is needed to start with a rigorous investigation of the behaviour of the process.

We will see very soon how *the existence* of a space $(\Omega, \mathcal{A}, \mathbb{P})$ with the desired properties can be established, Kolmogorov's theorem will provide the desired mathematical model. Before we are going to turn to this point we want to introduce a further definition which concerns certain special stochastic processes. Definition 1.2 is far too general, it covers *all*, even the most complicatedly defined "walks", and thus it is hardly to be expected that there are any interesting results. The situation will change considerably as soon as we are going to impose additional conditions, conditions which are not too general (this would lead to few interesting results) and not too special (then there would exist only few applications). There are several candidates with the desired properties. Here we will be concerned with *Markov processes*. They will be introduced next.

Markov processes

Consider rules for random walks on $\{1, \ldots, N\}$, say. There are incredibly many of them, all need some random mechanism to start, and one has to define rules by which one selects the positions at times $1, 2, \ldots$ We will speak of a *Markov process* if these rules are such that the choice of the next position for a walk which is at $i \in S$ after k steps only depends on k and i and *not* on the positions before k (i.e., one doesn't need a *memory* to determine the next step). In the remarks after definition 1.1 we have already given an example of a rule which obviously doesn't lead to a Markov process. Our other examples satisfy the condition, they are even more restrictive in that the rule to proceed always is *dependent only on i* and *not on k*; such processes will be called *homogeneous*.

We now will formulate the rigorous counterpart of the preceding discussion.

Definition 1.3 Let $X_0, X_1, \ldots$ be an S-valued stochastic process. It is called a *Markov process* if for every k and arbitrary $i_0, i_1, \ldots, i_{k-1}, i, j$ one has

$$\mathbb{P}(X_{k+1} = j \mid X_0 = i_0, X_1 = i_1, \ldots, X_{k-1} = i_{k-1}, X_k = i) = \mathbb{P}(X_{k+1} = j \mid X_k = i).$$

If in addition the numbers in this expression do not depend on k we will speak of a *homogeneous Markov process*.

Remark: It has to be noted that there is a little *technical difficulty* with this definition. The problem did not occur when we prescribed random walks by stochastic instructions, since there it will not cause confusion if certain rules never have to be applied. For example, everybody understands what is meant by:

> Start the walk on $\{-5, -4, \ldots, 4, 5\}$ at zero; if you are in position 5, return to zero next; if not, stay where you are or move one position to the right with equal probability.

Thus there *exists* a rule what to do next in position -4, but it is of *no interest* since no walk will ever reach this state. If one uses conditional probabilities as in definition 1.3, however, this leads to expressions of the form $\mathbb{P}(X_{k+1} = i \mid X_k = -4)$ which are not defined since $\mathbb{P}(X_k = -4) = 0$. The situation is even worse since one can easily imagine situations where the right hand side of the equation in 1.3 makes sense but not the left. Hence the more precise formulation of this definition would be: equality has to hold whenever both sides make sense, and "homogeneous" means that there are p_{ij} such that all conditional probabilities

$$\mathbb{P}(X_{k+1} = j \mid X_0 = i_0, X_1 = i_1, \ldots, X_{k-1} = i_{k-1}, X_k = i), \ \mathbb{P}(X_{k+1} = j \mid X_k = i)$$

which are defined coincide with this number[1].

Kolmogorov's theorem

By this theorem one can bridge the gap between the needs of the applications and rigorous probability theory. Let us return to the situation described at the beginning of our subsection on stochastic processes. There we have considered a (finite) set S, and after a sufficiently long observation of a particular class of random walks we *knew* – at least approximately – the probabilities of all events of the type

---

[1] From now on we will drop such remarks: we agree that equations containing conditional probabilities are considered only when they are defined.

"the walk starts at i_0, then it moves to $i_1,\ldots$, and after the k'th step it arrives at i_k"

which we will abbreviate by $q_{i_0,\ldots,i_k}$. One would like to have a probability space $(\Omega, \mathcal{A}, \mathbb{P})$ and random variables $X_k : \Omega \to S$, $k = 0, 1, \ldots$ such that always

$$\mathbb{P}(X_0 = i_0, X_1 = i_1, \ldots, X_k = i_k) = q_{i_0,\ldots,i_k}.$$

This can really be achieved under a rather mild fairness condition (see the next theorem). The problem does not arise if the q's are given as in our motivation, but for a general assertion one has to take care of this condition.

The rigorous formulation of the result which we have in mind reads as follows:

Theorem 1.4 *Let S be a finite set. Suppose that for any finite sequence $i_0, i_1, \ldots, i_k$ in S there is assigned a nonnegative number $q_{i_0,\ldots,i_k}$ such that $\sum_{i \in S} q_i = 1$, and $\sum_{i \in S} q_{i_0 \ldots i_{k-1} i}$ $= q_{i_0 \ldots i_{k-1}}$ for arbitrary $i_0, \ldots, i_{k-1}$ in S.*
Then there are a probability space $(\Omega, \mathcal{A}, \mathbb{P})$ and random variables $X_0, X_1, \ldots : \Omega \to S$ such that

$$\mathbb{P}(X_0 = i_0, X_1 = i_1, \ldots, X_k = i_k) = q_{i_0,\ldots,i_k}$$

holds for all $i_0, \ldots, i_k$.

Remark: This is true by *Kolmogorov's theorem* (see [12, p. 115], [16, p. 483] or any other textbook on measure theory for the general formulation). The *idea of the proof* is as follows. We define Ω to be the collection of all sequences $i_0, i_1, \ldots$ in S, i.e., $\Omega = S^{\mathbb{N}_0}$. For every k, let $X_k : \Omega \to S$ be the k'th projection: $(i_0, \ldots) \mapsto i_k$. It now "only" remains to provide Ω with a σ-algebra $\mathcal{A}$ and a measure such that the theorem holds. It is clear that, to achieve this aim, all sets

$$A_{i_0,\ldots,i_k} := \{(i_0, \ldots, i_k, i_{k+1}, \ldots) \mid i_{k+1}, \ldots \text{ arbitrary}\}$$

have to be in $\mathcal{A}$ with associated measure $q_{i_0,\ldots,i_k}$. Whereas it is easy to show that this definition can be extended to the ring generated by the $A_{i_0,\ldots,i_k}$ by a unique finitely additive measure it is rather cumbersome to check that one may apply Carathéodory's theorem to have an extension as a (σ-additive) measure to the generated σ-algebra.

Under the assumption of the existence of the Borel-Lebesgue measure on the real line one can give a rather elementary proof of theorem 1.4. Let $(\Omega, \mathcal{A}, \mathbb{P})$ be the unit interval together with the Borel subsets and the Borel-Lebesgue measure. We will assume that $S = \{1, \ldots, N\}$, the random variables $X_0, \ldots$ will be defined as follows:

- Write Ω as the disjoint union of intervals $I_1, \ldots, I_N$ such that I_i has length q_i; this is possible since the q_i sum up to one. $X_0 : \Omega \to S$ is defined such that the points in I_i are mapped to i. Then $\mathbb{P}(X_0 = i) = q_i$ for all i.

- For every i we now partition I_i into intervals $I_{i1}, \ldots, I_{iN}$ such that $\mathbb{P}(I_{ij}) = q_{ij}$; this can be done since $\sum_j q_{ij} = q_i$. And X_1 will be that mapping on Ω which maps the I_{ij} to j.
It is clear that then $\mathbb{P}(X_0 = i, X_1 = j) = q_{ij}$, and it also should be obvious how to proceed in order to define consecutively also $X_2, X_3, \ldots$.

Markov chains vs. Markov processes

We are now able to show that the definitions 1.1 and 1.3 are essentially equivalent. Start with a chain as in definition 1.1 by prescribing S, the $(p_i)_{i \in S}$ and the $(p_{ij})_{i,j \in S}$. With the help of Kolmogorov's theorem we want to construct an associated Markov process. The random variable X_0, i.e., the model for our starting position, should be such that $\mathbb{P}(X_0 = i) = p_i$, and hence we set $q_i := p_i$ in theorem 1.4. "The walk starts in i and next moves to j" is the event "$X_0 = i$ and $X_1 = j$"; by theorem 1.4 it *will have* probability q_{ij}, and according to our interpretation of 1.1 it *has* probability $p_i p_{ij}$; here we have used the identity $\mathbb{P}(A \cap B) = \mathbb{P}(A \mid B)\mathbb{P}(B)$ with $A =$ "the walk starts at i" and $B =$ "the position after step 1 is j". Thus we have no choice but to define $q_{ij} := p_i p_{ij}$, and similarly, for general k, we set

$$q_{i_0 \ldots i_k} := p_{i_0} p_{i_0 i_1} \cdots p_{i_{k-1} i_k}. \tag{1.1}$$

It is now routine to show that the q's satisfy the assumptions of Kolmogorov's theorem and that the X's provided by this theorem constitute a homogeneous Markov process with $\mathbb{P}(X_{k+1} = j \mid X_k = i) = p_{ij}$ whenever the left-hand side is defined.

Note that, by theorem 1.4, (1.1) implies that

$$P(X_0 = i_0, X_1 = i_1, \ldots, X_k = i_k) = p_{i_0} p_{i_0 i_1} \cdots p_{i_{k-1} i_k} \tag{1.2}$$

for all $i_0, \ldots, i_k$, this formula is often useful in concrete calculations.

Now suppose that, conversely, $X_0, X_1, \ldots$ is a homogeneous S-valued Markov process. In order to arrive at definition 1.1 we clearly have to set $p_i := \mathbb{P}(X_0 = i)$; then $p_i \geq 0$ and $\sum p_i = 1$ since $\mathbb{P}$ is a probability. The definition of the p_{ij} is not as easy since the natural approach $p_{ij} := \mathbb{P}(X_{k+1} = j \mid X_k = i)$ might fail: maybe the right-hand side is not defined. Thus we proceed as follows:

- Let i be such that there exists a k' with $\mathbb{P}(X_{k'} = i) > 0$; put

$$p_{ij} := \mathbb{P}(X_{k+1} = j \mid X_k = i)$$

for all j, where k is the smallest k' with $\mathbb{P}(X_{k'} = i) > 0$.

- Define the p_{ij} for the other i arbitrarily with $p_{ij} \geq 0, \sum_j p_{ij} = 1$.

This definition will be our candidate to come from definition 1.3 to definition 1.1. It follows easily from the assumptions in 1.3 – $\mathbb{P}$ is a probability measure, the process is homogeneous – that the p_{ij} satisfy the conditions in 1.1 and also that the "walk" given by the stochastic rules p_i, p_{ij} corresponds with $X_0, X_1, \ldots$. Also, (1.2) will hold with these p_i and p_{ij}.

Summing up, we arrive at the following conclusion:

Every chain defined by a probability vector and a stochastic matrix as in 1.1 gives rise to a homogeneous Markov process, and every such process is obtained in this way. There might be, however, chains which are formally different which generate the same process[2].

---

[2] As a simple example consider $S = \{1, 2\}$ with $p_1 = 1$, $p_2 = 0$. Here $\begin{pmatrix} 1 & 0 \\ 0 & 1 \end{pmatrix}$ and $\begin{pmatrix} 1 & 0 \\ 1 & 0 \end{pmatrix}$ determine the same Markov process.

Exercises

1.1: Let N be a fixed integer, we denote by K the collection of all probability vectors of length N:

$$K := \{(p_1, \ldots, p_N) \mid p_i \geq 0, \sum p_i = 1\}.$$

a) Prove that K is a compact convex subset of $\mathbb{R}^N$.

b) A point x of a convex set L is called an *extreme point (of L)* if $x = (y + z)/2$ can hold with $y, z \in L$ only if $y = z = x$.
What are the extreme points of K?

1.2: Let N be fixed and call K' the set of all stochastic $N \times N$-matrices.

a) K' is a compact convex subset of $\mathbb{R}^{N^2}$.

b) Which matrices P are extreme points of K'?

c) Let K'' be the collection of doubly stochastic $N \times N$-matrices. Prove that K'' is a closed subset of K' and identify the extreme points of this set.

d) A subset F of a convex set L is called a *face (of L)* if it is convex and if – for $y, z \in L$ – one has $(y + z)/2 \in F$ only if $y, z \in F$. Is K'' a face in K'?

1.3: Let P be a stochastic $N \times N$-matrix. We consider *two independent copies* of a Markov process on $\{1, \ldots, N\}$ with transition probabilities given by P, both are assumed to start deterministically at state 1.

The two processes can be thought of as a single process on $\{1, \ldots, N\}^2$. Is this a Markov process? What does the transition matrix look like?

1.4: Let S be a finite set and $\Phi : S \times S \to S$ a fixed function. Further we are given a sequence $(Y_k)_{k=0,1,\ldots}$ of independent and identically distributed S-valued random variables.

Define a process (X_k) by $X_0 := Y_0$ and $X_{k+1} := \Phi(X_k, Y_{k+1})$ for $k \geq 0$. Prove that (X_k) is a homogeneous Markov process and calculate the associated transition matrix in terms of Φ.

1.5: Let P be a stochastic matrix, we want to model a homogeneous Markov process $X_0, \ldots$ on a suitable probability space $(\Omega, \mathcal{A}, \mathbb{P})$. What precisely are the matrices P such that Ω can be chosen as a finite set?

1.6: Let (X_k) be the Markov process associated with the cyclic random walk on $\{0, \ldots, 9\}$ with a deterministic starting position at 5: the probability is $1/2$ for a step from i to $i+1 \bmod 10$ resp. to $i-1 \bmod 10$. Determine the following (conditional) probabilities:

a) $\mathbb{P}(X_3 \in \{2, 3, 4\})$.

b) $\mathbb{P}(X_3 = 6 \mid X_5 = 6)$.

c) $\mathbb{P}(X_5 = 6 \mid X_7 = 7, X_8 = 6)$.

1.7: Is it true that a stochastic matrix P has identical rows iff the following holds: regardless of how the starting distribution is chosen, for the associated Markov process (X_k) the X_k, $k \geq 1$, are independent and identically distributed random variables?

1.8: If $X_0, X_1, \ldots$ denotes a homogeneous Markov process with transition probabilities (p_{ij}) and starting distribution (p_i), then

$$P(X_0 = i_0, X_1 = i_1, \ldots, X_k = i_k) = p_{i_0} p_{i_0 i_1} \cdots p_{i_{k-1} i_k}$$

holds (see (1.2)). Use this to prove or to disprove that

a) $\mathbb{P}(X_k = i_k \mid X_{k-1} = i_{k-1},\, X_{k+1} = i_{k+1}) = \mathbb{P}(X_k = i_k \mid X_{k-1} = i_{k-1})$,

b) $\mathbb{P}(X_k = i_k \mid X_{k-1} = i_{k-1},\, X_{k+1} = i_{k+1},\, X_{k+2} = i_{k+2}) =$
$\mathbb{P}(X_k = i_k \mid X_{k-1} = i_{k-1},\, X_{k+1} = i_{k+1})$,

c) $\mathbb{P}(X_3 = i \mid X_2 = j) = \mathbb{P}(X_6 = i \mid X_5 = j)$.

2   Examples of Markov chains

By definition, a Markov chain is nothing but a probability vector (p_i) together with a stochastic matrix $P = (p_{ij})$. Mostly only P is given, and then it is tacitly assumed that one is interested in *all* starting distributions. Due to the law of total probability it suffices to study only the situations where one starts deterministically at a fixed but arbitrary point: if E denotes any event associated with the walk (e.g., "it visits four times state i_0 before it visits state j_0"), then the probability of E subject to the starting probability (p_i) is $\sum p_i \mathbb{P}_i(E)$, with $\mathbb{P}_i(E) :=$ "the probability of E when starting in i".

Because of the rather general setting examples of Markov chains abound. (Note that the identity matrix is also admissible, it gives rise to a particularly dull "random walk".) There are, however, some typical representatives which will be of some use later, mainly to motivate the concepts to be introduced and also to prepare our applications in part II and part III.

Example 1: The reflecting, the absorbing and the cyclic random walk

This is essentially a restatement of the walks of the introduction. One moves one step to the right or to the left on $\{1, \ldots, N\}$ (or stays at the same position) until one arrives at 1 or N. There, depending on the type of walk, some extra rule applies: 1 and N serve as reflecting walls, or the walk stays there forever at soon as it arrives at one of these positions, or $\{1, \ldots, N\}$ has to be be considered as a discrete circle, where the "neighbours" of 1 (resp. N) are 2 and N (resp. $N-1$ and 1).

The reflecting walk

Let a_i, b_i, c_i be nonnegative numbers for $i = 2, \ldots, N-1$ such that $a_i + b_i + c_i = 1$. Consider

$$P = \begin{pmatrix} 0 & 1 & 0 & \cdots & 0 & 0 & 0 \\ a_2 & b_2 & c_2 & \cdots & 0 & 0 & 0 \\ 0 & a_3 & b_3 & \cdots & 0 & 0 & 0 \\ \vdots & \vdots & \vdots & & \vdots & \vdots & \vdots \\ 0 & 0 & 0 & \cdots & a_{N-1} & b_{N-1} & c_{N-1} \\ 0 & 0 & 0 & \cdots & 0 & 1 & 0 \end{pmatrix}.$$

The absorbing walk

With a_i, b_i, c_i as above set

$$P = \begin{pmatrix} 1 & 0 & 0 & \cdots & 0 & 0 & 0 \\ a_2 & b_2 & c_2 & \cdots & 0 & 0 & 0 \\ 0 & a_3 & b_3 & \cdots & 0 & 0 & 0 \\ \vdots & \vdots & \vdots & & \vdots & \vdots & \vdots \\ 0 & 0 & 0 & \cdots & a_{N-1} & b_{N-1} & c_{N-1} \\ 0 & 0 & 0 & \cdots & 0 & 0 & 1 \end{pmatrix}.$$

The cyclic walk
This time the a_i, b_i, c_i are given with the above properties for $i = 1, \ldots, N$. The stochastic matrix is

$$
P = \begin{pmatrix}
b_1 & c_1 & 0 & \cdots & 0 & 0 & a_1 \\
a_2 & b_2 & c_2 & \cdots & 0 & 0 & 0 \\
0 & a_3 & b_3 & \cdots & 0 & 0 & 0 \\
\vdots & \vdots & \vdots & & \vdots & \vdots & \vdots \\
0 & 0 & 0 & \cdots & a_{N-1} & b_{N-1} & c_{N-1} \\
c_N & 0 & 0 & \cdots & 0 & a_N & b_N
\end{pmatrix}.
$$

Example 2: "Rules" vs. P

It is of crucial importance to be able to translate: what is the matrix P if the stochastic rules are given and, conversely, what type of walks is determined by a specific P?

For example, it should be clear that – for a walk on the states $1, 2, 3, 4, 5, 6$ – the rule

"if you are at 3, throw a fair die to determine the next position; in any other case, stay at i or go to $i + 1 \bmod 6$ with equal probability"

leads to

$$
P = \begin{pmatrix}
1/2 & 1/2 & 0 & 0 & 0 & 0 \\
0 & 1/2 & 1/2 & 0 & 0 & 0 \\
1/6 & 1/6 & 1/6 & 1/6 & 1/6 & 1/6 \\
0 & 0 & 0 & 1/2 & 1/2 & 0 \\
0 & 0 & 0 & 0 & 1/2 & 1/2 \\
1/2 & 0 & 0 & 0 & 0 & 1/2
\end{pmatrix}.
$$

Similarly the reader should try to treat further examples.

Conversely, one should "see" that

$$
P = \begin{pmatrix}
99/100 & 1/100 & 0 & 0 & 0 \\
0 & 99/100 & 1/100 & 0 & 0 \\
0 & 0 & 99/100 & 1/100 & 0 \\
0 & 0 & 0 & 99/100 & 1/100 \\
0 & 0 & 0 & 0 & 1
\end{pmatrix}
$$

describes a rather slow walk on $\{1, 2, 3, 4, 5\}$ from left to right until one arrives at state 5 (where one then stays forever).

What is going on when

$$
\begin{pmatrix}
0 & 0 & 1 & 0 \\
0 & 0 & 1 & 0 \\
1/4 & 1/4 & 1/4 & 1/4 \\
0 & 0 & 1/2 & 1/2
\end{pmatrix}
\quad \text{or} \quad
\begin{pmatrix}
0 & 1 & 0 & 0 \\
0 & 0 & 1 & 0 \\
0 & 0 & 0 & 1 \\
1/100 & 0 & 0 & 99/100
\end{pmatrix}
$$

are the relevant stochastic matrices?

Example 3: "Yes" or "no"?

Suppose you want to model a machine which can give the answers "yes" or "no" subject to the following rule: having answered "yes" the next reply will be "yes" or "no" with equal probability, but a "no" is always followed by a "yes". As soon as we have identified the states "yes" and "no" with 1 and 2, respectively, the associated matrix surely is

$$\begin{pmatrix} 1/2 & 1/2 \\ 1 & 0 \end{pmatrix}.$$

One also could use the more suggestive

$$\begin{array}{cc} & \boxed{\text{yes}} \quad \boxed{\text{no}} \\ \begin{array}{c} \boxed{\text{yes}} \\ \boxed{\text{no}} \end{array} & \begin{pmatrix} 1/2 & 1/2 \\ 1 & 0 \end{pmatrix}, \end{array}$$

in this book this will usually not be necessary.

Example 4: Markov chains as weighted graphs

In many cases the matrix P is *sparse*, i.e., there are few nonzero elements. Then it might be appropriate to visualize the chain as a graph: the *vertices* are the states of the chain, and we draw a directed *edge* (with weight p_{ij}) from i to j whenever $p_{ij} > 0$. Here you see an absorbing and a cyclic random walk on $\{1, 2, 3, 4, 5, 6\}$:

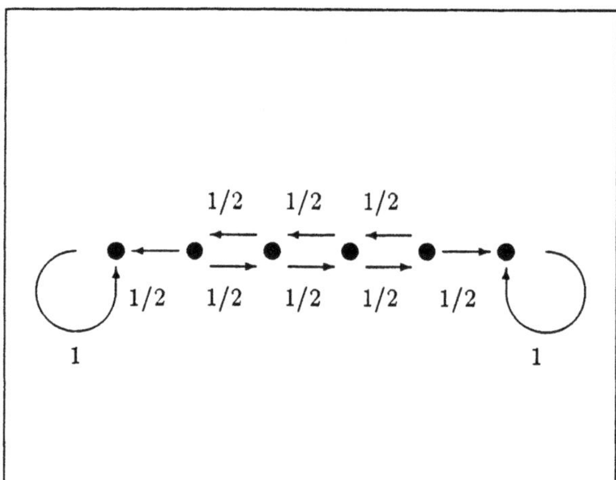

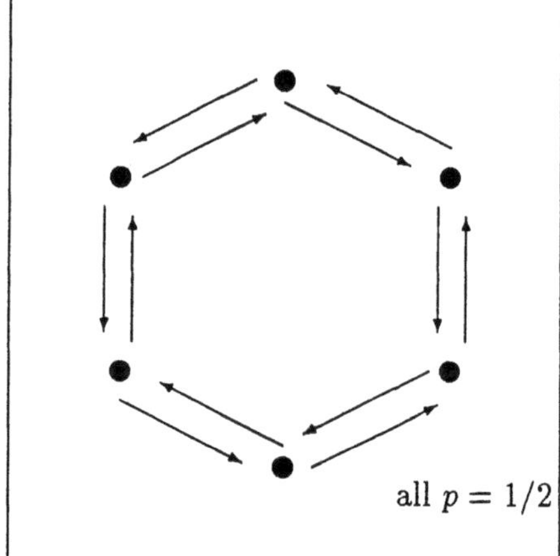

Example 5: Random permutations

In this example the state space S might be incredibly large, so large that even the most powerful computers are unable to provide an array of length N (= the cardinality of S), not to say the possibility to store all coefficients of the matrix P. It will thus *not* be possible to work with P as in our theoretical investigations to come. On the other hand it is rather simple to write computer programs which simulate "typical" walks on S and – as it will turn out later – to apply rigorously proved results successfully, even if N is that huge.

Imagine a situation where there are r books on a bookshelf. Someone uses this library in such a way that he or she picks up a book at random and – having had a look to it – puts it back randomly. How will the arrangement of the books develop?

To phrase it in the language of Markov chains we have as our *state space* the collection of the $r!$ permutations of $\{1, \ldots, r\}$. Denote permutations by $\tau = (i_1, \ldots, i_r)$ with $1 \le i_k \le r$ (1 is mapped to i_1, 2 to i_2 and so on).

A typical transition is defined by two random decisions: first one chooses the k'th book ($1 \le k \le r$) and then it is put back at position l among the remaining $k-1$ books ($1 \le l \le r$); it is assumed that k and l are selected independently according to the uniform distribution. Hence a transition from τ to itself will occur with probability $1/r$, this happens iff the k'th book is reshelved where it had been before. Also it may happen that one passes from $(i_1, i_2, \ldots, i_r)$ to $(i_2, i_1, i_3, \ldots, i_r)$ or to any other permutation which arises from τ by interchanging two adjacent entries. There are *two* possibilities to arrive at such a transition[1], hence the associated probability is $2/r^2$. There remain $r^2 - r - 2(r-1)$ transitions to permutations which only occur with precisely one choice of k, l so that their probability is $1/r^2$.

Summing up, there are $r^2 - r - 2(r-1) + 1 + (r-1) = r^2 - 2r + 2$ states which one can reach being in state τ. One transition, that from τ to τ, will occur with probability $1/r$, and there are $r - 1$ resp. $r^2 - 3r + 2$ states with transition probabilities $2/r^2$ resp. $1/r^2$.

Remark: Another way to look at this example is to consider a deck of r cards and to "shuffle" it in such a way that a randomly chosen card is restored randomly. This is abbreviated as a *random-to-random shuffle*. It should be clear what is meant by the similarly defined *top-to-random shuffle* or the *random-to-bottom shuffle*.

Example 6: Processes with short memory

By definition, a Markov process has *no* memory, the choice of the next position only depends on the present state and not on those occupied in the past.

Here we want to point out that it is sometimes possible to use Markov chain results even if the Markov condition is violated. To explain the simple idea consider a "walk" on $\{a, b\}$ defined by

- The first two positions are a.

- If you have been at a (resp. b) for the last two consecutive steps, move next to b (resp. a) with probability 0.8 and stay where you are with probability 0.2; in all other situations stay or move with equal probability 0.5.

This is surely *not* a Markov process on $\{a, b\}$. However, if we consider as states the pairs (a, a), (a, b), (b, a), (b, b) of all possible consecutive positions – which will be identified with $1, 2, 3, 4$ – then the rules give rise to a Markov chain: the chain will start deterministically in state 1, and the transition matrix is

$$P = \begin{pmatrix} 0.2 & 0.8 & 0 & 0 \\ 0 & 0 & 0.5 & 0.5 \\ 0.5 & 0.5 & 0 & 0 \\ 0 & 0 & 0.8 & 0.2 \end{pmatrix}.$$

(Why, for example, is $p_{23} = 0.5$? We are in state $2 = (a, b)$, i.e., the walk is now at b coming from a. Hence the next pair of consecutive positions will be $(b, \text{something})$, and with equal probability the states $3 = (b, a)$ and $4 = (b, b)$ will occur.)

---

[1] The transition $(i_1, i_2, \ldots, i_r) \mapsto (i_2, i_1, i_3, \ldots, i_r)$, for example, occurs if $k = 1$ and $l = 2$ or vice versa.

Similarly, any process on $\{1, \ldots, N\}$ for which the transition probabilities possibly depend on the last k_0 steps can be considered as a Markov chain with N^{k_0} states. Note that questions concerning the original process can be answered by considering the first component of the walk on $\{1, \ldots, N\}^{k_0}$.

Example 7: A diffusion model

Imagine two boxes each containing r balls. When we begin, there are only white balls in the left box, and the right one contains only red balls. Now a "move" means to choose randomly one ball in each box and to exchange their positions. Consider as a state a possible distribution of balls. Since the state is uniquely determined once we know the number of red balls in the left box, we may and will identify the states with the elements of $S = \{0, 1, \ldots, r\}$. The start is deterministic at state 0, and by elementary probability the entries of the transition matrix are

$$
p_{ij} = \begin{cases}
\frac{2i(r-i)}{r^2} & : \quad j = i \\
\frac{(r-i)^2}{r^2} & : \quad j = i+1 \\
\frac{i^2}{r^2} & : \quad j = i-1 \\
0 & : \quad \text{otherwise.}
\end{cases}
$$

This chain is called the *Bernoulli-Laplace model of diffusion* or the *Ehrenfest model*. Formally it is a special example of a reflecting random walk (see page 12).

Example 8: 0-1-sequences

Let r be a fixed integer and $S = \{0, 1\}^r$ the collection of all 0-1-sequences of length r. Transitions from $(\varepsilon_1, \ldots, \varepsilon_r)$ to $(\eta_1, \ldots, \eta_r)$ are possible if and only if precisely one of the numbers $\varepsilon_i - \tau_i$ is different from zero, and each of these r transitions have the same probability $1/r$.

One can think of S as the set of vertices of an r-dimensional hypercube, and transitions are admissible along an edge to a neighbour.

Example 9: Random feedback

(This example is due to the young composer *Orm Finnendahl* who lives and works in Berlin. He uses Markov chains to provide sequences of numbers which are transformed to musical events.)

Let m and r be integers, our state space will be $S = \{1, \ldots, m\}^r$. Consider any state $\tau = (\eta_1, \ldots, \eta_r)$. If $\eta := \eta_r$ does *not* occur among the numbers $\eta_1, \ldots, \eta_{r-1}$, only one transition, namely to τ, is admissible. Otherwise define I to be the nonvoid set of indices i such that $1 \leq i \leq r-1$ and $\eta_i = \eta$; then choose a random $i \in I$ (with respect to the uniform distribution) and put $\eta' := \eta_{i+1}$. The next state will be $(\eta', \eta_1, \ldots, \eta_{r-1})$.

> Suppose that, for example, $m = 3, r = 12$, and $\tau = (213321221311)$. Since $\eta = 1$ is followed once by 1, once by 2 and twice by 3 we will have $\eta' = 1, 2,$ or 3 – and thus transitions to $(121332122131), (221332122131),$ or (321332122131) – with the respective probabilities $1/4, 1/4, 2/4$.

The composer was rather surprised by the phenomenon that the walk usually visits only few states: after starting the chain with a randomly chosen τ one arrives rather soon at a situation where η is not among the $\eta_1, \ldots, \eta_{r-1}$ so that the chain produces no new elements. The problem to find an explanation was communicated to the German mathematical community in [14]. This, however, did not lead to any convincing solutions.

Exercises

2.1: What will a "typical" walk look like if the transition matrix is

$$\frac{1}{1000}\begin{pmatrix} 0 & 1 & 999 \\ 1 & 0 & 999 \\ 1 & 1 & 998 \end{pmatrix}, \text{ or } \frac{1}{1000}\begin{pmatrix} 1 & 999 & 0 \\ 0 & 1 & 999 \\ 999 & 0 & 1 \end{pmatrix}, \text{ or } \frac{1}{1000}\begin{pmatrix} 999 & 0 & 1 \\ 500 & 0 & 500 \\ 0 & 1 & 999 \end{pmatrix}?$$

2.2: Let $S = \{1, \ldots, 10\}$, we consider the following walk:

Start at 1, transitions in the k'th step are according to the following rule:

– if you are at 10, stay where you are;

– otherwise, throw a fair die, let the result be d; move d units to the right if this is possible, if not, stay where you are.

What is the stochastic matrix associated with this walk?

2.3: At the end of chapter 1 we have given a counterexample: it is in general not possible to reconstruct P from the Markov process induced by P and a fixed starting position. Prove that P can be found if one has access to *all* Markov processes with transition matrix P and *arbitrary* starting distributions.

2.4: Consider a deck of r cards which is in its natural order. A number $\rho \in \{1, \ldots, r\}$ is chosen uniformly at random and then one cuts ρ times a single card from the top of the deck to the bottom. If this procedure is thought of as a single "step" of a random walk, which positions are possible? What are the transition probabilities?

2.5: Prove rigorously that the walk introduced in example 6 is not Markov on $\{a, b\}$.

2.6: Consider the following walk on $\{0, 1, 2\}$: the first three positions are $X_0 = 0$, $X_1 = 1$, $X_2 = 2$, and – for $k \geq 2$ – the $(k+1)$'th position is

$$(X_k + X_{k-1} + X_{k-2} + d) \bmod 3,$$

where d is found by throwing a fair die.

a) Prove that (X_k) is *not* a Markov process.

b) Show that it is nevertheless possible to associate with (X_k) a Markov process on a set with 27 states; cf. example 6.

2.7: Verify the formula for the transition probabilities in example 7.

2.8: Let (X_k) be a homogeneous Markov process on a finite set S with transition matrix P. Fix a map $\Phi : S \to S'$, where S' is another finite set. Is $(\Phi(X_k))$ a Markov process on S'? If not, what conditions on P and Φ have to be fulfilled in order to arrive at a Markov process?

2.9: Fix a state space S together with a stochastic matrix P. A bijection $\rho : S \to S$ is called a *symmetry* if

$$p_{ij} = p_{\rho(i)\rho(j)}$$

holds for all i, j.

a) Prove that the collection of all symmetries is a group with respect to composition. We will call this group G_P in the sequel.

b) Calculate G_P for the symmetric, the absorbing and the cyclic random walk (see example 1), the answer will depend on a_i, b_i, c_i.

c) Determine G_P for

$$P = \frac{1}{10} \begin{pmatrix} 8 & 1 & 1 & 0 \\ 8 & 0 & 1 & 1 \\ 8 & 1 & 0 & 1 \\ 8 & 1 & 1 & 0 \end{pmatrix}.$$

d) Provide an example where G_P is trivial (i.e., only the identity is a symmetry).

e) What can be concluded if *all* bijections are symmetries?

2.10: We will say that two states i, j of a Markov chain are *equivalent* if there exists a symmetry ρ with $\rho(i) = j$; we will write $i \sim j$ in this case.

a) Prove that "$\sim$" is an equivalence relation.

b) Verify that for every disjoint partition $S = S_1 \cup \cdots \cup S_r$ of a finite set S there exists a stochastic matrix P such that the $S_1, \ldots, S_r$ are the equivalence classes with respect to "$\sim$".

3 How linear algebra comes into play

Let a Markov chain be given by a probability vector $(p_1, \ldots, p_N)$ and a stochastic matrix $P = (p_{ij})$ as in definition 1.1. On page 9 we saw how one may associate a Markov process $X_0, X_1, \ldots$ defined on some space $(\Omega, \mathcal{A}, \mathbb{P})$ with values in $S := \{1, \ldots, N\}$.
We now apply the "law of total probability", i.e.,

$$\mathbb{P}(A) = \sum_i \mathbb{P}(A \mid B_i)\mathbb{P}(B_i), \tag{3.1}$$

whenever Ω is the disjoint union of the $B_1, B_2, \ldots$. We want to use this elementary fact to calculate the numbers

$$p_i^{(k)} := \mathbb{P}(X_k = i),$$

that is the probabilities that the walk occupies position i after k steps; we write the "k" in brackets since it is not an exponent here.
One has $p_i^{(0)} = p_i$ and, by (3.1),

$$p_i^{(k+1)} = \mathbb{P}(X_{k+1} = i) = \sum_j \mathbb{P}(X_{k+1} = i \mid X_k = j)\mathbb{P}(X_k = j) = \sum_j p_{ji} p_j^{(k)}.$$

The *crucial observation* is that this equation just means that $(p_1^{(k+1)}, \ldots, p_N^{(k+1)})$, the *row vector* associated with the $p_i^{(k+1)}$, is nothing but the matrix product of the row vector $(p_1^{(k)}, \ldots, p_N^{(k)})$ with our matrix P.

> It has to be stressed that we have to multiply *row vectors from the left* and not *column vectors from the right* due to our decision to denote transition probabilities by p_{ij} and not p_{ji}. This is suggestive if one writes from left to right, but it is nothing but a convention.

By induction we get immediately

$$(p_1^{(k)}, \ldots, p_N^{(k)}) = (p_1, \ldots, p_n) \begin{pmatrix} p_{11} & p_{12} & \cdots & p_{1N} \\ p_{21} & p_{22} & \cdots & p_{2N} \\ \vdots & \vdots & & \vdots \\ p_{N1} & p_{N2} & \cdots & p_{NN} \end{pmatrix}^k, \tag{3.2}$$

and in this way we may hope that it will be possible to calculate probabilities in connection with Markov chains by using matrix algebra. In fact it will turn out that this machinery can be applied successfully. We will see why and how eigenvalues, eigenvectors and inverse matrices are of importance here.

We will use the *convention* that elements of $\mathbb{R}^N$ are *column vectors*. It is, however, typographically more convenient to deal with row vectors which can easily be achieved by passing from an $x \in \mathbb{R}^N$ to the *transposed vector* $x^\top$. For example, if we want to define a vector e having the entries $1, \ldots, 1$ we can do this by putting $e^\top := (1, \ldots, 1)$ (or by $e := (1, \ldots, 1)^\top$).

As an example let us consider *the role of certain left eigenvectors*. Suppose that you have found such an eigenvector with associated eigenvalue 1, that is a $\pi = (\pi_1, \ldots, \pi_N)^\top$ with $(\pi_1, \ldots, \pi_N) = (\pi_1, \ldots, \pi_N)P$. Suppose that the entries of π can be thought of as probabilities, that is $\pi_i \geq 0, \sum \pi_i = 1$. Then, by choosing $p_i := \pi_i$, it follows from (3.2) that

$$(p_1^{(k)}, \ldots, p_N^{(k)}) = (\pi_1, \ldots, \pi_N)$$

for all k, and for this reason such a π will be called an *equilibrium distribution*.

Let us try to understand the consequences. Imagine, for example, the simplest reflecting walk on $\{1, 2, 3, 4\}$:

$$P = \begin{pmatrix} 0 & 1 & 0 & 0 \\ 1/2 & 0 & 1/2 & 0 \\ 0 & 1/2 & 0 & 1/2 \\ 0 & 0 & 1 & 0 \end{pmatrix}. \tag{3.3}$$

If you know, for example, that the walk starts at 2, that is if $(p_1^{(0)}, p_2^{(0)}, p_3^{(0)}, p_4^{(0)}) = (0, 1, 0, 0)$, then (3.2) enables you to calculate the $(p_1^{(k)}, p_2^{(k)}, p_3^{(k)}, p_4^{(k)})$ in a simple way. All these vectors are different, and all carry non-trivial information. For example, whenever k is even, the walk will *not* occupy one of the states 1 or 3.

Now let π be defined by $\pi^\top := (1/6, 1/3, 1/3, 1/6)$. We have $\pi^\top = \pi^\top P$ so that, if we choose the components of π as starting probabilities, our information will not change with k. We can, e.g., assert that the walk occupies state 3 with probability 1/3, regardless of how huge k is.

As another illlustration consider

$$P = \begin{pmatrix} 0 & 1/2 & 0 & 1/2 \\ 1/2 & 0 & 1/2 & 0 \\ 0 & 1/2 & 0 & 1/2 \\ 1/2 & 0 & 1/2 & 0 \end{pmatrix},$$

the cyclic random walk on $\{1, 2, 3, 4\}$. This time $(1/4, 1/4, 1/4, 1/4)$ is a left eigenvector, that is the *uniform distribution* is an equilibrium distribution in this case. Note that this always happens if P is *a doubly stochastic matrix*, i.e., if $\sum_i p_{ij} = 1$ holds for every j.

There is *a related phenomenon*. Consider once more the above example (3.3). Suppose that you know that the walk was started either deterministically at 1 or deterministically at 2. In the first (resp. in the second) case the walk will occupy a state in $\{1, 3\}$ (resp. in $\{2, 4\}$) after k steps for every even k. Thus if you know that the 100,000'th position is 4 you are sure that the walk started at 2. Loosely speaking one could say that, concerning its starting position, *the walk keeps some memory*.

However, if we pass from (3.3) to

$$P = \begin{pmatrix} 0 & 1 & 0 & 0 \\ 1/3 & 1/3 & 1/3 & 0 \\ 0 & 1/3 & 1/3 & 1/3 \\ 0 & 0 & 1 & 0 \end{pmatrix},$$

then this is not to be expected. Since now the walk can pause at states 2 and 3, there is no obvious way to decide, knowing the position after $k = 100{,}000$ steps, whether it was started at 1 or 2.

Surprisingly the information concerning the starting position is lost rather rapidly here. To check this, let us calculate some powers of P:

$$P^2 = \frac{1}{9} \begin{pmatrix} 3 & 3 & 3 & 0 \\ 1 & 5 & 2 & 1 \\ 1 & 2 & 5 & 1 \\ 0 & 3 & 3 & 3 \end{pmatrix},$$

$$P^4 = \frac{1}{81} \begin{pmatrix} 15 & 30 & 30 & 6 \\ 10 & 35 & 26 & 10 \\ 10 & 26 & 35 & 10 \\ 6 & 30 & 30 & 15 \end{pmatrix},$$

$$P^8 = \frac{1}{6561} \begin{pmatrix} 861 & 2460 & 2460 & 780 \\ 820 & 2501 & 2420 & 820 \\ 820 & 2420 & 2501 & 820 \\ 780 & 2460 & 2460 & 861 \end{pmatrix}.$$

Thus even after a rather small number of steps it turns out that the entries in the first and the second row are pretty close together. Since these by (3.2) represent the probability distribution after 8 steps when starting at 1 or at 2 our calculation justifies the intuition.

In fact, *all* rows are close to each other (*every* starting positions leads after 8 steps to roughly the same distribution), and an evaluation of further powers of P would indicate that they converge rapidly to $(1/8, 3/8, 3/8, 1/8)$. It is of course not by chance that this is a left eigenvector of P with associated eigenvalue 1. A great part of this book will be devoted to understand and to apply this loss-of-memory phenomenon[1].

Exercises

3.1: Prove that the collection of all stochastic $N \times N$-matrices such that 1 is a simple eigenvalue is an open subset of the set K' of exercise 1.2.

3.2: Let P be a stochastic matrix with associated process $X_0, X_1, \ldots$, and k_0 an integer. Then $\hat{P} := P^{k_0}$ is that stochastic matrix which gives rise to the walk $X_0, X_{k_0}, X_{2k_0}, \ldots$. Which matrices arise in this way:

a) Determine all stochastic 2×2-matrices P such that there is a stochastic matrix Q with $P = Q^2$.

---

[1] cf. chapter 7, in particular theorem 7.4.

b) Prove that for every $N > 1$ and $k_0 > 1$ there is a stochastic $N \times N$-matrix P such that P is *not* of the form Q^{k_0} for a stochastic matrix Q.

3.3: Let a walk start on $S = \{1, \ldots, N\}$ in accordance with an initial distribution $p_1, \ldots, p_N$, the transitions in the k'th step are governed by a stochastic matrix P_k. Prove that the probability $p_i^{(k)}$ to find the walk after k steps in state i is the i'th component of the vector

$$(p_1, \ldots, p_N) P_1 \cdots P_k.$$

3.4: Prove that – similarly to the case of homogeneous Markov chains – the inhomogeneous chain of the preceding exercise can be modelled rigorously by a Markov process $X_0, X_1, \ldots : \Omega \to S$ on a suitable probability space $(\Omega, \mathcal{A}, \mathbb{P})$. (More precisely: given $(p_1, \ldots, p_N)$ and the P_k there is a Markov process (X_k) with $\mathbb{P}(X_0 = i) = p_i$ and

$$\mathbb{P}(X_k = j \mid X_{k-1} = i) = (P_k)_{ij}$$

for $k > 0$ and all $i \in S$.)

3.5: Recall that a probability vector $(\pi_1, \ldots, \pi_N)$ is called an equilibrium distribution of a stochastic matrix P provided that $(\pi_1, \ldots, \pi_N) P = (\pi_1, \ldots, \pi_N)$.

a) Verify that for a doubly stochastic matrix the uniform distribution $(1/N, \ldots, 1/N)$ is an equilibrium distribution. Is it possible that there are other equilibria in this case?

b) Prove that the collection of all equilibrium distributions of a fixed P is a nonvoid compact convex subset K of $\mathbb{R}^N$.

c) Find an explicit expression of an equilibrium distribution of an arbitrary stochastic 2×2-matrix.

d) Determine an equilibrium distribution of a stochastic matrix P where all rows are identical. Is it unique?

3.6: Convince yourself of the loss-of-memory phenomenon by considering a stochastic matrix P with strictly positive entries and calculating the matrix products P, P^2, P^4, $P^8, \ldots$.

3.7: With the notation of exercise 2.9 let ρ be a symmetry. Prove that $(\pi_{\rho(1)}, \ldots, \pi_{\rho(N)})$ is an equilibrium for every equilibrium distribution $(\pi_1, \ldots, \pi_N)$.

4   The fundamental notions in connection with Markov chains

What is essential? It is one thing to fix a set of axioms as the starting point of a hopefully interesting new theory. However, it is usually much more difficult to find the relevant notions which enable one to study – sometimes even to completely classify – the new objects. Examples are very rare where this has been achieved by a single mathematician. Generally it takes years or decades where many suggestions are under consideration, where it turns out to be necessary to modify the axioms and where many researchers are involved.

The relevant notions for Markov chains have mostly been found in the first half of the twentieth century (cf. also the historical comments in chapter 8), here we will be concerned with

- closed subsets of a Markov chain,

- states which communicate,

- the period of a state, and

- recurrent and transient states.

It will turn out that the study of a general Markov chain can be split up into the investigation of certain special states (transient states) and Markov chains of a particular type (irreducible chains). Transient chains will be studied in some detail in chapter 5. We then will need a little digression to prove an important analytical lemma (chapter 6). This will enable us to continue our study of irreducible chains in chapter 7. Part I ends with a summary of our results and some notes and remarks in chapter 8.

Closed subsets of the state space

As in chapter 2 we fix a finite set S and a stochastic matrix $P = (p_{ij})_{i,j\in S}$, a particular starting distribution is *not* prescribed.

Definition 4.1

   (i) A nonvoid subset C of S is called *closed* (or *invariant*) provided that

$$p_{ij} = 0 \text{ whenever } i \in C \text{ and } j \notin C.$$

  (ii) By an *absorbing state* we mean a state i_0 such that $\{i_0\}$ is closed (i.e., one for which $p_{i_0 i_0} = 1$ holds).

 (iii) The chain is called *irreducible* if S itself is the only closed subset.

For concrete chains it is usually not hard to identify the closed sets: the cyclic walk

$$P = \begin{pmatrix} 0 & 1/2 & 0 & \cdots & 0 & 1/2 \\ 1/2 & 0 & 1/2 & \cdots & 0 & 0 \\ \vdots & \vdots & \vdots & & \vdots & \vdots \\ 1/2 & 0 & 0 & \cdots & 1/2 & 0 \end{pmatrix}$$

on $\{1,\ldots,N\}$ is irreducible, for the absorbing walk

$$P = \begin{pmatrix} 1 & 0 & 0 & \cdots & 0 & 0 & 0 \\ 1/2 & 0 & 1/2 & \cdots & 0 & 0 & 0 \\ \vdots & \vdots & \vdots & & \vdots & \vdots & \vdots \\ 0 & 0 & 0 & \cdots & 1/2 & 0 & 1/2 \\ 0 & 0 & 0 & \cdots & 0 & 0 & 1 \end{pmatrix}$$

only $\{1\}, \{N\}, \{1,N\}$ and $\{1,\ldots,N\}$ are closed, and for

$$P = \begin{pmatrix} 1 & 0 & 0 & \cdots & 0 & 0 \\ 1/2 & 1/2 & 0 & \cdots & 0 & 0 \\ 0 & 1/2 & 1/2 & \cdots & 0 & 0 \\ \vdots & \vdots & \vdots & & \vdots & \vdots \\ 0 & 0 & 0 & \cdots & 1/2 & 1/2 \end{pmatrix} \tag{4.1}$$

precisely the N sets $\{1,\ldots,r\}, r = 1,\ldots,N$ have this property.

(Of course, the identification of the closed subsets is not always *that* simple. For example, in order to prove that the random-to random shuffle from chapter 2 (page 15) is irreducible one has to remember how permutations can be built up from transpositions.)

In the presence of closed subsets one might hope to simplify the study of the chain: if C is closed, then the "restriction" of P to C can be defined reasonably, and for "small" C the reduced chain should be much simpler. Also it is to be expected that an essential step in the classification of arbitrary chains will be the understanding of the behaviour of irreducible chains. We will see in chapter 6 and chapter 7 how this can be made precise.

All assertions in the following lemma are easy to check:

Lemma 4.2
- (i) *Unions and nonvoid intersections of closed sets are closed.*
- (ii) *Let C_1, C_2 be different closed sets which are minimal with respect to " $\subset$ ". Then $C_1 \cap C_2 = \emptyset$.*
- (iii) *Every closed set contains a minimal one.*

Remark: For assertion (iii) to hold it is essential that we restrict ourselves to *finite* state spaces. (iii) is not true for arbitrary S, the simplest counterexample seems to be the deterministic walk on the integers which is defined by $p_{i,i+1} = 1$ for all i.

Suppose that C is closed in $S = \{1,\ldots,N\}$. If C is the set $\{1,\ldots,N'\}$ – which of course can easily be achieved by passing to another enumeration of the states –, then P will have the form

$$\begin{pmatrix} * & \cdots & * & 0 & \cdots & 0 \\ \vdots & & \vdots & \vdots & & \vdots \\ * & \cdots & * & 0 & \cdots & 0 \\ * & \cdots & * & * & \cdots & * \\ \vdots & & \vdots & \vdots & & \vdots \\ * & \cdots & * & * & \cdots & * \end{pmatrix}.$$

More generally, if $C_1, \ldots, C_r$ are the minimal closed sets, we may assume that

$$C_1 = \{1, \ldots, N_1\}, \; C_2 = \{N_1 + 1, \ldots, N_2\}, \ldots, \; C_r = \{N_{r-1} + 1, \ldots, N_r\}.$$

Then, as a consequence of lemma 4.2(ii), P can be written as

$$\begin{pmatrix} P_1 & 0 & 0 & \cdots & 0 & 0 & 0 \\ 0 & P_2 & 0 & \cdots & 0 & 0 & 0 \\ \vdots & \vdots & \vdots & & \vdots & \vdots & \vdots \\ 0 & 0 & 0 & \cdots & 0 & P_r & 0 \\ * & * & * & \cdots & * & * & * \end{pmatrix}, \tag{4.2}$$

where P_ρ is the stochastic matrix which corresponds to the restriction of the chain to C_ρ ($\rho = 1, \ldots, r$). "0" here denotes a matrix with zero entries, and the " $*$ " stand for further matrices. (Note that always some C_ρ will exist, the $*$-matrices, however, might be absent.)

(4.2) will be referred to as *a standard form of the chain.*

For example, a standard form of the absorbing random walk on $\{1, 2, 3, 4\}$ is

$$\begin{pmatrix} 1 & 0 & 0 & 0 \\ 0 & 1 & 0 & 0 \\ 1/2 & 0 & 0 & 1/2 \\ 0 & 1/2 & 1/2 & 0 \end{pmatrix},$$

where we have renumbered the states $1, 2, 3, 4$ as $1, 3, 4, 2$.

It is plain by this example that it is not always natural to pass from the original P to the form (4.2).

States which communicate

Let C be a proper closed subset of S. Then a walk starting in C will never visit any $j \notin C$. In order to examine more closely what can happen it is convenient to introduce

Definition 4.3 Let i, j be arbitrary states.

(i) If the probability is positive that a walk which starts at i will visit j we will write $i \to j$. This happens precisely if one has $p_{ij}^{(k)} > 0$ for some k, where we have written the k'th power P^k of our matrix P as $P^k = (p_{ij}^{(k)})$. In this definition $k = 0$ is admissible (with $P^0 =$ the identity matrix). Hence $i \to i$ for all i.

(ii) We will say that i and j *communicate* if $i \to j$ and $j \to i$. In this case we will write $i \leftrightarrow j$.

Usually it is rather easy to decide by inspection whether or not one has $i \to j$, in particular if the chain is given as a weighted graph as in chapter 2, example 4 (page 14). Then "$i \to j$" is nothing but "there is a directed path from i to j".

Proposition 4.4

(i) " $\leftrightarrow$ " *is an equivalence relation.*

(ii) *Let C be a closed minimal subset of S. Then each two states in C communicate.*

(iii) *Every minimal closed subset is an equivalence class with respect to* " $\leftrightarrow$ ".

(iv) *A chain is irreducible iff $i \leftrightarrow j$ for all i, j.*

Proof. (i) It is clear from the definition that $i \leftrightarrow i$ and that $i \leftrightarrow j$ yields $j \leftrightarrow i$. Transitivity is plain if one thinks of a chain as a walk subject to stochastic rules or as a weighted directed graph.

A formal proof is also easy: let i, j, l be states such that $i \to j$ and $j \to l$. Then there are k, k' with $p_{ij}^{(k)}, p_{jl}^{(k')} > 0$, and we need a k'' with $p_{il}^{(k'')} > 0$. $k'' := k + k'$ has this property since $P^{k+k'} = P^k P^{k'}$, and thus

$$
\begin{aligned}
p_{il}^{(k+k')} \;&=\; p_{ij}^{(k)} p_{jl}^{(k')} + \text{(something} \geq 0) \\
&\geq\; p_{ij}^{(k)} p_{jl}^{(k')} \\
&>\; 0.
\end{aligned}
$$

It follows that " $\to$ " and consequently also " $\leftrightarrow$ " are transitive.

(ii) We know that $C = C'$ whenever C' is a closed subset of C, and this has to suffice to prove that $i \to j$ for arbitrary $i, j \in C$.

The proof uses a little trick. Fix $j \in C$ and define $C_{[j]}$ to be the collection of all starting positions $i \in C$ which never reach j:

$$
C_{[j]} := \{ i \mid i \in C,\ p_{ij}^{(k)} = 0 \text{ for all } k \}.
$$

We are done once we know that $C_{[j]} = \emptyset$, and this will follow as soon as we have shown that $C_{[j]}$ is a *proper* subset of C and that $C_{[j]}$ is closed.

The first part is clear since, by definition, $j \notin C_{[j]}$. For the second part, let $i \in C_{[j]}$, $j' \notin C_{[j]}$ be given.

Case 1: $j' \notin C$.
Then $p_{ij'} = 0$ since C is closed.

Case 2: $j' \in C$.
By the definition of $C_{[j]}$ we know that $j' \to j$. Thus necessarily $p_{ij'} = 0$ since otherwise $i \to j'$ and thus $i \to j$, a contradiction.

(iii) That equivalence classes associated with elements of a closed C are subsets of C follows from the definition. And (ii) says that the class of i is at least as large as C if i belongs to the minimal closed set C.

(iv) This is a special case of (iii). $\square$

The period of a state

As we have noted in chapter 3 the reflecting walk (3.3) on $\{1, 2, 3, 4\}$ has the property that after an *odd* number of steps a state in $\{2, 4\}$ will be occupied if the walk was started at 1, say. In particular, only after an *even* number of steps the walk can be back at 1 again. This section is devoted to provide an appropriate definition in order to investigate such when-might-the-walk-return questions.

First we remind the reader of a *definition from elementary number theory*. If $\mathbb{M}$ is a subset of $\{0, 1, 2, \ldots\}$ which strictly contains $\{0\}$, then *the greatest common divisor of* $\mathbb{M}$ is the number d which satisfies

- $d|n$ for all $n \in \mathbb{M}$ ("$d|n$" means "d divides n"),

- if $d'|n$ for all $n \in \mathbb{M}$, then $d'|d$.

This number always exists and is uniquely determined, it is usually denoted by $\gcd \mathbb{M}$.

> Those who know only the greatest common divisor of *finite* sets of integers might argue as follows: consider the collection M of all numbers which are of the form $\gcd\{n_1, \ldots, n_r\}$, where r is arbitrary and the $n_1, \ldots, n_r$ are in $\mathbb{M}$. Since M is a nonvoid subset of $\mathbb{N}$ and $\mathbb{N}$ is well-ordered there exists a minimal element d in M. It is easy to see that d has the claimed properties.

Definition 4.5 Let i be a state such that $p_{ii}^{(k)} > 0$ for some $k > 0$, that is there is a positive probability that a walk which starts at i returns. Then the *period of i* is the greatest common divisor of the set $N_i := \{k \mid k \geq 0, \ p_{ii}^{(k)} > 0\}$.
If i has period 1 it will be called *aperiodic*.

In the above example of the reflecting walk all states obviously have period 2 since $N_i = \{0, 2, 4, \ldots\}$ for all i. Here is a more interesting chain:

> Let S be the set $\{-a, -a+1, \ldots, -1, 0, 1, 2, \ldots, b\}$, where a and b are arbitrary integers. Define the transition probabilities by
>
> - $p_{i,i+1} = 1$ for $i = -a, -a+1, \ldots, -1, 1, 2, 3, \ldots, b-1$,
> - $p_{b,0} = 1$,
> - $p_{0,1} = p_{0,-a} = 1/2$.
>
> Thus the chain consists of the "cycles" $0 \to -a \to -a+1 \to \cdots \to -1 \to 0$ and $0 \to 1 \to 2 \to \cdots \to b \to 0$; only at zero a random decision is necessary, otherwise the walk is deterministic.
>
> Now consider the state 0. Since the "cycles" have length $a+1$ and $b+1$, the k where $p_{0,0}^{(k)} > 0$ are precisely the $k = r(a+1) + s(b+1)$ with $r, s = 0, 1, \ldots$. Therefore the period of 0 is $\gcd((a+1)\mathbb{N}_0 + (b+1)\mathbb{N}_0) = \gcd(a+1, b+1)$. Hence – surprisingly – 0 can have a small period or even be aperiodic even for huge a, b.

Thus an assertion "i has period d" always should be interpreted very carefully. It *does not mean* that a walk starting in i will be back in i after $d, 2d, 3d, \ldots$ steps with a positive probability; it rather implies that *it is for sure that the walk does not occupy position i again after k steps* whenever k is not in $\{d, 2d, \ldots\}$ (and also that d is maximal with respect to this property).

A chain can have states with different periods, the absorbing random walk on $\{1, 2, 3, 4\}$, e.g., contains states with periods 1 and 2. If one checks similar examples it turns out that this phenomenon seems never to occur in the situation of irreducible chains. Here is the explanation:

Proposition 4.6 *Let i, j be different states such that $i \leftrightarrow j$. Then the periods for i and j are defined and coincide. Consequently, by lemma 4.4, all states in a minimal closed subset have the same period.*

Proof. Choose positive k', k'' with $p_{ij}^{(k')}, p_{ji}^{(k'')} > 0$. Then $p_{ii}^{(k'+k'')}, p_{jj}^{(k'+k'')} > 0$ by the argument from the proof of proposition 4.4 so that both i and j have a period.

Denote the period of i (resp. j) by d (resp. d'). With N_i, N_j as in definition 4.5 we have $k' + k'' \in N_i$, and this yields $d | (k' + k'')$. For arbitrary $k \in N_j$ we know that

$$p_{ii}^{(k+k'+k'')} \geq p_{ij}^{(k')} p_{jj}^{(k)} p_{ji}^{(k'')} > 0,$$

hence $k + k' + k'' \in N_i$ and thus $d | (k + k' + k'')$ as well. Consequently, since d divides all $k \in N_j$, we have $d | d'$.
$d' | d$ follows by symmetry, and this shows that $d = d'$.   □

We close this section with a *simple observation*. Suppose that we have a situation where all states have the same period $d > 0$ and that we now pass from the original chain to a new one where we have replaced the stochastic matrix P by its d'th power $Q := P^d$. (This can be thought of as an abbreviated version of the original walk, we only pay attention to the steps $0, d, 2d, \ldots$) Denote for arbitrary i the sets N_i of definition 4.5 by N_i^P or N_i^Q depending on whether they are calculated with respect to the old or the new stochastic matrix. It is clear that $N_i^P = dN_i^Q$ and hence *all states i will be aperiodic* now.

This procedure particularly can be applied if the original chain is irreducible. The new chain will possibly *not* have this property, but in the minimal closed subsets now all states are aperiodic.

The simplest example is provided by

$$P = \begin{pmatrix} 0 & 1 \\ 1 & 0 \end{pmatrix};$$

the associated chain is irreducible with period $d = 2$. The matrix Q is the identity matrix, and the Q-chain is aperiodic on its minimal closed sets $\{1\}$ and $\{2\}$.

Recurrent and transient states

Now we want to investigate a *quantitative aspect* of the notion $i \to j$. We will denote, for $k = 1, 2, \ldots$, by $f_{ij}^{(k)}$ the *probability that a walk which starts at i visits j for the first time in the k'th step.* In the language of chapter 1 this just means that

$$f_{ij}^{(k)} := \mathbb{P}(X_1 \neq j, \ldots, X_{k-1} \neq j, \ X_k = j), \tag{4.3}$$

where $X_0, X_1, X_2, \ldots$ stands for the Markov process with transition matrix P and a deterministic start at i.

With this notation the number $f_{ij}^* := \sum_{k=1}^{\infty} f_{ij}^{(k)}$ is the probability that a walk starting at i will occupy state j at some later step, and therefore $f_{ij}^* > 0$ implies $i \to j$. (Note, however, that the converse only holds if $i \neq j$ since, by definition, one *always* has $i \to i$.)

How long will it take to come from i to j? With $X_0 = i, X_1, \ldots : \Omega \longrightarrow S$ as in (4.3) a reasonable measure will be the "expectation" $\int W_{ij} \, d\mathbb{P}$, with $W_{ij} : \Omega \longrightarrow [1, \infty]$ defined by

$$W_{ij}(\omega) := \min\{k \geq 1 \mid X_k(\omega) = j\}$$

(here $\min \emptyset := \infty$). In terms of the $f_{ij}^{(k)}$ this integral is easily calculated as

$$\mu_{ij} := \sum_{k=1}^{\infty} k f_{ij}^{(k)} + (1 - f_{ij}^*)\infty. \tag{4.4}$$

The μ_{ij} don't carry interesting information if $f_{ij}^* < 1$, in many books they are defined only if $f_{ij}^* = 1$. Also then it is a priori not clear that μ_{ij} is finite (which would allow us to consider this number as the expectation of the running time from i to j). In fact this will be shown later in proposition 7.2.

In order to prepare these investigations we introduce the

Definition 4.7

   (i) A state i will be called *recurrent*, if $f_{ii}^* = 1$. If $f_{ii}^* < 1$, then we will speak of a *transient* state.

   (ii) A *positive recurrent* (resp. a *null recurrent*) state is a recurrent state such that $\mu_{ii} < \infty$ (resp. $\mu_{ii} = \infty$).

As an *example*, defined by the graph notation which we have already met on page 14, consider

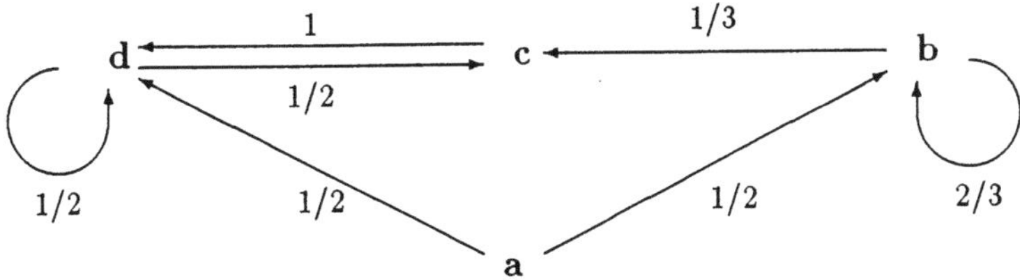

It is plain that state $\mathbf{a}$ is transient since a walk starting there will *never* return to this position. For the other states one has to check the graph to determine the probabilities for a first return in precisely the k'th step:

- $f_{\mathbf{bb}}^{(1)} = 2/3$, $f_{\mathbf{bb}}^{(k)} = 0$ for $k > 1$; hence $f_{\mathbf{bb}}^{*} = 2/3$, i.e., $\mathbf{b}$ is also transient.

- $f_{\mathbf{cc}}^{(1)} = 0$, $f_{\mathbf{cc}}^{(k)} = 1/2^{k-1}$ for $k \geq 2$, hence $\mathbf{c}$ is recurrent. The expectation value of the return time is (with the help of the formula $\sum_{1}^{\infty} kq^{k-1} = 1/(1-q)^2$) determined as $\mu_{\mathbf{cc}} = \sum_{k=2}^{\infty} k/2^{k-1} = 3$, and therefore $\mathbf{c}$ is positive recurrent.

- Finally, $\mathbf{d}$ is positive recurrent as well: $f_{\mathbf{dd}}^{(1)} = f_{\mathbf{dd}}^{(2)} = 1/2$, hence $f_{\mathbf{dd}}^{*} = 1$ and $\mu_{\mathbf{dd}} = 3/2$.

We want to *characterize* recurrent and transient states. To this end, fix a state i and model a walk starting at i by a homogeneous Markov process $X_0 = i, X_1, \ldots$ defined on some probability space $(\Omega, \mathcal{A}, \mathbb{P})$ (see page 9). By homogeneity we have

$$
\begin{aligned}
\mathbb{P}(X_{k+1} = i_1, \ldots, X_{k+k'} = i_{k'} \mid X_k = i) &= \mathbb{P}(X_1 = i_1, \ldots, X_{k'} = i_{k'} \mid X_0 = i) \\
&= p_{ii_1} p_{i_1 i_2} \cdots p_{i_{k'-1}, k'}.
\end{aligned}
$$

In other words: a walk which is at position i at "time" k will behave precisely as one which starts deterministically at i.

This is intuitively clear since the random generators (= the rows of P) are the same for all times. The result also can be easily proved rigorously with the help of formula (1.2) in chapter 1.

In chapter 12 we will learn that what is important here is a special case of the *strong Markov property*.

Hence

$$
\begin{aligned}
f_{ii}^{(k')} &= \sum_{i_1, \ldots, i_{k'-1} \neq i} \mathbb{P}(X_1 = i_1, \ldots, X_{k'-1} = i_{k'-1}, X_{k'} = i \mid X_0 = i) \\
&= \mathbb{P}(X_{k+1} \neq i, \ldots, X_{k+k'-1} \neq i, X_{k+k'} = i \mid X_k = i),
\end{aligned}
$$

and thus the probability that our walk occupies i at times $k, k + k'$ but *not* at times $1, \ldots, k-1, k+1, \ldots, k+k'-1$ is precisely $f_{ii}^{(k)} f_{ii}^{(k')}$. It follows that the walk returns to i *at least two times* with probability $\sum_{k,k' \geq 1} f_{ii}^{(k)} f_{ii}^{(k')} = (f_{ii}^{*})^2$. Similarly one obtains the more general equation $\mathbb{P}(B_s) = (f_{ii}^{*})^s$, where B_s stands for the event "the walk returns to i at least s times". The B_s are decreasing, and therefore the probability of their intersection (that is the probability of infinitely many returns) is

$$
\lim_s (f_{ii}^{*})^s = \begin{cases} 0 &: f_{ii}^{*} < 1 \\ 1 &: f_{ii}^{*} = 1. \end{cases}
$$

These calculations have led us to

Proposition 4.8 *A state i is recurrent iff a walk starting at i returns infinitely often with probability one. It is transient iff infinitely many returns for such walks occur only with probability zero.*

Remark: Note that this 0–1-law for the events $E_k :=$ "the walk occupies position i at the k'th step" holds although the E_k are not independent in general.

Surely E_k has probability $p_{ii}^{(k)}$ (recall that we have written P^k as $(p_{ij}^{(k)})_{ij}$ and that $\mathbb{P}(E_k) = (0, \ldots, 0, 1, 0, \ldots, 0)P^k$ with the 1 at the i'th position). E_k is the disjoint union of the events

$$
\begin{aligned}
F_{1,k} &:= \{X_1 = X_k = i\}, \\
F_{2,k} &:= \{X_1 \neq i, X_2 = X_k = i\}, \\
&\;\;\vdots \quad \vdots \quad \vdots \\
F_{k,k} &:= \{X_1 \neq i, \ldots, X_{k-1} \neq i, X_k = i\}.
\end{aligned}
$$

Similarly to the calculations leading to proposition 4.8 one gets $\mathbb{P}(F_{k-t,k}) = f_{ii}^{(k-t)} p_{ii}^{(t)}$ for $t = 0, \ldots, k-1$ (recall that $p_{ii}^{(0)} = 1$) and thus $p_{ii}^{(k)} = \sum_{t=0}^{k-1} f_{ii}^{(k-t)} p_{ii}^{(t)}$.

Now let k_0 be arbitrary:

$$
\begin{aligned}
\sum_{k=1}^{k_0} p_{ii}^{(k)} &= \sum_{k=1}^{k_0} \sum_{t=0}^{k-1} f_{ii}^{(k-t)} p_{ii}^{(t)} \\
&= \sum_{t=0}^{k_0-1} p_{ii}^{(t)} \sum_{k=t+1}^{k_0} f_{ii}^{(k-t)} \\
&\leq \Big(\sum_{t=0}^{k_0} p_{ii}^{(t)}\Big) f_{ii}^* \\
&= \Big(1 + \sum_{t=1}^{k_0} p_{ii}^{(t)}\Big) f_{ii}^*
\end{aligned}
$$

Hence $(1 - f_{ii}^*) \sum_{t=1}^{k_0} p_{ii}^{(t)} \leq f_{ii}^*$, and we are ready for the proof of

Proposition 4.9 *A state i is recurrent iff $\sum_{t=0}^{\infty} p_{ii}^{(t)} = \infty$, and consequently it is transient iff $\sum_{t=0}^{\infty} p_{ii}^{(t)} < \infty$.*

Proof. By the preceding inequality and proposition 4.8 we already know that $\sum_{t=0}^{\infty} p_{ii}^{(t)}$ must be finite for transient states and that $\sum_{t=0}^{\infty} p_{ii}^{(t)} = \infty$ implies that i is recurrent.

Now the (easy part of) the Borel-Cantelli lemma comes into play. Suppose that $\sum_{t=0}^{\infty} p_{ii}^{(t)} < \infty$. Then – with the above notation – $\sum \mathbb{P}(E_k) < \infty$ and thus the walk will occupy state i infinitely often only with probability zero. Hence, by proposition 4.8, i is transient. This completes our proof since the remaining statement "i recurrent $\Rightarrow \sum_{t=0}^{\infty} p_{ii}^{(t)} = \infty$" follows by logical transposition. $\qquad\square$

Some first remarkable consequences of these characterizations are contained in

Proposition 4.10

   (i)   *Let i, j be states such that i is recurrent and $i \leftrightarrow j$. Then j is also recurrent.*

  (ii)   *There always exists at least one recurrent state[1].*

 (iii)   *All states in a minimal invariant set are recurrent.*

Proof. (i) Choose k', k'' such that $p_{ij}^{(k')}, p_{ji}^{(k'')} > 0$. Then, by the calculation from the proof of proposition 4.4(i), we have $p_{jj}^{(k+k'+k'')} \geq p_{ii}^{(k)} p_{ij}^{(k')} p_{ji}^{(k'')}$ and thus $\sum_k p_{jj}^{(k)} \geq (\sum_k p_{ii}^{(k)}) p_{ij}^{(k')} p_{ji}^{(k'')}$. This inequality implies the result by our characterization 4.9.

(ii) Suppose that the chain is irreducible (if this is not the case, pass to a minimal closed set and consider the restricted chain). Therefore, by proposition 4.4(iv), all states communicate.

Fix any state i_0, we claim that it is recurrent.

First we note that $\sum_j p_{i_0 j}^{(k)} = 1$ for all k since P^k is a stochastic matrix. Therefore $\sum_{j,k} p_{i_0 j}^{(k)} = \infty$ and thus there must be a j_0 with $\sum_k p_{i_0 j_0}^{(k)} = \infty$; here – for the second time – we have used the fact that S is finite. Similarly to the proof of part (i) it follows that $\sum_k p_{i_0 i_0}^{(k)} = \infty$ (there is a k' such that $p_{j_0 i_0}^{(k')} > 0$, and $p_{i_0 i_0}^{(k+k')} \geq p_{i_0 j_0}^{(k)} p_{j_0 i_0}^{(k')}$). Hence, by proposition 4.9, i_0 is recurrent.

(iii) This follows from (i), (ii), and proposition 4.4(ii). $\square$

Recall that we have denoted by $C_1, ..., C_r$ the minimal closed subsets of our chain (see page 25). We will define $T := S \setminus (C_1 \cup \cdots \cup C_r)$, note that T might be empty. Whereas the preceding proposition tells us that the i in $C_1 \cup \cdots \cup C_r$ are recurrent nothing is known up to now for the $i \in T$. The fact that we have decided to denote this set by "T" is far from being accidental:

Proposition 4.11 *All i in T are transient.*

Proof. Fix any $i \in T$ and consider the set $C' := \{j \mid i \to j\}$. It is plain that it is closed and hence it must meet $C := C_1 \cup \cdots \cup C_r$: if the intersection were empty this would contradict proposition 4.2 and the definition of T.

Denote by $p_{iC}^{(k)}$ the sum $\sum_{j \in C} p_{ij}^{(k)}$. That $C \cap C' \neq \emptyset$ may be rephrased by saying that for every $i \in T$ there is a k such that $p_{iC}^{(k)} > 0$. But $p_{iC}^{(k)} \leq p_{iC}^{(k+1)}$ since there are no transitions from C to T, and this allows us to reverse the quantifiers: we find a k_0 such that $p_{iC}^{(k_0)} > 0$ for *all* $i \in T$. (Note that this argument is only possible since T is finite.) Let p be the (positive) minimum of these numbers.

Fix again an $i \in T$ and model a walk starting at i by a homogeneous process (X_k) with $X_0 = i$. Since $p_{iC}^{(k_0)} \geq p$ it follows that

$$p_{iT}^{(k_0)} := \mathbb{P}(X_{k_0} \in T) = \sum_{j \in T} p_{ij}^{(k_0)} = 1 - p_{iC}^{(k_0)} \leq 1 - p.$$

The homogeneity of the walk and the fact that there are no transitions from C to T imply that

---

[1] We emphasize once more that we are dealing only with *finite* chains in this book. The result does *not hold* for infinite chains as is easily seen by considering once again the deterministic walk to the right on the integers (see page 24).

$$
\begin{aligned}
p_{iT}^{(2k_0)} &= \mathbb{P}(X_{2k_0} \in T) \\
&= \sum_{j \in S} p_{ij}^{(k_0)} \mathbb{P}(X_{2k_0} \in T \mid X_{k_0} = j) \\
&= \sum_{j \in T} p_{ij}^{(k_0)} \mathbb{P}(X_{2k_0} \in T \mid X_{k_0} = j) \\
&\leq \sum_{j \in T} p_{ij}^{(k_0)} (1 - p) \\
&= p_{iT}^{(k_0)} (1 - p) \\
&\leq (1 - p)^2;
\end{aligned}
$$

that the conditional probabilities $\mathbb{P}(X_{2k_0} \in T \mid X_{k_0} = j)$ are bounded by $1 - p$ since they can be treated as the $\mathbb{P}(X_{k_0} \in T)$ above can be justified like the corresponding formula for $\mathbb{P}(X_{k+1} = i_1, \ldots, X_{k+k'} = i_{k'} \mid X_k = i)$ on page 30.

Similarly one gets $p_{iT}^{(rk_0)} \leq (1 - p)^r$ for arbitrary r. To finish the proof it suffices to note that the $(p_{iT}^{(k)})_{k=1,\ldots}$ are *decreasing* (since $p_{iT}^{(k)} = 1 - p_{iC}^{(k)}$ and the $(p_{iC}^{(k)})$ are increasing): it follows that

$$
\begin{aligned}
p_{ii}^{(1)} + p_{ii}^{(2)} + \cdots &\leq p_{iT}^{(1)} + p_{iT}^{(2)} \cdots \\
&\leq k_0 p_{iT}^{(0)} + k_0 p_{iT}^{(k_0)} + k_0 p_{iT}^{(2k_0)} + \cdots \\
&\leq k_0 + k_0(1 - p) + k_0(1 - p)^2 + \cdots \\
&< \infty,
\end{aligned}
$$

and therefore, by proposition 4.9, the proof is complete. $\qquad\Box$

The main results of the present section can be summarized as follows:
The state space S of a finite Markov chain can always be written as a disjoint union $S = C_1 \dot{\cup} \cdots \dot{\cup} C_r \dot{\cup} T$, where

- $r \geq 1$, the C_ρ are minimal closed sets, and the restriction of the chain to C_ρ is irreducible;

- any i in any C_ρ is recurrent;

- the $i \in T$ are transient (note, however, that T might be empty).

In particular, the problem to characterize finite Markov chains is reduced to

1. the answer to the question "What happens with transient states?" and
2. the study of irreducible chains.

Transient states will be studied in the next chapter, the more detailed inspection of irreducible chains will be postponed until chapter 7: it needs a preparation with which we will be concerned in chapter 6.

Exercises

4.1: Consider a deck of r cards such that $r > 3$. Determine the closed subsets of all permutations of these cards with respect to the 3-to-random shuffle (the name should be self-explaining).

4.2: A stochastic matrix is called *deterministic* if it contains in each row only one nonzero entry. Prove that the associated chain cannot be irreducible and aperiodic in this case. Can it be irreducible? Is it possible that aperiodic states exist?

4.3: Provide a canonical form of the matrix

$$\frac{1}{10} \begin{pmatrix} 1 & 0 & 0 & 0 & 9 \\ 0 & 4 & 6 & 0 & 0 \\ 0 & 10 & 0 & 0 & 0 \\ 2 & 2 & 2 & 2 & 2 \\ 8 & 0 & 0 & 0 & 2 \end{pmatrix}.$$

4.4: Let S be the symmetric group $\mathcal{S}_r$, i.e., the group of all permutations over $\{1, \ldots, r\}$. Transitions are defined as follows: if the chain is now in state $(i_1, \ldots, i_r)$, choose a $\rho \in \{2, \ldots, r\}$ uniformly at random and pass to

$$(i_1, \ldots, i_{\rho-2}, i_\rho, i_{\rho-1}, i_{\rho+1}, \ldots, i_N).$$

Is this chain irreducible and aperiodic?

4.5: On page 24 it was claimed that the chain defined on the nonnegative integers by $p_{i,i+1} = 1$ admits no minimal closed subsets. Prove that the following more general result holds: if a chain is defined on $\mathbb{N}_0$ by a stochastic matrix $P = (p_{ij})$ such that $p_{i,i+1} > 0$ for every i, then there are no minimal closed subsets. Is the converse also true?

4.6: It has been mentioned that the definitions in this chapter can also be considered for countable state spaces, and it has been emphasized that some of our results do not hold in this more general setting.

a) A state i is called *transient* if a walk which starts at i will return infinitely often only with probability zero. Give an example of a chain on the integers where all i are transient. (For finite state spaces this is not possible, cf. proposition 4.10.)

b) Use results from elementary probability to verify that 0 is a recurrent state for the symmetric random walk on the integers (where *"recurrent"* in the general setting means that the walk returns infinitely often with probability 1). Prove also that 0 is null recurrent. (For finite state spaces, all recurrent states are positive recurrent; see proposition 7.2 below.)

4.7: Assume that the stochastic matrix P admits a strictly positive equilibrium distribution (see exercise 3.5). Prove that there cannot exist any transient states.

4.8: Let $i \in C$, where C is a closed subset of S having r elements. Prove that $d \leq r$ if i has period d.

4.9: Let d and N be integers with $d \leq N$. Under what conditions on d and N does there exist a Markov chain on $\{1, \ldots, N\}$ such that all states have period d?

4.10: Fix N and denote by K_{aper} the collection of stochastic $N \times N$-matrices which give rise to irreducible and aperiodic chains. Prove that K_{aper} is a convex, dense and open subset of the collection of all stochastic matrices.

4.11: Let i be a state for which the period with respect to a stochastic matrix P is d. What is the period of i with respect to P^{k_0}?

4.12: For integers N, N_0, N_1, $\ldots$, N_d such that $N = N_0 + \cdots + N_d$ there exists a Markov chain on $\{1, \ldots, N\}$ with the following properties:
- there are precisely N_0 transient states;
- there are, for $j = 1, \ldots, d$, subsets C_j with N_j elements such that the chain acts irreducibly on $C_1 \cup \cdots \cup C_d$ and each state in this union has period d;
- only transitions from C_j to C_{j+1} are possible (with $d + 1 := 1$).

4.13: Let a Markov chain be given by using the graph notation (ε and δ denote arbitrary numbers between 0 and 1):

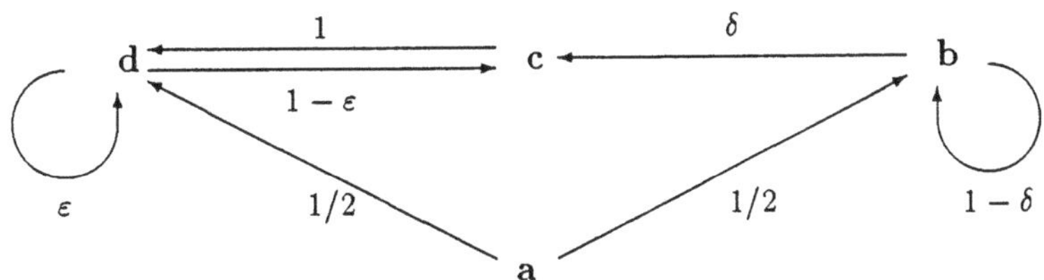

Depending on ε, δ, which of the states **a**, **b**, **c**, **d** is transient or recurrent, which of the recurrent states is positive recurrent, what are the values of the $f_{aa}^{(k)}$ etc.?

4.14: In this exercise we consider Markov chains on $\{1, \ldots, N\}$ given by a stochastic $N \times N$-matrix P. We call a property of such a chain *robust* if it holds simultaneously for all cases where the matrices P have their non-zero entries at the same positions. (As a simple illustration consider the property "a transition from 1 to 3 is possible".) Which of the following properties are robust?

a) $i \to j$, for fixed states i, j;

b) i has period d;

c) the chain is irreducible;

d) P is doubly stochastic;

e) there is a transient state;

f) find further examples which are robust and others which fail to have this property.

4.15: Let $\mathbb{M}$ be a subset of $\mathbb{N}_0$ such that $\mathbb{M} + \mathbb{M} \subset \mathbb{M}$. Prove that there are a Markov chain and a suitable state i such that the set $\mathbb{N}_i$ from definition 4.5 coincides with $\mathbb{M}$.

5   Transient states

As in the preceding chapter we fix a Markov chain on a finite state space S. The notation
will be as before: $C_1, \ldots, C_r$ are the minimal closed sets, and $T := S \setminus (C_1 \cup \cdots \cup C_r)$.

The $i \in T$ are the transient states. We already know that an i is transient iff $f_{ii}^* < 1$,
and also iff $\sum_k p_{ii}^{(k)} < \infty$. In particular it follows that $p_{ii}^{(k)} \to 0$, but more is true:

Proposition 5.1 *Let i be a transient state. Then:*

   (i) *For every state j the probabilities $p_{ji}^{(k)}$ tend to zero for $k \to \infty$.*

  (ii) *Almost surely a walk which starts at i will be in some C_ρ after finitely many
      steps.*

 (iii) *If j is also transient, then $e_{ij} := \sum_{k \geq 0} p_{ij}^{(k)}$ is finite. e_{ij} is, for a walk starting at
      i, just the expectation of the number of j-visits before entering $C := C_1 \cup \cdots \cup C_r$.
      Consequently $\sum_j e_{ij}$ is the expectation of the total number of steps in T.*

Proof. (i) We start with a slight generalization of the calculations which have preceded
proposition 4.9. Fix states i and j (i does not need to be transient here) and model a
walk starting at j by a homogeneous Markov process $X_0 = j$, $X_1, \ldots$ defined on some
probability space $(\Omega, \mathcal{A}, \mathbb{P})$. Put

$$
\begin{aligned}
E_k &:= \{X_k = i\}, \\
F_{1,k} &:= \{X_1 = X_k = i\}, \\
F_{2,k} &:= \{X_1 \neq i, X_2 = X_k = i\}, \\
\cdots \quad &\cdots \quad \cdots \\
F_{k,k} &:= \{X_1 \neq i, \ldots, X_{k-1} \neq i, X_k = i\}.
\end{aligned}
$$

Then E_k is the disjoint union of the $F_{\kappa,k}$ so that $\mathbb{P}(E_k) = \sum_\kappa \mathbb{P}(F_{\kappa,k})$. But $\mathbb{P}(F_{k-t,k}) = f_{ji}^{(k-t)} p_{ii}^{(t)}$ by the homogeneity of the chain, and therefore

$$
\mathbb{P}(E_k) = p_{ji}^{(k)} = f_{ji}^{(k)} + f_{ji}^{(k-1)} p_{ii}^{(1)} + \cdots + f_{ji}^{(1)} p_{ii}^{(k-1)}, k = 1, 2, \ldots \tag{5.1}
$$

holds.

The end the proof is easy, it will need – of course – the assumption that i is transient. It
suffices to note that $\sum_k f_{ji}^{(k)} \leq 1$; then (5.1) together with $p_{ii}^{(k)} \to 0$ immediately imply
that $p_{ji}^{(k)} \to 0$.

(ii) This time we need a Markov process $X_0 = i$, $X_1, \ldots : \Omega \longrightarrow S$ which starts at i. Let
F_k be the event $\{X_k \in T\}$. Surely

$$
\mathbb{P}(F_k) = \sum_{j \in T} \mathbb{P}(X_k = j) = \sum_{j \in T} p_{ij}^{(k)}.
$$

By part (i) – with the roles of i and j interchanged – the $p_{ij}^{(k)}$ tend to zero so that
$\mathbb{P}(F_k) \to 0$.

Since $S \setminus T$ is a closed set we know that $X_{k+1} \notin T$ whenever $X_k \notin T$, and this implies $F_1 \supset F_2 \supset \cdots$. Therefore the probability of $\bigcap_k F_k$, being the limit of the $\mathbb{P}(F_k)$, is zero. This finishes the proof: $\bigcap F_k$ contains precisely the ω where the walk never leaves T.

(iii) We will need the equations (5.1), this time with the roles of i and j reversed:

$$\begin{aligned}
p_{ij}^{(1)} &= f_{ij}^{(1)}, \\
p_{ij}^{(2)} &= f_{ij}^{(2)} + f_{ij}^{(1)} p_{jj}^{(1)}, \\
p_{ij}^{(3)} &= f_{ij}^{(3)} + f_{ij}^{(2)} p_{jj}^{(1)} + f_{ij}^{(1)} p_{jj}^{(2)}, \\
&\quad \vdots \quad \vdots \quad \vdots
\end{aligned}$$

Summation leads to

$$\sum_{k=1}^{\infty} p_{ij}^{(k)} = \Big(\sum_{k=1}^{\infty} f_{ij}^{(k)} \Big) \Big(1 + \sum_{k=1}^{\infty} p_{jj}^{(k)} \Big), \tag{5.2}$$

and we may conclude that $\sum_k p_{ij}^{(k)} < \infty$ (since $\sum_k f_{ij}^{(k)} \le 1, \sum_k p_{jj}^{(k)} < \infty$).

Now let – with the notation of the preceding part (ii) of this proof – E_k be the event $\{X_k = j\}$ and $Y_k : \Omega \longrightarrow \mathbb{R}$ its indicator function:

$$Y_k(\omega) := \begin{cases} 1 & : \quad \omega \in E_k \\ 0 & : \quad \text{otherwise.} \end{cases}$$

Then the function $Y : \Omega \to [0, \infty]$, $Y := \sum_k Y_k$, counts the number of the j-visits so that $\int Y d\mathbb{P}$ is the number we are interested in. By the monotone convergence theorem (see [12, p. 50], [16, p. 208]) we may interchange summation and integration, i.e., $\int Y d\mathbb{P} = \sum_k \int Y_k d\mathbb{P}$. These integrals are easy to evaluate: $\int Y_k d\mathbb{P} = \mathbb{P}(E_k) = p_{ij}^{(k)}$, and this completes the proof. $\qquad\square$

Fix any transient state i and consider a walk starting there. With $X_0 = i$, $X_1, \ldots :$ $\Omega \longrightarrow S$ as in the preceding proof (part (ii)) we put

$$w_{ij}^{(k)} := \mathbb{P}(X_1, X_2, \ldots, X_{k-1} \in T, X_k = j), \text{ and } w_{ij} := \sum_{k \ge 1} w_{ij}^{(k)}$$

for every $j \in C$; then part (ii) of proposition 5.1 may be rephrased by $\sum_{j \in C} w_{ij} = 1$.

Note the difference between $f_{ij}^{(k)}$ and $w_{ij}^{(k)}$: in the definition of $f_{ij}^{(k)}$ we ask for the probability that the first visit of j occurs in the k'th step whereas in the case of $w_{ij}^{(k)}$ this first visit has to coincide with a transition from T to C.

Here is *an example*. Put

$$P := \begin{pmatrix} 1 & 0 & 0 & 0 \\ 0 & 1/2 & 1/2 & 0 \\ 0 & 1 & 0 & 0 \\ 1/8 & 1/8 & 1/2 & 1/4 \end{pmatrix}. \tag{5.3}$$

With respect to this transition matrix the Markov chain has $C_1 = \{1\}$ and $C_2 = \{2,3\}$ as its minimal irreducible sets, and there is only one transient state: $T = \{4\}$. Let a walk start at 4. It will stay in T for $k - 1$ steps and then jump – for example – to state 2 with probability $w_{42}^{(k)} = (1/4)^{k-1}(1/8)$, and consequently $w_{42} = 1/8 + (1/8)(1/4) + (1/8)(1/4)^2 + \cdots = 1/6$. Similarly one gets $w_{41} = 1/6$ and $w_{43} = 4/6$.

For more complicated situations it is not obvious how to manage the necessary calculations. We will use *linear algebra* to reduce them to the evaluation of a certain matrix inverse.

Let us first fix notation. We resume what has been done in chapter 4, page 25, that is we renumber – if necessary – the states such that the $C_1, \ldots, C_r$ are of the form $\{1, \ldots, N_1\}, \{N_1 + 1, \ldots, N_2\}, \ldots, \{N_{r-1} + 1, \ldots, N_r\}$. The stochastic matrix P then looks as follows:

$$\begin{pmatrix} P_1 & 0 & 0 & \cdots & 0 & 0 & 0 \\ 0 & P_2 & 0 & \cdots & 0 & 0 & 0 \\ \vdots & \vdots & \vdots & & \vdots & \vdots & \vdots \\ 0 & 0 & 0 & \cdots & P_{r-1} & 0 & 0 \\ 0 & 0 & 0 & \cdots & 0 & P_r & 0 \\ R_1 & R_2 & R_3 & \cdots & R_{r-1} & R_r & Q \end{pmatrix}, \tag{5.4}$$

Let N stand, as usual, for the cardinality of S. Then $t := N - N_r$ is the number of transient states (which will be assumed to be nonzero in this chapter from now on), Q is a $t \times t$-matrix, and R_ρ is a matrix with t rows and $N_\rho - N_{\rho-1}$ columns[1] for $\rho = 1, \ldots, r$. The matrix Q will play a particularly important role, as a preparation of the proof of the following theorem we show

Lemma 5.2 *Consider the sequence of matrices* $(\tilde{Q}^{(k)})_k := (Id + Q + Q^2 + \cdots + Q^k)_k$. *If* $\tilde{Q}^{(k)}$ *is written as* $(\tilde{q}_{ij}^{(k)})_{i,j \in T}$, *then* $\lim_{k \to \infty} \tilde{q}_{ij}^{(k)}$ *exists for all* $i, j \in T$ *and equals* e_{ij} *(as defined in proposition 5.1(iii)). The matrix* $F := (e_{ij})_{i,j \in T}$ *is the inverse of* $Id - Q$:

$$(Id - Q)F = F(Id - Q) = Id.$$

F is called the fundamental matrix *associated with the chain.*

Proof. The first part is obvious since the entries of Q^k are just the $p_{ij}^{(k)}$. Also, by simple matrix manipulation, one has $(Id - Q)(Id + Q + Q^2 + \cdots + Q^k) = Id - Q^{k+1}$. Thus, since multiplication is continuous and since the entries of Q^k tend to zero with $k \to \infty$ we get $(Id - Q)F = Id$. $F(Id - Q) = Id$ is proved similarly. $\square$

Remark: The lemma should remind you of the well-known formula $1 + q + q^2 + \cdots = 1/(1-q)$ (for $|q| < 1$) of the geometric series, it states that $Id + Q + Q^2 + \cdots = (Id - Q)^{-1}$. What we have shown is a special case of the following more general fact: whenever Q is an element of a Banach algebra (e.g., a space of continuous functions or of linear continuous operators on a Banach space), then $Id + Q + Q^2 + \cdots$ is an inverse of $Id - Q$ as soon as one knows that $Id + Q + Q^2 + \cdots$ converges (here Id denotes the neutral element of multiplication which is assumed to exist). $Id + Q + Q^2 + \cdots$ is called the *Neumann series* associated with Q.

---

[1] with $N_0 := 0$.

Once it is guaranteed that $(Id - Q)^{-1}$ exists, all numbers which we have introduced so far are easy to evaluate:

Theorem 5.3 *Let F be the fundamental matrix of the chain. Then*

   (i) *F has as its entries the numbers e_{ij}, $i, j \in T$, from proposition 5.1(iii). Consequently the vector $F(1, 1, \ldots, 1)^\top$ contains at its i'th position the expectation of the number of steps[2] of a walk starting at $i \in T$ before it is absorbed in C.*

   (ii) *f_{ii}^*, the probability that a walk returns to i, equals $(e_{ii} - 1)/e_{ii}$ for $i \in T$.*

   (iii) *The numbers w_{ij}, $i \in T$, $j \in C$, are the components of the matrix FR, where R denotes the $t \times N_r$-matrix $(R_1 R_2 \ldots R_r)$. Therefore the probability that $i \in T$ will land in some particular C_ρ is just the i'th component of $FR(0, 0, \ldots, 0, 1, \ldots, 1, 0, \ldots$ where the vector $(0, 0, \ldots, 0, 1, \ldots, 1, 0, \ldots, 0)^\top$ contains a 1 precisely at the positions $N_{\rho-1}+1, \ldots, N_\rho$.*

Proof. (i) is part of the assertion of the preceding lemma, and (ii) is a special case of (5.2).

To prove (iii), fix $i \in T, j \in C$ and consider a process $X_0 = i, X_1, \ldots : \Omega \to S$ as a model for a walk starting at i. We want to condition on X_1, we put

$$E := \{X_1, \ldots, X_{k-1} \in T, X_k = j \text{ for some } k\},$$

and we define $E_l := \{X_1 = l\}$ for $l \in T \cup \{j\}$.

Then E is the disjoint union of E_j and the $E \cap E_l$, $l \in T$. Since the chain is homogeneous, we know that $\mathbb{P}(E \cap E_l)/\mathbb{P}(E_l) = \mathbb{P}(X_1, \ldots, X_{k-1} \in T, \ X_k = j \text{ for some } k \mid X_1 = l) = w_{lj}$. Therefore $w_{ij} = \mathbb{P}(E) = p_{ij} + \sum_{l \in T} p_{il} w_{lj}$ which is nothing but the matrix equality $V = R + QV$, where $V := (w_{ij})$. It follows that $(Id - Q)V = R$ and thus $V = FR$ as claimed. $\qquad\square$

Let us treat some *examples* to apply our results.

1) First we consider the chain with the transition matrix (5.3) above. Here Q is the 1×1-matrix $(1/4)$ so that $F = (1 - 1/4)^{-1} = (4/3)$. Thus $e_{44} = 4/3$, $f_{44}^* = 1/4$ and

$$(w_{41} \ w_{42} \ w_{43}) = FR = (4/3)(1/8 \ 1/8 \ 1/2) = (1/6 \ 1/6 \ 4/6)$$

as in our previous calculation.

2) Let P be given in standard form as

$$P = \begin{pmatrix} 1/3 & 1/3 & 1/3 & 0 & 0 & 0 & 0 \\ 1/3 & 1/3 & 1/3 & 0 & 0 & 0 & 0 \\ 1/3 & 1/3 & 1/3 & 0 & 0 & 0 & 0 \\ 0 & 0 & 0 & 1/2 & 1/2 & 0 & 0 \\ 0 & 0 & 0 & 1/2 & 1/2 & 0 & 0 \\ 1/6 & 1/6 & 1/6 & 0 & 1/6 & 0 & 1/3 \\ 0 & 1/2 & 0 & 0 & 1/4 & 1/4 & 0 \end{pmatrix}.$$

The set of transient states is $\{6, 7\}$, and since $Q = \begin{pmatrix} 0 & 1/3 \\ 1/4 & 0 \end{pmatrix}$ it follows that

---

[2] Note that this number includes the starting position: even if the walk jumps immediately from i to C it will be one.

$F = \frac{1}{11} \begin{pmatrix} 12 & 4 \\ 3 & 12 \end{pmatrix}$. A walk starting at 6, for example, will on the average be 12/11 times at 6 before it leaves T (since here the starting position is included one has to subtract 1 if the expectation of returns is of interest); the total number of steps in T for such a walk has an average of $12/11 + 4/11 = 16/11$.

The matrix R here equals $\begin{pmatrix} 1/6 & 1/6 & 1/6 & 0 & 1/6 \\ 0 & 1/2 & 0 & 0 & 1/4 \end{pmatrix}$, and therefore

$$(w_{ij})_{i=6,7, j=1,\ldots,5} = FR = \frac{1}{22} \begin{pmatrix} 4 & 8 & 4 & 0 & 6 \\ 1 & 13 & 1 & 0 & 7 \end{pmatrix}.$$

The absorption probabilities of state 6 with respect to the invariant sets $\{1,2,3\}$ and $\{4,5\}$ are $4/22 + 8/22 + 4/22 = 8/11$ and $0 + 6/22 = 3/11$, respectively, and for state 7 we obtain the values 15/22 and 7/22.

These examples should suffice to illustrate the usefulness of our preceding results, everyone is invited to produce more impressing ones with the help of suitable matrix calculation packages.

Exercises

5.1: Let P be a stochastic matrix, we suppose that a certain state i_0 is transient. Now let k_0 be an arbitrary integer. Prove that i_0 is transient also with respect to the chain P^{k_0}.

5.2: Let i,j be states such that $i \to j$. Prove that with j also i is transient.

5.3: If P is a doubly stochastic matrix, then the associated chain admits no transient states.

5.4: Consider an $N \times N$-matrix $Q = (q_{ij})$ all entries of which are nonnegative such that $\sum_j q_{ij} < 1$ for all i. Prove that $Id - Q$ is invertible.

5.5: In lemma 5.2 we have shown that, under the assumption that the series $F := Id + Q + Q^2 + \cdots$ converges, F is the inverse of $Id - Q$.

a) Is the converse also true: does $Id + Q + Q^2 + \cdots$ converge if $(Id - Q)^{-1}$ exists? (Here Q denotes an arbitrary $N \times N$-matrix.)

b) Does the existence of $(Id - Q)^{-1}$ follow from $Q^k \to 0$?

5.6: The following stochastic matrix is already in standard form:

$$P = \frac{1}{10} \begin{pmatrix} 0 & 10 & 0 & 0 & 0 & 0 \\ 5 & 5 & 0 & 0 & 0 & 0 \\ 0 & 0 & 10 & 0 & 0 & 0 \\ 1 & 4 & 2 & 2 & 1 & 0 \\ 0 & 0 & 1 & 3 & 3 & 3 \\ 0 & 0 & 0 & 0 & 10 & 0 \end{pmatrix}.$$

Calculate the fundamental matrix F and the matrix of the (w_{ij}). How long will the transient state 5 survive on the average before it enters the union $C = \{1,2,3\}$ of the minimal closed subsets? For state 6, is it more likely to be absorbed at state 1 or at state 2?

5.7: Suppose that a chain has precisely two transient states, $N-1$ and N, say, and that the Q-matrix has the form

$$Q = \begin{pmatrix} a & b \\ 1/2 & 1/2 \end{pmatrix}.$$

What is the expectation of the number A of steps of a walk starting at $N-1$ before it is absorbed in the subset of recurrent states? What values of A are possible?

5.8: Consider the chain defined by the stochastic matrix

$$P = \frac{1}{8} \begin{pmatrix} 1 & 2 & 0 & 5 \\ 3 & 2 & 2 & 1 \\ 0 & 3 & 3 & 2 \\ 1 & 1 & 3 & 3 \end{pmatrix}.$$

What is the probability that a walk starting at state 2 is in position 1 before it is in position 3?

(Hint: Pass to a suitable modification of P which transforms the problem into an absorption problem.)

5.9: Which matrices Q can be the Q-matrix of the transient states of a Markov chain; cf. (5.4)? Is the collection of these Q convex, is it open in the set of all stochastic matrices?

5.10: In theorem 5.3 we have derived a formula for the f_{ii}^* in terms of the entries of the fundamental matrix. Find, more generally, expressions for the f_{ij}^*, where i, j are arbitrary transient states.

5.11: Let i, j be states such that i is transient and $i \sim j$ (see exercise 2.10). Prove that j also is transient.

6   An analytical lemma

Let $x = (x_0, x_1, \ldots)$ and $y = (y_0, y_1, \ldots)$ be real sequences. Their *convolution* is defined to be the sequence

$$x * y = ((x * y)_k) = (x_0 y_0, x_0 y_1 + x_1 y_0, \ldots, \sum_{i=0}^{k} x_i y_{k-i}, \ldots).$$

In the applications we have in mind the x_i, y_i will stand for certain probabilities, and convolutions arise when applying the law of total probability: $\mathbb{P}(E) = \sum_i \mathbb{P}(E \mid A_i)\mathbb{P}(A_i)$. We will be concerned with sequences x, y which satisfy the convolution equation

$$x_k = (x * y)_k \text{ for } k = 1, 2, \ldots. \tag{6.1}$$

Under mild additional conditions this has far-reaching consequences for the behaviour of the sequence x, and this fact will play an important role in our further investigations.

We have already met sequences for which (6.1) is satisfied, for example in the investigations following proposition 4.8 (with $x_k = p_{ii}^{(k)}, y_k = f_{ii}^{(k)}$) or again in the proof of proposition 5.1. In fact, these are particularly typical examples, and therefore we prefer to switch our notation from x_i, y_i to the more suggestive p_i, f_i.

The fundamental analytical result we have in mind reads as follows:

Lemma 6.1 *Let* $p = (p_0, p_1, \ldots)$ *and* $f = (f_0, f_1, \ldots)$ *be sequences of nonnegative real numbers such that* $p_0 = 1, f_0 = 0$. *Suppose that the following conditions are satisfied:*

(i) $p_k = (p * f)_k$ *for* $k \geq 1$; *explicitly this means that*

$$\begin{aligned}
p_1 &= f_1, \\
p_2 &= f_2 + p_1 f_1, \\
&\vdots \quad \vdots \quad \vdots \\
p_k &= f_k + p_1 f_{k-1} + \cdots + p_{k-1} f_1, \\
&\vdots \quad \vdots \quad \vdots
\end{aligned}$$

(ii) $f_1 + f_2 + \cdots = 1$;

(iii) *the greatest common divisor of the indices* k *such that* $f_k > 0$ *is one.*

Then the sequence $(p_k)_k$ *converges to* $1/\sum k f_k$ *(with the convention* $1/\infty := 0$).

The rest of this chapter will be devoted to the proof of this lemma. Since it is rather involved we will split it up into *several parts*.

As a *first step* we introduce the numbers $r_k := f_{k+1} + f_{k+2} + \cdots$ for $k = 0, 1, \ldots$. Then (i) can be rewritten as

$$p_k = p_0(r_{k-1} - r_k) + p_1(r_{k-2} - r_{k-1}) + \cdots + p_{k-1}(r_0 - r_1),$$

and this gives rise to

$$r_0 p_k + r_1 p_{k-1} + \cdots + r_k p_0 = r_0 p_{k-1} + r_1 p_{k-2} + \cdots + r_{k-1} p_0.$$

(Note that, by (ii), $r_0 = 1$.) Call the left-hand side of this equation A_k. Since the right-hand side has the same structure (with k replaced by $k-1$), we know that

$$A_k = A_{k-1} = \cdots = A_0 = 1,$$

and we thus arrive at

$$1 = r_0 p_k + r_1 p_{k-1} + \cdots + r_k p_0 \text{ for all } k. \tag{6.2}$$

Since all summands are nonnegative we may conclude that $1 \geq r_0 p_k = p_k$ for all k. (This assertion, however, could also have been derived in a simpler way: use induction and note that p_k is a convex combination of $p_0 = 1, p_1, \ldots, p_{k-1}, 0$.)

The lemma claims that $\lim p_k = 1/\mu$, where $\mu := f_1 + 2f_2 + 3f_3 + \cdots = r_0 + r_1 + r_2 + \cdots$. This will be established as soon as we have proved that

$$A := \limsup p_k \leq 1/\mu \text{ and } B := \liminf p_k \geq 1/\mu.$$

These two inequalities will be treated in step 3 and step 4 below, as a preparation we need an elementary number theoretical fact which will be essential to understand the role of condition (iii).

Step 2: We claim that every subset $\mathbb{M}$ of $\{0, 1, 2, \ldots\}$ which contains a nonzero element and which satisfies $\mathbb{M} + \mathbb{M} \subset \mathbb{M}$ and $\gcd \mathbb{M} = 1$ has the property that there is an integer k_0 such that $\{k_0, k_0+1, k_0+2, \ldots\} \subset \mathbb{M}$; here $\mathbb{M} + \mathbb{M}$ is the usual abbreviation of $\{k+l \mid k, l \in \mathbb{M}\}$. To prove this fact we first choose $k_1, \ldots, k_r \in \mathbb{M}$ with $\gcd\{k_1, \ldots, k_r\} = 1$ (such k_ρ exist by the argument presented on page 27). That the greatest common divisor is 1 may be rephrased as $k_1 \mathbb{Z} + \cdots + k_r \mathbb{Z} = \mathbb{Z}$, and consequently there are $a_1, \ldots, a_r \in \mathbb{Z}$ with $a_1 k_1 + \cdots + a_r k_r = 1$.

Let $K \in \mathbb{N}$ be any number which dominates all $|a_\rho|$, we will show that every k such that $k \geq k_0 := K k_1 (k_1 + \cdots + k_r)$ lies in $\mathbb{M}$. Write k as $k = K k_1 (k_1 + \cdots + k_r) + b k_1 + c$ with $b, c \in \mathbb{N}_0$ and $c < k_1$. Then

$$\begin{aligned}
k &= K k_1 (k_1 + \cdots + k_r) + b k_1 + c(a_1 k_1 + \cdots + a_r k_r) \\
&= k_1(K k_1 + b + c a_1) + k_2(K k_1 k_2 + c a_2) + \cdots + k_r(K k_1 k_r + c a_r),
\end{aligned}$$

where – due to $K + a_\rho \geq 0$ and $c < k_1$ – the factors at $k_1, \ldots, k_r$ are in $\mathbb{N}_0$. But $\mathbb{M} + \mathbb{M} \subset \mathbb{M}$ together with $k_1, \ldots, k_r \in \mathbb{M}$ yield $k_1 \mathbb{N}_0 + \cdots + k_r \mathbb{N}_0 \subset \mathbb{M}$, and thus we have proved that $k \in \mathbb{M}$ as claimed.

Step 3: We are ready to show that $A \leq 1/\mu$. Since this is trivially true if $A = 0$ we may assume that $A > 0$.

The *crucial idea* is to apply step 2 to the set $\mathbb{M}$ of all numbers k_0 such that

$$\lim_{s \to \infty} p_{k_s - k_0} = A$$

whenever $(p_{k_s})_s$ is a subsequence of (p_k) with $\lim_{s \to \infty} p_{k_s} = A$.

WHY MAY WE APPLY STEP 2? $\mathbb{M} + \mathbb{M} \subset \mathbb{M}$ follows easily from the definition, $\gcd \mathbb{M} = 1$ will be shown by verifying that $\{j \mid f_j > 0\} \subset \mathbb{M}$; *here* condition (iii) is needed.

Let (p_{k_s}) be arbitrary with $p_{k_s} \to A$ and fix j_0 with $f_{j_0} > 0$. The idea is to conclude from conditions (i) and (ii),

$$p_k = f_0 p_k + f_1 p_{k-1} + \cdots + f_{j_0} p_{k-j_0} + \cdots + f_k p_0,$$

that p_{k-j_0} is necessarily "large" (= close to A) if p_k is (since p_k is "nearly" a convex combination of $p_0, \ldots, p_k$ with the positive factor f_{j_0} at p_{k-j_0}[1].)

This will now be made precise. Fix an arbitrary $\delta > 0$ and choose

- k' such that $f_{k'+1} + \cdots < \delta$,

- k'' with $p_k \leq A + \delta$ for $k \geq k''$.

If then $k \geq k''' := \max\{k' + k'', j_0\}$ is arbitrary we have

$$R := f_{k'+1} p_{k-k'-1} + f_{k'+2} p_{k-k'-2} + \cdots + f_k p_0 \leq \delta$$

and also $p_k, p_{k-1}, \ldots, p_{k-k'} \leq A + \delta$. Consequently

$$\begin{aligned}
p_k &= f_0 p_k + \cdots + f_{j_0} p_{k-j_0} + \cdots + f_{k'} p_{k-k'} + R \\
&\leq f_{j_0} p_{k-j_0} + (f_0 + \cdots + f_{j_0-1} + \cdots + f_{j_0+1} + \cdots)(A + \delta) + \delta \\
&= f_{j_0} p_{k-j_0} + (1 - f_{j_0})A + 2\delta.
\end{aligned}$$

Now consider the particular case when $k = k_s$, and k_s satisfies $k_s \geq k'''$ as well as $p_{k_s} \geq A - \delta$. Then $A - \delta \leq f_{j_0} p_{k_s-j_0} + (1 - f_{j_0})A + 2\delta$, hence $p_{k_s-j_0} \geq A - (3/f_{j_0})\delta$. In this way we have shown that $A - (3/f_{j_0})\delta \leq p_{k_s j_0} \leq A + \delta$ for large s, hence $\lim_{s\to\infty} p_{k_s-j_0} = A$ and thus $j_0 \in \mathbb{M}$.

WHY DOES IT HELP? Let $\hat{k}$ and $\varepsilon > 0$ be arbitrary. Step 2 provides a k_0 with $k_0, k_0+1, \ldots, k_0 + \hat{k} \in \mathbb{M}$. Fix an arbitrary sequence (p_{k_s}) converging to A and consider the $\hat{k} + 1$ sequences $(p_{k_s-k_0})_s, (p_{k_s-k_0-1})_s, \ldots, (p_{k_s-k_0-\hat{k}})_s$. They all converge to A, and thus we may choose a sufficiently large s with $p_{k_s-k_0}, p_{k_s-1-k_0}, \ldots, p_{k_s-\hat{k}-k_0} \geq A - \varepsilon$.

We set $k := k_s - k_0$, and we will finish the proof of step 3 with the help of (6.2):

$$\begin{aligned}
1 &= r_0 p_k + \cdots + r_k p_0 \\
&\geq r_0 p_k + \cdots + r_{\hat{k}} p_{k-\hat{k}} \\
&\geq (r_0 + \cdots + r_{\hat{k}})(A - \varepsilon).
\end{aligned}$$

This is true for *every* $\hat{k}$ and *every* ε, and therefore $A \leq 1/(r_0 + r_1 + \cdots) = 1/\mu$ as claimed.

Step 4: To show that $B \geq 1/\mu$ we argue similarly. This time we may suppose that $\mu < \infty$, and now we consider as our set $\mathbb{M}$ the collection of all k_0 such that the $p_{k_s-k_0}$ tend to B with $s \to \infty$ whenever $p_{k_s} \to B$. The proof that $\{j \mid f_j > 0\}$ lies in $\mathbb{M}$ parallels the one above, and with step 2 we arrive at a k such that $p_k, p_{k-1}, \ldots, p_{k-\hat{k}} \leq B + \varepsilon$, where $\hat{k}$ and $\varepsilon > 0$ are prescribed arbitrarily.

The end of the proof is slightly different. We choose – for given $\varepsilon > 0$ – the number $\hat{k}$ such that $\sum_{j>\hat{k}} r_j < \varepsilon$. Then

---

[1] As a more evident example consider unknown $x, y \in [0, 1]$ such that $\lambda x + (1 - \lambda)y$ is close to one. If it is known that λ is "not too small" one may conclude that x itself is close to one as well.

ract

$$
\begin{aligned}
1 &= r_0 p_k + \cdots + r_k p_0 \\
&\leq r_0 p_k + \cdots + r_{\hat{k}} p_{k-\hat{k}} + \varepsilon \\
&\leq (r_0 + \cdots + r_{\hat{k}})(B + \varepsilon) + \varepsilon \\
&\leq (\mu - \varepsilon)(B + \varepsilon) + \varepsilon,
\end{aligned}
$$

and this can happen only if $B \geq 1/\mu$. $\square$

Remarks:

1. The proof has an essential drawback in that it is *not constructive*: we know that $p_k \to 1/\mu$, but it is hardly possible from an inspection of the preceding argument to find a k_0 for a given $\varepsilon > 0$ such that $|p_k - 1/\mu| < \varepsilon$ for $k \geq k_0$. Only with such an information at hand, however, one could decide in concrete applications when a calculation should terminate in order to have the result with prescribed accuracy.

In particular this would be desirable in the case when we are going to apply our lemma to prove the convergence of an irreducible and aperiodic Markov chain to its equilibrium. Part II of this book will contain techniques which provide concrete bounds for this situation. Lemma 6.1 is contained here since it is of independent interest. Its consequences are very far-reaching, a first application can be found in the following remark.

2. By the equations in lemma 6.1(i) the sequences p and f determine each other. Thus it would suffice to start with nonnegative $f_1, f_2, \ldots$ which satisfy (ii) and (iii) and then to *define* $p_1, p_2, \ldots$ recursively by (i).

This can be used to derive *a first probabilistic interpretation of the lemma*. Let a probability measure $\tilde{\mathbb{P}}$ on $\mathbb{N}$ be given which satisfies $\gcd\{k \mid \tilde{\mathbb{P}}(\{k\}) > 0\} = 1$. The $f_k := \tilde{\mathbb{P}}(\{k\})$ satisfy 6.1(ii) and (iii), but what is the meaning of the associated p_k?

Imagine a random walk on $\{0, 1, 2, \ldots\}$ which starts at 0 and, whenever the position in the k'th step is n, passes next to $n + m$, where $m \in \mathbb{N}$ is chosen at random with probability f_m. This, of course, corresponds to the Markov chain with state space $\mathbb{N}_0$ and the doubly infinite transition matrix

$$
\begin{pmatrix}
0 & f_1 & f_2 & f_3 & \cdots \\
0 & 0 & f_1 & f_2 & \cdots \\
0 & 0 & 0 & f_1 & \cdots \\
\vdots & \vdots & \vdots & \vdots & \vdots
\end{pmatrix} .
$$

Choose as a probabilistic model a probability space $(\Omega, \mathcal{A}, \mathbb{P})$ and random variables $X_0 = 0, X_1, \ldots : \Omega \longrightarrow \mathbb{N}_0$ (cf. exercise 6.4). It might be interesting to know whether or not a walk visits a particular state n. The probability that this happens is

$$
\mathbb{P}\{\omega \mid \text{there exists } k \text{ such that } X_k(\omega) = n\},
$$

a number which will be called ρ_n for the moment.

To calculate the ρ_n we condition on the position *just before* going to n. More precisely, put

$$
\begin{aligned}
E_n &:= \{\omega \mid \text{there exists } k > 0 \text{ such that } X_k(\omega) = n\} \\
E_{n,m} &:= \{\omega \mid \text{there exists } k > 0 \text{ such that } X_{k-1}(\omega) = m, X_k(\omega) = n\}
\end{aligned}
$$

for $0 \leq m < n$. Then E_n is the disjoint union of the $E_{m,0}, \ldots, E_{n,n-1}$, and the probability of $E_{n,m}$ is $\mathbb{P}(E_m)$ times the probability to pass in one step from m to n, i.e., $\rho_m f_{n-m}$. Therefore the ρ_n satisfy the set of equations in lemma 6.1(i):

$$
\begin{aligned}
\rho_1 &= f_1, \\
\rho_2 &= f_2 + \rho_1 f_1, \\
&\vdots \quad \vdots \quad \vdots \\
\rho_k &= f_k + \rho_1 f_{k-1} + \cdots + \rho_{k-1} f_1, \\
&\vdots \quad \vdots \quad \vdots
\end{aligned}
$$

and thus – by uniqueness – they are precisely the p_k of the preceding calculations. In particular they converge to $1/\mu$, where $\mu = \sum n f_n$ is the expectation of the stepsize of our random walk (this assertion is called the *renewal theorem*).

> As an illustration consider a game where a player starts at zero, and the number of units to proceed on $\mathbb{N}_0$ is determined by throwing two dice. The expected value of the step size is 7, and thus the probability that the walk touches a particular "large" n is roughly $1/7$. As already mentioned, the proof of lemma 6.1 does not provide information what "roughly" and "large" here mean precisely. We will return to this question later, concrete estimates will be obtained as a by-product of the results in part II (see the end of chapter 10).

We must resist the temptation to proceed further along these lines. What we just have developed are the very beginnings of *discrete renewal theory* (see [33] for a more extensive introduction to this field). The name stems from the fact that it is possible to model simple renewal situations in just this way.

> Imagine that one deals with machines/bulbs/transistors/… each working for some time of which only the probability distribution is known (lifetime n days with probability f_n). One starts with a new machine/bulb/transistor, and it is replaced as soon it is defect. Then one will have to renew an item on a particular day in the future with a probability which is roughly 1 divided by the expectation of the lifetime, at least if the assumptions of lemma 6.1 are met.

It is this connection why our lemma is sometimes called the *fundamental lemma of discrete renewal theory* (or *the discrete renewal theorem*).

Exercises

6.1: The main result of the present chapter states that under three assumptions a certain sequence converges. Is each of these assumptions essential? (Try to find suitable counterexamples).

6.2: For a fixed real sequence (a_k) and a real A put

$$
\mathbb{M} := \{ k_0 \mid \lim_s a_{k_s - k_0} = A \text{ whenever } \lim_s a_{k_s} = A \}.
$$

a) Prove that $\mathbb{M} + \mathbb{M} \subset \mathbb{M}$.

b) Give an example where $\mathbb{M} = \{0, 2, 4, \ldots\}$.

6.3: Let $\mathbb{M}$ be a nonempty subset of the positive integers such that $\mathbb{M} + \mathbb{M} \subset \mathbb{M}$. Prove that there are integers $d > 0$ and k_0 such that $\mathbb{M}$ contains all kd for $k \geq k_0$.

6.4: Prove the existence of $\mathbb{N}_0$-valued random variables $X_1, \ldots$ on $[0, 1]$ with the Borel-Lebesgue measure which model the random walk given by the renewal probabilities $f_1, f_2, \ldots$. (Hint: cf. our second proof of theorem 1.4.)

6.5: We resume the application to renewal theory, this time we start with a probability vector $f_0, f_1, \ldots$. This means – if $f_0 > 0$ – that we allow a positive probability that the walk pauses. Try to find an expression for $\lim \rho_k$ in this more general situation.

7   Irreducible Markov chains

The results of the preceding chapter will enable us to complete our picture of finite Markov chains. We will proceed as follows:

- Investigation of the behaviour of the $p_{ii}^{(k)}$ for "large" k in the case of recurrent states i.

- Proof of the existence of an equilibrium distribution for irreducible aperiodic chains.

- Discussion of irreducible chains with an arbitrary period.

- Calculation of the first passage time matrix.

Long-time behaviour of recurrent and positive recurrent states

We know that all states in an irreducible chain are recurrent (proposition 4.10) and that $\sum_k p_{ii}^{(k)} = \infty$ for such i (proposition 4.9). In contrast to the corresponding characterization for transient states (where one might conclude from $\sum_k p_{ii}^{(k)} < \infty$ that $p_{ii}^{(k)} \to 0$) this does not contain much information on the limit behaviour of the $p_{ii}^{(k)}$. This is provided by part (i) of the following proposition:

Proposition 7.1 *Let i be an arbitrary recurrent state.*

  (i) *Suppose that i has period d. Then $p_{ii}^{(k')} = 0$ whenever k' is not of the form kd, and $\lim_{k \to \infty} p_{ii}^{(kd)} = d/\mu_{ii}$; (recall that $\mu_{ii} = \sum_k k f_{ii}^{(k)}$ denotes the expectation[1] of the number of steps to come from i to i).*

  (ii) *Let j be another state such that $i \leftrightarrow j$. If i is positive recurrent, then so is j.*

Proof. (i) Suppose first that $d = 1$. Put $p_k := p_{ii}^{(k)}$, $f_k := f_{ii}^{(k)}$. Since i is recurrent, we know that $\sum f_k = 1$, and we have already shown that the p_k, f_k satisfy the recurrence relation 6.1(i) (see the calculations after proposition 4.8). Thus it only remains to show that condition (iii) of lemma 6.1 holds since then the assertion to be proved is true by this lemma.

Put $G_p := \gcd\{k \mid p_k \neq 0\}$ and $G_f := \gcd\{k \mid f_k \neq 0\}$. We have to show that $G_p = 1$ (our assumption) yields $G_f = 1$. To this end let m be a number such that m divides all k such that $f_k > 0$. We claim that m divides all k with $p_k > 0$ as well which would imply $G_f \leq G_p$ and thus $G_f = 1$.

We proceed by induction. The case $k = 1$ is clear since $p_1 = f_1$. Now let k be any number, we assume that $m|k'$ whenever k' is such that $k' < k$ and $p_{k'} > 0$ and also that $p_k > 0$. We know that

---

[1] Cf. (4.4) in chapter 4; we will prove in the next proposition that the μ_{ii} are finite so that they really can be thought of as expectations.

$$p_k = f_k + p_1 f_{k-1} + \cdots + p_{k-1} f_1,$$

and we consider two possibilities. Either $f_k > 0$ in which case $m|k$ trivially holds. Or $f_k = 0$, but then there has to exist a $k' < k$ with $p_{k'} > 0$, $f_{k-k'} > 0$. m thus divides both k' and $k - k'$, hence $m|k$ and the proof of (i) in the case $d = 1$ is complete.

Now we consider the case of an arbitrary period d. That only the numbers $p_{ii}^{(kd)}$ among the $p_{ii}^{(1)}, p_{ii}^{(2)}, p_{ii}^{(3)}, \ldots$ can have non-zero values follows from the definition of d. Their convergence is proved by reduction to the case $d = 1$ as follows: pass from the original chain (with transition matrix P) to the chain with transition matrix $\tilde{P} := P^d$. Whereas in the original chain i had period d it now is aperiodic; this is nothing but the elementary relation $\gcd d\mathbb{M} = d \gcd \mathbb{M}$. Thus, by the first part of the proof, the $\tilde{p}_{ii}^{(k)}$ converge to $1/\tilde{\mu}_{ii}$, where the ~-notation is used to remind us that we are now dealing with the chain defined by the transition matrix $\tilde{P}$. However, the relation between the numbers with or without the ~ is simple, namely

$$\tilde{p}_{ii}^{(k)} = p_{ii}^{(kd)},$$
$$\tilde{f}_{ii}^{(k)} = f_{ii}^{(kd)}.$$

Therefore $\tilde{\mu}_{ii} = \sum k \tilde{f}_{ii}^{(k)} = \mu_{ii}/d$, and this completes the proof of (i).

(ii) Choose k', k'' as in the proof of proposition 4.10: the numbers $p_{ij}^{(k')}, p_{ji}^{(k'')}$ are positive, and

$$p_{jj}^{(k+k'+k)} \geq p_{ii}^{(k)} p_{ij}^{(k')} p_{ji}^{(k'')}$$

for all k. Since the $p_{ii}^{(kd)}$ tend to d/μ_{ii} and this number is strictly positive by assumption, it follows that $p_{jj}^{(k)}$ is larger than $d p_{ij}^{(k')} p_{ji}^{(k'')}/2\mu_{ii}$ for infinitely many k, i.e., these numbers do *not* converge to zero. But we know from proposition 4.6 and proposition 4.10 that j is recurrent and also has period d. Hence the $p_{jj}^{(kd)}$ converge to d/μ_{jj} by (i), and it follows that $\mu_{jj} < \infty$.

$\square$

Proposition 7.2 *All states i which lie in some minimal closed set C_ρ are positive recurrent. In particular all states in an irreducible chain have this property.*[2]

Proof. By passing to a minimal closed set we may assume that i communicates with all other states. They all are recurrent by proposition 4.10, and by the preceding proposition there are only two possibilities: all are null recurrent or all are positive recurrent.

Suppose that $\mu_{ii} = \infty$ would hold for all i. This would imply not only $p_{ii}^{(k)} \to 0$ (by proposition 7.1) but also $p_{ji}^{(k)} \to 0$ for all i, j (see the proof of proposition 5.1). As a consequence the numbers $\sum_j p_{ij}^{(k)}$ would tend to zero with $k \to \infty$ since there are only finitely many j. But this is surely a contradiction, since $\sum_j p_{ij}^{(k)} = 1$.

$\square$

---

[2] It is stressed again here that we deal with finite chains only; the result does *not* hold in the general case.

The equilibrium distribution

Chains which are irreducible and where all states are aperiodic will play an important role, they will be called *irreducible and aperiodic* chains for short[3]; occasionally we will also speak of an irreducible and aperiodic transition matrix P if the associated chain has this property.

First we prove a simple characterization:

Lemma 7.3

(i) *A chain is irreducible iff there exists a k such that all entries of the matrix $P + P^2 + \cdots + P^k$ are strictly positive.*

(ii) *It is irreducible and aperiodic iff P^k has strictly positive components for a suitable integer k.*

Proof. (i) Under the assumption of irreducibility we have $i \to j$ for all i, j, and therefore we find $k(i,j)$ with $p_{ij}^{(k(i,j))} > 0$. Then $P + P^2 + \cdots + P^k$ will be strictly positive as soon as k majorizes all $k(i,j)$. Conversely, if the (i,j)-component of $P + P^2 + \cdots + P^k$ is greater than zero, there must be a $k' \le k$ with $p_{ij}^{(k')} > 0$ so that $i \to j$.

(ii) Let all components of P^k be strictly positive (which we will abbreviate by $P^k > 0$). It is clear that then the chain is irreducible.

From $P^k > 0$ it follows that $P^{k+1} > 0$ since P is a stochastic matrix. Conclusion: $p_{ii}^{(k')} > 0$ for $k' \ge k$ so that all i are aperiodic.

Conversely, let the chain be irreducible and aperiodic. Since $\lim_{k \to \infty} p_{ii}^{(k)} = 1/\mu_{ii} > 0$ by proposition 7.1 and proposition 7.2 we may choose a k' such that $p_{ii}^{(k')} > 0$ for all i. Then – as can easily be calculated – $P^{k'+k''} > 0$ for any k'' which majorizes all $k(i,j)$ (which have the same meaning as in the first part of this proof). $\square$

We are now able to understand the phenomenon described at the end of chapter 3:

Theorem 7.4 *Consider a Markov chain with state space $\{1, \ldots, N\}$ which is assumed to be irreducible and aperiodic.*

(i) *The powers P^k of the transition matrix P converge componentwise to a stochastic matrix W in which all rows are equal. If we denote a typical row by $\pi^{\mathsf{T}} = (\pi_1, \ldots, \pi_N)$, then we have $\pi_i > 0$ for all i and $\sum_i \pi_i = 1$.*

(ii) *π is the unique vector such that $\pi^{\mathsf{T}} P = \pi^{\mathsf{T}}$ and $\sum_i \pi_i = 1$.*

(iii) *For every i, the i'th component of π is just $1/\mu_{ii}$.*

This unique π is called the equilibrium distribution *associated with the chain[4].*

Proof. (i) All i are positive recurrent so that $p_{ii}^{(k)} \to 1/\mu_{ii} > 0$. From this and the identity

$$p_{ji}^{(k)} = f_{ji}^{(k)} + f_{ji}^{(k-1)} p_{ii}^{(1)} + \cdots + f_{ji}^{(1)} p_{ii}^{(k-1)}$$

it follows immediately that the $p_{ji}^{(k)}$ also converge and that the limit is f_{ji}^* / μ_{ii}.

---

[3] Sometimes the word *"ergodic"* is used instead of "irreducible and aperiodic".

[4] Cf. the following remark 1, there it is explained why this notion is appropriate.

We claim that $f_{ji}^* = 1$. For the proof of this fact fix the smallest k' with $p_{ij}^{(k')} > 0$. Then, since $1 - f_{ji}^*$ is the probability that a walk starting at j will never arrive at i, the product $p_{ij}^{(k')}(1 - f_{ji}^*)$ is the probability that a walk which starts at i is at j in step number k' and is later never seen at its starting position[5]. In particular $p_{ij}^{(k')}(1 - f_{ji}^*) \leq 1 - f_{ii}^*$, which is the probability of no return to i for a walk starting there. But *this* probability is zero since i is recurrent, hence $f_{ji}^* = 1$.

So far we have shown that $p_{ji}^{(k)} \to 1/\mu_{ii} =: \pi_i$ with $k \to \infty$. That the π_i sum up to one follows from the fact that all matrices P^k are stochastic and that this property is preserved under coordinate-wise limits. The proof of (i) is now complete, and the assertion (iii) was established as a by-product.

(ii) If π is as in (i), then $\pi^\top P = \pi^\top$ is equivalent with $WP = W$, and this identity follows easily from the continuity of matrix multiplication:

$$WP = (\lim P^k)P = \lim P^{k+1} = W.$$

Conversely, let $\tilde{\pi}$ be given such that $\tilde{\pi}^\top P = \tilde{\pi}^\top$ and $\sum \tilde{\pi}_i = 1$. $\tilde{\pi}^\top$ is also an eigenvector with associated eigenvalue 1 for all P^k and thus – by continuity – of W. We consequently get

$$\tilde{\pi}^\top = \tilde{\pi}^\top W = \pi^\top,$$

where the last equality follows from $\sum \tilde{\pi}_i = 1$. $\qquad\square$

Equilibrium distributions will be very important in the sequel. Some additional **remarks** concerning the theorem are in order:

1. The *convergence* of the numbers $p_{ij}^{(k)}$ means that for large k it is more and more difficult to identify k given the $p_{ij}^{(k)}$: the chain "forgets" the length of its history. The fact that *the limit does not depend on i* implies that – for large k – it is hardly possible to assert where the starting position was, even if all $p_{ij}^{(k)}$ have been estimated rather carefully: the chain "forgets" the initial position.

As we have already mentioned in chapter 3, an interesting phenomenon occurs if we take $\pi^\top$ as the initial distribution: the chain forgets the length of its history completely and immediately. *This* explains why π is called the equilibrium distribution.

2. Whereas there does not seem not to exist any simple way to determine the μ_{ii} directly from the definition, the eigenvector equation $\pi^\top P = \pi^\top$ (together with the condition $\sum \pi_i = 1$) is rather simple to solve, in particular since one can guarantee that there exists precisely one solution. And once the π_i are calculated, the $\mu_{ii} = 1/\pi_i$ are also known.

3. The theorem depends essentially on proposition 7.1 which in turn was a corollary to the analytical lemma 6.1. Since no explicit information on the rate of convergence was provided there it has to be admitted that theorem 7.4 in the present form will be of little practical interest. Explicit bounds, however, would be interesting in view of the applications we have in mind in part III. Part II of this book will be mainly devoted to the development of techniques by which it is possible to remedy this drawback.

---

[5] Once more a special case of the strong Markov property is used here: a walk starting at j behaves like one which after k' steps is restarted there. Cf. the discussion leading to proposition 4.8 and – for a more thorough treatment – chapter 12.

The reader is invited to have a look at proposition 10.5 or at proposition 10.8. They provide – together with concrete bounds – two independent possibilities to prove convergence directly.

Here are some **examples**:

1. Suppose that our stochastic matrix P is not only irreducible and aperiodic but also doubly stochastic: $\sum_i p_{ij} = 1$ for all j; this is, e.g., the case if P is symmetric. Then $\pi^\top := (1/N, \dots, 1/N)$ surely *is* a solution of $\pi^\top P = \pi^\top$ with $\sum_i \pi_i = 1$, and since this is unique it follows that the equilibrium distribution is the uniform distribution. Proposition 7.4(iii) here means that it takes a walk on the average N steps between a visit of and a return to a particular state i.

2. Let the transition matrix P be such that all rows coincide (in order to have an irreducible and aperiodic chain we assume that all entries are strictly positive). Then all P^k are equal so that $W = P$ in this case. This simple example also shows that *every* probability distribution on a finite set which is strictly positive everywhere can occur as an equilibrium distribution of an irreducible and aperiodic chain.

3. Consider, with positive a, b such that $2a + b = 1$, the following reflecting random walk on $\{1, 2, 3, 4\}$:

$$P := \begin{pmatrix} 2a & b & 0 & 0 \\ a & b & a & 0 \\ 0 & a & b & a \\ 0 & 0 & b & 2a \end{pmatrix}.$$

The equilibrium distribution is easily calculated:

$$\pi^\top = \left(\frac{a}{1+b}, \frac{b}{1+b}, \frac{b}{1+b}, \frac{a}{1+b} \right).$$

This is the uniform distribution precisely if $a = b\, (= 1/3)$.

Irreducible chains with an arbitrary period

We consider now an irreducible chain with transition matrix P which is not necessarily aperiodic. By proposition 4.6 we know that all states have the same period, say d (sometimes such a chain is called *an irreducible chain with period d*). Now we pass – as in the proof of proposition 7.1 – to the chain with transition matrix P^d. This will *not* be irreducible in general. However, on the minimal closed sets G the chain behaves like an irreducible aperiodic chain, and this idea gives rise to a complete description:

Theorem 7.5 *Let a Markov chain on a finite state space S be given by an irreducible stochastic matrix and denote by d its period.*

    (i) *There is a partition of S into disjoint non-empty subsets $G_0, \dots, G_{d-1}$ such that:*

        • *The G_δ are minimal closed sets with respect to P^d; the restriction of the chain associated with P^d to each G_δ is irreducible and aperiodic.*

- *For the original chain there are transitions only from G_δ to $G_{\delta+1}$ (with the convention $G_d := G_0$). For the transition probabilities this means that*

$$\sum_{j \in G_{\delta+1}} p_{ij} = 1$$

for every $i \in G_\delta$.

(ii) *The Cesàro limit W of the matrix sequence $P, P^2, \ldots$ exists:*

$$W = \lim_{k \to \infty} (P + \cdots + P^k)/k$$

(which, as usual, is meant to hold for every component). W is a stochastic matrix in which all rows are identical. This row vector, which we will denote by $\pi^\top$, is the unique solution of the eigenvector equation $\pi^\top P = \pi^\top$ such that $\sum_i \pi_i = 1$. As in theorem 7.4 also here $\pi_i = 1/\mu_{ii}$ holds, and again $\pi^\top$ is called the equilibrium distribution *associated with the chain.*

Proof. (i) Consider the chain with transition matrix P^d and denote by $\mathcal{G}$ the collection of its minimal closed subsets. Since all states of the original chain are recurrent, the numbers $p_{ii}^{(dk)}$ converge with $k \to \infty$ to a non-zero value (7.1(i)) und thus the i are *not* transient with respect to the new chain (by 4.9). This shows that the state space S is the disjoint union of the $G \in \mathcal{G}$.

Now fix any element of $\mathcal{G}$ and call it G_0. Define subsets G_δ of S for $\delta = 1, \ldots, d-1$ by

$$G_\delta := \{j \mid \text{there are } i \in G_0 \text{ and } k \in \{\delta, \delta + d, \delta + 2d, \ldots\} \text{ such that } p_{ij}^{(k)} > 0\}.$$

(Note that it would be admissible to use this definition also with $\delta = 0$: the set G_0 which is defined in this way is precisely the set G_0 we started with. The reason is that all $i \in G_0$ communicate with respect to the new chain. However, it is not clear a priori that the G_δ lie in $\mathcal{G}$.)

Claim 1: S is the disjoint union of the $G_0, \ldots, G_{d-1}$.

Proof: It is clear that $S = G_0 \cup \cdots \cup G_{d-1}$ since $i \to j$ for all i, j. Now fix two different δ', δ'' and arbitrary $j' \in G_{\delta'}, j'' \in G_{\delta''}$. There are – for suitable $i', i'' \in G_0$ and integers $k', k'', k, \tilde{k}$ – transitions with positive probabilities

$$\begin{array}{lllllll}
\text{from} & i' & \text{to} & j' & \text{in} & \delta' + k'd & \text{steps,} \\
\text{from} & i'' & \text{to} & j'' & \text{in} & \delta'' + k''d & \text{steps,} \\
\text{from} & i' & \text{to} & i'' & \text{in} & kd & \text{steps,} \\
\text{from} & j' & \text{to} & i' & \text{in} & \tilde{k} & \text{steps.}
\end{array}$$

This follows from the definition of the G_δ and the fact that the original (resp. the new) chain acts irreducibly on S (resp. on G_0).

Now suppose that $j' = j''$ would hold. Then the preceding observations would give rise to two possible ways from i' to i' with positive probability, namely one of length $k_1 := \delta' + k'd + \tilde{k}$ and another of length $k_2 := kd + \delta'' + k''d + \tilde{k}$. Since i' has period d, both k_1 and k_2 are divisible by d. In particular we would have $d|\delta' - \delta''$, a contradiction. This proves that $G_{\delta'} \cap G_{\delta''} = \emptyset$.

Claim 2: For any δ there are only transitions (of the original chain) from G_δ to $G_{\delta+1}$ (with $G_d := G_0$).

Proof: This is proved similarly: the existence of other transitions would violate the d-periodicity of all states.

Claim 3: $\mathcal{G}$ is just the collection $G_0, \ldots, G_{d-1}$.

Proof: We claim that – with respect to P^d – each G_δ is invariant and that each two j', j'' in G_δ communicate. The invariance is clear from the definition. For the second part choose i'' in G_0 and $k'', k \geq 0$ such that the probabilities for a transition

$$
\begin{array}{ccccccc}
\text{from} & i'' & \text{to} & j'' & \text{in} & \delta + k''d & \text{steps,} \\
\text{and from} & j' & \text{to} & i'' & \text{in} & d - \delta + dk & \text{steps}
\end{array}
$$

are positive (for the existence of k one has to combine claim 2 with the fact that $j' \to i''$). This shows that one may pass from j' to j'' under P^d.

(ii) Denote the entries of $(P + \cdots + P^k)/k$ by $q_{ij}^{(k)}$. We consider first the case where i, j lie in the same G_δ. Since P^d acts as an irreducible and aperiodic chain on G_δ we know from proposition 7.4 that the $p_{ij}^{(dk')}$ tend to d/μ_{jj} with $k' \to \infty$. Since the $p_{ij}^{(k)}$ vanish for $k \neq dk'$ it follows that $q_{ij}^{(k)} \to 1/\mu_{jj}$.

Now fix δ and consider $i \in G_\delta, j \in G_{\delta+r}$ for some $1 \leq r < d$ (with $\delta + r := (\delta + r)$ mod d). Then $p_{ij}^{(r+dk)} = \sum_{j' \in G_{\delta+r}} p_{ij'}^{(r)} p_{j'j}^{(dk)}$, and the other $p_{ij}^{(k)}$ vanish. Therefore $q_{ij}^{(k)} \to 1/\mu_{jj}$ since $q_{j'j}^{(k)} \to 1/\mu_{jj}$ and the $p_{ij'}^{(r)}$ sum up to one.

The rest of the proof parallels that of theorem 7.4: $(P + \cdots + P^k)/k \to W$ implies

$$
WP = \lim_k (P^2 + \cdots + P^{k+1})/k = \lim_k (P + P^2 + \cdots + P^{k+1})/(k+1) = W,
$$

and this means that $\pi^\top P = \pi^\top$ if we set $\pi_i := 1/\mu_{ii}$. $\sum \pi_i = 1$ is a consequence of the fact that W is stochastic as the limit of stochastic matrices.
Conversely: $\tilde{\pi}^\top P = \tilde{\pi}^\top$ yields $\tilde{\pi}^\top (P + \cdots + P^k)/k = \tilde{\pi}^\top$ and thus $\tilde{\pi}^\top W = \tilde{\pi}^\top$. Hence $\tilde{\pi}^\top = \pi^\top$ if the components of $\tilde{\pi}^\top$ sum up to one. $\qquad\square$

Most of the **remarks** following the proof of theorem 7.4 could be repeated here. Note, however, that an irreducible chain with $d > 1$ does *not* completely forget the length of its history: if the walk starts at a state in G_0 and you find it for some – arbitrarily large – k in G_3 you know that k mod $d = 3$. But this is essentially all that can be said.
A similar remark applies to a guess of the starting position given the position in the k'th step.

Note that the theorem in particular implies that the transition matrix P has the form

$$
\begin{pmatrix}
0 & * & 0 & \cdots & 0 \\
0 & 0 & * & \cdots & 0 \\
\vdots & \vdots & \vdots & & \vdots \\
0 & 0 & 0 & \cdots & * \\
* & 0 & 0 & \cdots & 0
\end{pmatrix},
$$

if $S = \{1, \ldots, N\}$ and the states are labelled such that the first ones belong to G_0, the next ones to G_1 etc.; here the 0 denote matrices with zero entries and the $*$ stand for stochastic – not necessarily square – matrices.

The first passage time matrix

Now we will restrict ourselves again to the case of irreducible and aperiodic chains. Similarly to the end of chapter 5 where we used linear algebra to calculate certain numbers in connection with transient states we now want to apply the same techniques to determine the expectations of running times until return. The μ_{ij} as a measure of the average number of steps to pass from i to j have been introduced already in (4.4) of chapter 4. These numbers, however, have played a role so far only in the case $i = j$. Under this condition we know that they are finite by proposition 7.2, we already have shown that $f_{ij}^{*} = 1$ for all i, j (see the proof of 7.4(i)), but this is not sufficient to guarantee that all μ_{ij} are finite. A little trick is necessary:

Lemma 7.6 *For an irreducible and aperiodic chain all μ_{ij} are finite (so that these numbers can be considered as the expectation of the number of steps to come from i to j). The matrix $M = (\mu_{ij})_{ij}$ is called the* first passage time matrix *of our chain.*

Proof. To be specific we assume that the state space is $\{1, \ldots, N\}$, and we will show that $\mu_{ij} < \infty$ for, e.g., $j = 1$ and all $i \neq 1$. The trick is to declare $\{1\}$ as a minimal closed set by passing from the original transition matrix $P = (p_{ij})$ to

$$
\tilde{P} := \begin{pmatrix}
1 & 0 & 0 & \cdots & 0 \\
p_{21} & p_{22} & p_{23} & \cdots & p_{2N} \\
p_{31} & p_{32} & p_{33} & \cdots & p_{3N} \\
\vdots & \vdots & \vdots & & \vdots \\
p_{N1} & p_{N2} & p_{N3} & \cdots & p_{NN}
\end{pmatrix}.
$$

For the chain defined by $\tilde{P}$ the states $2, \ldots, N$ are transient: the original chain is irreducible, and consequently $j \to 1$ for all j. Therefore the number of steps to come from state $i \in \{2, \ldots, N\}$ to 1 (subject to P) is precisely the number of steps which is needed for the transient state i to be absorbed in the collection of minimal closed sets – which is just $\{1\}$ – if we argue with respect to $\tilde{P}$.
In this way the assertion of the lemma is reduced to proposition 5.1(iii). $\square$

By the next theorem the matrix M can be determined by simple matrix calculations. The following notation will be used: if $R = (r_{ij})_{ij}$ is any square matrix, then R_{diag} is defined to be the matrix $(r_{ij}\delta_{ij})_{ij}$ (with $\delta_{ij} =$ the Kronecker delta), i.e., R_{diag} has the same diagonal as R, but all other entries vanish; and the symbol E will stand for an $N \times N$-matrix where all entries are 1.

Theorem 7.7 *Consider an irreducible aperiodic chain given by a stochastic matrix P. In proposition 7.4 we have shown that $W = \lim P^{k}$ exists, and we already know that the rows of W are the normalized solutions of an eigenvalue equation.*

    (i)   *The matrix $Id - (P - W)$ is invertible, its inverse will be called Z.*

    (ii) *The first passage time matrix can be calculated as*

$$
M = (Id - Z + E Z_{\mathrm{diag}}) M_{\mathrm{diag}}.
$$

Explicitly written this means that $\mu_{ij} = (\delta_{ij} - z_{ij} + z_{jj})/\pi_{j}$.

Note: Part (ii) seems to be circular since M appears on both sides of the equation. However, on the right hand side only M_{diag} is of importance, and this matrix is known since the elements μ_{ii} on the diagonal are the inverses of the components π_i of the equilibrium distribution.

Proof. (i) We have $(Id - Q)(Id + Q + Q^2 + \cdots + Q^k) = Id - Q^{k+1}$ for any square matrix Q. Suppose that we know for some reason that $Q^k \to 0$. Then the determinant of $Id - Q^{k+1}$, being a continuous function of the entries, will tend to 1, the determinant of Id. In particular this determinant will be nonzero for large k. But then $Id - Q$ and also $Id + Q + \cdots + Q^k$ have a nonzero determinant as well by the determinant product formula, and thus both of them are invertible. If we now let k tend to infinity in the equation $Id + Q + \cdots + Q^k = (Id - Q)^{-1}(Id - Q^{k+1})$, it follows from $Q^k \to 0$ that $Id + Q + Q^2 + \cdots$ exists and is the inverse of $Id - Q$. (Note that the argument is a little bit more involved than that in the proof of lemma 5.2: there the existence of $Id + Q + \cdots$ could be assumed.)

To prove the assertion (i) we will apply the preceding argument with $Q = P - W$. From $W = \lim P^k$ it follows at once – as in the proof of theorem 7.4(ii) above – that $WP = W = W^2 = PW$. And from this we get by induction that $(P-W)^k = P^k - W \to 0$, and hence it is justified to use the preceding argument.

(ii) First we claim that the first passage time matrix M is the unique matrix R which coincides on the diagonal with M and satisfies the matrix equation $R = P(R - R_{\text{diag}}) + E$.

The matrix M is a solution.

The condition concerning the diagonal is trivially satisfied, and the matrix equation means

$$\mu_{ij} = \sum_{l \neq i} p_{il}\mu_{lj} + 1 = \sum_{l \neq i} p_{il}\mu_{lj} + \sum_{l} p_{il} = p_{ij} + \sum_{l \neq i} p_{il}(\mu_{lj} + 1).$$

That this holds follows by a calculation involving *conditional expectations*.

For any random variable Y on a probability space $(\Omega, \mathcal{A}, \mathbb{P})$ whose expectation $\mathbb{E}(Y)$ exists one has $\mathbb{E}(Y) = \sum_{\sigma} \mathbb{P}(B_\sigma)\mathbb{E}(Y_{B_\sigma})$, whenever $B_1, \ldots, B_s$ is a disjoint partition of Ω; here Y_{B_σ} means the restriction of the random variable Y to the probability space $(B_\sigma, \mathcal{A}|_{B_\sigma}, \mathbb{P}/\mathbb{P}(B_\sigma))$.

We will apply this fact as follows. Fix i and j and consider a Markov process $X_0, X_1, \ldots$, defined on some $(\Omega, \mathcal{A}, \mathbb{P})$, with transition probabilities given by P, and $X_0 = i$. Then, with $Y := \inf\{k \mid X_k = j\}$ we have $\mu_{ij} = \mathbb{E}(Y)$ by definition. Ω is the disjoint union of the $B_l := \{X_1 = l\}$, and with the notation of the preceding paragraph we know that $\mathbb{E}(Y_{B_j}) = 1$ and $\mathbb{E}(Y_{B_l}) = \mu_{lj} + 1$ for $l \neq j$ ("+1", since we must not forget the first step from i to l in our calculation; it should be stressed that here again it is crucial that we deal with *homogeneous* chains). It now suffices to note that $\mathbb{P}(B_l) = p_{il}$.

M is the only solution.

Let $\tilde{M}$ be another solution, we put $R := \tilde{M} - M$. The diagonal of R vanishes, and also $R = P(R - R_{\text{diag}}) = PR$ holds. Let z be any column of R. The matrix equation means that $Pz = z$. Thus $P^k z = z$ for all k and consequently $Wz = z$. But all rows in W equal the same vector $\pi^\top$, hence all components of z must be identical (with the value $\pi^\top z$). Since R has a vanishing diagonal, at least one of these components is zero. Thus z is zero and we have shown that $R = 0$.

With this preparation in mind it remains to show that

$$R := (Id - Z + EZ_{\text{diag}})M_{\text{diag}}$$

satisfies $R_{\text{diag}} = M_{\text{diag}}$ as well as $R = P(R - R_{\text{diag}}) + E$. The first condition can be verified by simple calculation, it is a consequence of the fact that every entry on the diagonal of $Id - Z + EZ_{\text{diag}}$ is one.

The matrix equation needs further preparations; we claim that

$$
\begin{aligned}
Id - Z &= W - PZ, & (7.1)\\
WM_{\text{diag}} &= E, & (7.2)\\
PEZ_{\text{diag}} &= EZ_{\text{diag}}. & (7.3)
\end{aligned}
$$

The *first equation* follows from

$$
Z = Id + \sum_{k=1}^{\infty}(P - W)^k = Id + \sum_{k=1}^{\infty}(P^k - W)
$$

since this series expansion together with $W = PW$ show that

$$
Id - Z = \sum_{k=1}^{\infty}(W - P^k) = W - PZ.
$$

The *second one* is clear by theorem 7.4(iii), and for the proof of the *third one* one only has to use the fact that P is stochastic.

Here is the end of the proof:

$$
\begin{aligned}
P(R - R_{\text{diag}}) + E &= P(-Z + EZ_{\text{diag}})M_{\text{diag}} + E \\
&= (-PZ + EZ_{\text{diag}})M_{\text{diag}} + E \\
&= R + (-Id + Z - PZ)M_{\text{diag}} + E \\
&= R - WM_{\text{diag}} + E \\
&= R;
\end{aligned}
$$

the first transformation is justified since $R_{\text{diag}} = M_{\text{diag}}$, the second is clear by (7.3), in the next one only the definition of R is reproduced, and finally (7.1) and (7.2) are used.
$\qquad\qquad\qquad\qquad\qquad\qquad\qquad\qquad\qquad\qquad\qquad\qquad\qquad\qquad\qquad\quad\Box$

We close this chapter with some **examples:**

1) Let P be such that $p_{ij} = 1/N$ for all i, j; here as usual N denotes the cardinality of the state space. This example has the advantage that we know the result beforehand since the number of steps to walk from i to j corresponds to the number of independent trials until the first "success" occurs, where "success" means that the random generator produces just j; the probability of this to happen is $1/N$, and hence the expected number of trials should be N. In fact this turns out to be true here: the equilibrium distribution is the uniform distribution, and it follows that $Z = Id$. Consequently, as it was to be expected, M is the matrix $EZ_{\text{diag}}M_{\text{diag}} = NE$, that is all entries equal the number N.

2) The same argument applies in the more general situation where P is a transition matrix such that all rows are identical. If we denote the entries of a typical row by $p_1, \ldots, p_N$, then the formula for M in theorem 7.7 in fact produces

$$M = \begin{pmatrix} 1/p_1 & 1/p_2 & \cdots & 1/p_N \\ 1/p_1 & 1/p_2 & \cdots & 1/p_N \\ \vdots & \vdots & & \vdots \\ 1/p_1 & 1/p_2 & \cdots & 1/p_N \end{pmatrix},$$

as it should be.

3) As an example where one really has to do some calculations consider

$$P = \frac{1}{10} \begin{pmatrix} 8 & 1 & 1 & 0 \\ 8 & 0 & 1 & 1 \\ 8 & 1 & 0 & 1 \\ 8 & 1 & 1 & 0 \end{pmatrix}.$$

This describes a chain on $\{1, 2, 3, 4\}$ which obviously is irreducible and aperiodic. Some qualitative aspects can be read from the matrix: there is a strong tendency of the walks towards state 1, also note that the states 2 and 3 play a completely symmetric role.

The equilibrium is easily calculated as the normalized solution of a system of linear equations:

$$\pi^T = \frac{1}{55}(44, 5, 5, 1).$$

And this leads to

$$Id - (P - W) = \frac{1}{110} \begin{pmatrix} 110 & -1 & -1 & 2 \\ 0 & 120 & -1 & -9 \\ 0 & -1 & 120 & -9 \\ 0 & -1 & -1 & 112 \end{pmatrix}$$

and thus to

$$Z = (Id - (P - W))^{-1} = \frac{1}{121} \begin{pmatrix} 121 & 1 & 1 & -2 \\ 0 & 111 & 1 & 9 \\ 0 & 1 & 111 & 9 \\ 0 & 1 & 1 & 119 \end{pmatrix}.$$

Finally we arrive at

$$M = \frac{1}{4} \begin{pmatrix} 5 & 40 & 40 & 220 \\ 5 & 44 & 40 & 200 \\ 5 & 40 & 44 & 200 \\ 5 & 40 & 40 & 220 \end{pmatrix}.$$

Notice the obvious symmetry between states 2 and 3 in all these calculations (see also exercise 7.12 below).

Exercises

7.1: In example 5 of chapter 2 we have introduced various shuffles. Prove that in all these cases the associated chains are irreducible and aperiodic and that the equilibrium distribution is the uniform distribution.

7.2: For a chain on a state space with N elements which is given by a stochastic matrix the following are equivalent:

a) the chain is irreducible;

b) $Id + P + \cdots + P^N$ is strictly positive.

7.3: Consider once more the product chain of example 1.3. Prove or disprove: if the original chain is irreducible (resp. irreducible and aperiodic) then so is the product chain.

7.4: Let P be an irreducible stochastic matrix such that $p_{11} > 0$. Prove that all states are aperiodic.

7.5: For a general stochastic $N \times N$-matrix P a probability vector $(\pi_1, \ldots, \pi_N)$ is called an equilibrium distribution if $(\pi_1, \ldots, \pi_N)P = (\pi_1, \ldots, \pi_N)$ holds (cf. exercise 3.5). As before we will denote the collection of these $\pi^\top$ by K.

a) K is always nonempty, and for every extreme point $(\pi_1, \ldots, \pi_n)$ of K there is a minimal closed subset C such that $\pi_i = 0$ for $i \notin C$.

b) For every $(\pi_1, \ldots, \pi_N) \in K$ and every transient i the number π_i vanishes.

7.6: Suppose that a chain admits more than one minimal closed subset. Then there are at least two different equilibria.

7.7: In theorem 7.5 we have used Cesàro limits. These limits play an important role in various fields, those readers who have never met them before are invited to investigate some basic properties.

We will say that a sequence (a_k) of real numbers is C-convergent if $\lim_k (a_1 + \cdots + a_k)/k$ exists in $\mathbb{R}$. This limit will be called the C-limit of (a_k) and written $C\text{-}\lim a_k$.

a) If (a_k) is convergent, then it is C-convergent. The converse does not hold.

b) Sums of C-convergent sequences (a_k) and (b_k) are also C-convergent. In this case one has

$$C\text{-}\lim(a_k + b_k) = C\text{-}\lim a_k + C\text{-}\lim b_k.$$

c) Does every sequence (a_k) admit a subsequence which is C-convergent?

d) Are subsequences of C-convergent sequences also C-convergent?

e) Are there unbounded C-convergent sequences?

7.8: Let P be an irreducible matrix. With $E=$ "the matrix where all entries are 1" prove that $Id - P + E$ is invertible and that the equilibrium distribution $\pi^\top$ of P is the unique solution of

$$(\pi_1, \ldots, \pi_N)(Id - P + E) = (1, \ldots, 1).$$

7.9: In this exercise P is an arbitrary irreducible 2×2-matrix. Calculate explicitly the matrix of running times.

7.10: The chain given by

$$P = \frac{1}{8} \begin{pmatrix} 0 & 0 & 4 & 4 & 0 \\ 0 & 0 & 2 & 6 & 0 \\ 0 & 0 & 0 & 0 & 8 \\ 0 & 0 & 0 & 0 & 8 \\ 3 & 5 & 0 & 0 & 0 \end{pmatrix}$$

is irreducible with period 3. Analyse this chain (equilibrium, behaviour of P^3 on the irreducible subsets, etc.)

7.11: Let P be irreducible, π the unique equilibrium and i, j states such that $i \sim j$ (cf. exercise 2.10). Prove that $\pi_i = \pi_j$.

7.12: How can the symmetry between the states 2 and 3 in the last example of the present chapter be explained by general properties of the relation "$\sim$"?

7.13: Let P and P' be stochastic $N \times N$-matrices such that P is irreducible and aperiodic. Then $\lambda P + (1 - \lambda)P'$ has the same property for every $\lambda \in]0, 1]$.

8   Notes and remarks

Part I was intended to introduce Markov chains, to define the notions relevant for the further investigations and to reveal the structure of an arbitrary chain on a finite state space. Here is a short **summary**:

What is a Markov chain?

Formally a Markov chain is nothing but a finite set S plus a stochastic matrix P plus a probability vector p. The latter encodes the distribution how to start, the rows of P contain the information how to move to the next position. Usually, however, one is interested in the behaviour of the chain with respect to *arbitrary starting distributions*, and therefore the matrix P is much more important than the vector p.

Rather than to consider a chain as something static it should be thought of as *an instruction to perform a random walk* on S; the parameters of the random generators to be used can be found in p and P.

And finally, there is the model in the *framework of probability theory*. Using this language we have to consider an S-valued *stochastic process* which is defined on some probability space, a process which is in a certain sense memoryless and homogeneous. One can show that such a model always can be constructed, its existence is necessary to build the proofs on the safe ground of probability spaces.

Markov chains can be defined in various ways. Most common is the description by defining P and p directly, but sometimes it is more convenient to use *weighted directed graphs* with the elements of S as vertices or even to fix the stochastic transition rules verbally.

The steps to analyse a chain

A first crucial step is to find the *minimal closed subsets* $C_1, \ldots, C_r$; once they are known also the set of transient states $S \setminus (C_1 \cup \cdots \cup C_r)$ is identified. In most cases the C_ρ are easily determined once one has understood the "dynamics" of the chain.

If this is not possible for some reason one could proceed as follows. First find the nonzero elements in $Q := P + P^2 + \cdots + P^N$ (see exercise 7.2); this can considerably be facilitated by replacing in every step of the calculation the nonzero entries by 1. Then the nonzero elements in the i'th row of Q are precisely at those positions j where $i \to j$. The proof of this fact is easy, see exercise 8.1.

In this way we know the relation "$\to$" and thus "$\leftrightarrow$" and now it suffices to recall that the $C_1, \ldots, C_r$ can be found among the equivalence classes with respect to "$\leftrightarrow$" (see proposition 4.4 and exercise 8.2).

The next step will be an investigation of the *transient states*: calculate the fundamental matrix $(Id - Q)^{-1}$, then theorem 5.3 will enable you to determine the absorption probabilities and the expected running times until absorption for each transient i.

And finally the behaviour within the C_ρ can be studied. To this end, fix any minimal invariant C and pass from P to the restriction of the chain to this set (which for simplicity we will continue to call P here). Calculate the period d (e.g., by having a look at the nonzero elements of the diagonal of the $P, P^2, \ldots$) and identify the minimal closed subsets of C with respect to P^d. *There* the chain works as an aperiodic and irreducible chain, and a thorough analysis now necessitates the determination of the equilibrium distribution (theorem 7.4) and the matrix of first passage times (theorem 7.7).

An example

As an example we consider a chain on $\{1, 2, \ldots, 8\}$ defined by

$$
P = \frac{1}{10}
\begin{pmatrix}
1 & 9 & 0 & 0 & 0 & 0 & 0 & 0 \\
5 & 5 & 0 & 0 & 0 & 0 & 0 & 0 \\
0 & 0 & 0 & 2 & 8 & 0 & 0 & 0 \\
0 & 0 & 10 & 0 & 0 & 0 & 0 & 0 \\
0 & 0 & 10 & 0 & 0 & 0 & 0 & 0 \\
2 & 1 & 1 & 2 & 1 & 1 & 1 & 1 \\
0 & 0 & 0 & 0 & 0 & 5 & 0 & 5 \\
1 & 1 & 0 & 0 & 0 & 0 & 0 & 8
\end{pmatrix}.
$$

(In order to save space we have skipped the very first step where one has to identify the minimal closed sets. Our chain is already in canonical form.)

One can see from the matrix that $C_1 := \{1, 2\}$ and $C_2 := \{3, 4, 5\}$ are the minimal closed sets and that the states $6, 7, 8$ are transient. Now we start with the **matrix calculations.**

a) The matrices which govern the behaviour of the transient states
Here are the matrix Q, the fundamental matrix F and the matrices R and FR (notation as in theorem 5.3):

$$
\frac{1}{10}
\begin{pmatrix}
1 & 1 & 1 \\
5 & 0 & 5 \\
0 & 0 & 8
\end{pmatrix},
\quad
\frac{1}{17}
\begin{pmatrix}
20 & 2 & 15 \\
10 & 18 & 50 \\
0 & 0 & 85
\end{pmatrix},
$$

$$
\frac{1}{10}
\begin{pmatrix}
2 & 1 & 1 & 2 & 1 \\
0 & 0 & 0 & 0 & 0 \\
1 & 1 & 0 & 0 & 0
\end{pmatrix},
\quad
\frac{1}{170}
\begin{pmatrix}
55 & 35 & 20 & 40 & 20 \\
70 & 60 & 10 & 20 & 10 \\
85 & 85 & 0 & 0 & 0
\end{pmatrix}.
$$

By theorem 5.3 these matrices contain the relevant information concerning the transient states. E.g.:

- A walk starting at state 5 will on the average need $(20 + 2 + 15)/17 = 37/17$ steps until it will be absorbed forever in $C_1 \cup C_2$; recall that this number includes the starting position so that it surely would be more realistic to deal with $37/17 - 1 = 20/17$.

- The absorption of such a walk will take place at state 2 with probability $35/170$ $(= 7/34)$, and the probabilities to be absorbed in C_1 or in C_2 are

$$
\frac{55 + 35}{170} = \frac{9}{17} \quad \text{and} \quad \frac{20 + 40 + 20}{170} = \frac{8}{17},
$$

respectively.

- The return probability for state 8 is

$$f_{88}^* = \frac{85/17 - 1}{85/17} = \frac{68}{85}.$$

b) The period of C_1
Since all entries of the restriction of the transition matrix P to C_1 are strictly positive, the restricted chain is aperiodic (and, of course, irreducible).

c) The equilibrium distribution of C_1
This necessitates to solve the matrix equation

$$(\pi_1, \pi_2) = (\pi_1, \pi_2) \begin{pmatrix} 1/10 & 9/10 \\ 5/10 & 5/10 \end{pmatrix}, \quad \pi_1 + \pi_2 = 1.$$

One easily finds $(\pi_1, \pi_2) = (5/14, 9/14)$ as the unique solution.

d) The matrix of running times for C_1
We adopt the notation of proposition 7.7. W is already known, both rows equal $5/14, 9/14$. We obtain consecutively $Id - (P - W)$, its inverse Z and finally the running time matrix M by the formula derived in theorem 7.7:

$$\frac{1}{35} \begin{pmatrix} 44 & -9 \\ -5 & 40 \end{pmatrix}, \frac{1}{49} \begin{pmatrix} 40 & 9 \\ 5 & 44 \end{pmatrix}, \frac{1}{45} \begin{pmatrix} 126 & 50 \\ 90 & 70 \end{pmatrix}.$$

e) The period of C_2
On C_2 the chain oscillates between $C_{21} := \{3\}$ and $C_{22} := \{4, 5\}$, and the square of the restriction of P to C_2 is

$$\begin{pmatrix} 1 & 0 & 0 \\ 0 & 1/5 & 4/5 \\ 0 & 1/5 & 4/5 \end{pmatrix}.$$

Hence the period of C_2 is 2, and with respect to P^2 the set C_2 splits into the minimal closed subset C_{21} and C_{22}.

f) The equilibrium distributions of P^2 on C_{21} and C_{22} and the associated matrices of running times
These are particularly easy to determine: the associated transition matrices are (1) and $\begin{pmatrix} 1/5 & 4/5 \\ 1/5 & 4/5 \end{pmatrix}$, and we get the unit mass on $\{3\}$ and the distribution $(1/5, 4/5)$ on $\{4, 5\}$ as equilibrium distributions. The running time matrices are (1) and $\begin{pmatrix} 5 & 5/4 \\ 5 & 5/4 \end{pmatrix}$, respectively.

(Note that these running times refer to P^2. For state 4, e.g., one needs $2 \cdot 5$ steps to return on the average if one counts with respect to the original chain.)

We continue with some **notes and remarks**. First we want to emphasize that all results presented in part I were found in the early decades of Markov chain theory at the beginning of the twentieth century, also the examples are standard (with the exception of example 9 and – maybe – example 5: the connection between Markov chains and card shuffling is rather recent[1]). We omit to try to associate the various results with certain mathematicians; those readers who are interested in the *general history* of this field are referred to [28] (chapter XII), the notes and remarks in [50], and [68] (chapter 10).

The material is selected and presented according to the author's taste, there are numerous monographs where other approaches have been chosen (two recommendable recent references are [20] and [60], more advanced introductions are [17] or [22]).

Whereas the chains considered here give rise to many interesting applications they are surely the most elementary representatives of mathematical models for "memoryless" stochastic phenomena; they are, in a sense, the simplest situation after sequences of independent identically distributed random variables which can assume only finitely many values. They have the advantage that elementary probability suffices to develop the theory rigorously, also nearly all of the abstract existence results can be complemented by recipes by which the probabilities or expectations can be determined explicitly with the help of easy *matrix calculations*. We have tried here to give typical examples of this interplay between probability and linear algebra. However, what has been shown is far from being exhaustive (further results into this direction can be found in chapter XVI of [30], [42], [43], or [49]).

Several generalizations have been studied. One of these has already been mentioned, the case of *countable state spaces*. The theory can still be developed in an elementary way, expectedly some of the results fail to be true (cf. exercise 4.6). For a systematic development of the countable theory the reader is referred to [20], [50], or [60].

Whereas the step from finite to countable state spaces does not lead to conceptual difficulties it is not as easy to pass from discrete to *continuous time*. Let us try to understand the underlying idea. We start with a finite state space S, and as before we want to fix rules for a "walk" on S. In this book we have done this by prescribing an initial distribution and stochastic rules what to do at "times" $k = 1, 2, \ldots$. Now we want to refine this, the "position of the walk at time t" shall have a meaning for *all real* $t \geq 0$. This is essentially done as before, the next position in S is chosen according to a probability distribution on S which only depends on the present position. But there is a new feature: the walk will pause after the occupation of a new state i for some time T before it continues, where T has an exponential distribution the parameter of which might depend on i. It should be clear that it is again rather simple to simulate such generalized walks. Also, as in the case considered in this book, all that has to be known about such a walk is *encoded in a single matrix*. There is a matrix Q with the following property: if a walk is started according to an initial distribution given by a probability vector $(p_i)_{i \in S}$, then the probability to find the walk at a state i_0 at time t is just the i_0-component of the vector

$$(p_i)e^{tQ};$$

here e^{tQ} means the matrix exponential of Q, i.e., the matrix $Id + (tQ)/1! + (tQ)^2/2! + \cdots$.

---

[1] The reader will find further investigations in the following chapters. Standard references are [1], [3] and [24]. For the "deterministic" theory of card shuffling cf. [59].

It is considerably more difficult than in the case of discrete time to transform this idea into a family $(X_t)_{t\geq 0}$ of S-valued random variables on a suitable probability space, and further severe problems have to be overcome if one passes from finite to countable state spaces. We refer the reader to [50] or [60] for details.

Even more advanced is the theory of *Markov processes in continuous time on uncountable state spaces*. A number of nontrivial technical difficulties have to be taken into account (measurability of the paths, the definition of conditional probabilities as a special case of conditional expectations, ...), this general approach is beyond the scope of this book. See [48] and the literature cited there.

One of the most surprising properties of Markov chains is the fact that – under the mild assumption of irreducibility and aperiodicity of the transition matrix P – they tend to forget their history. All rows of P^k converge to the same vector $\pi^\top$, i.e., neither k nor the starting position can be read from the probability to find the walk at state j in the k'th step. A positive formulation of this phenomenon could be the statement that one knows all probabilities to find the walk in the various states i provided that it has run for sufficiently many steps. But *how often* will a special state be visited? The naive answer would be the following: if the probability to find the walk at j is π_j, then it is to be expected that roughly for $\pi_j \cdot k$ times out of k steps the walk will occupy this state. That this is in fact true is the *ergodic theorem for Markov chains*, here is the rigorous formulation:

> Let P be an irreducible stochastic $N \times N$-matrix and $X_0, X_1, \ldots$ a homogeneous $\{1, \ldots, N\}$-valued Markov process with the transition probabilities given by P (the initial distribution (p_i) might be arbitrary here). Then, if $\pi^\top$ denotes the unique equilibrium, one has with probability one that
>
> $$\frac{\operatorname{card}\{k' \mid 0 \leq k' \leq k, X_{k'} = i\}}{k}$$

tends to π_i for every i.

(For a proof, see chapter 1.10 in [60] or chapter 3.4 in [20].)

Exercises

8.1: Prove that the entry at the i-j-position of $Id + P + \cdots + P^N$ is strictly positive iff $i \to j$.

8.2: Which of the equivalence classes with respect to "$\leftrightarrow$" correspond to the minimal closed subsets of the chain?

8.3: Analyse, similarly to the example in this chapter, the chain given by the matrix

$$P := \frac{1}{10}\begin{pmatrix} 1 & 9 & 0 & 0 & 0 & 0 & 0 & 0 \\ 10 & 0 & 0 & 0 & 0 & 0 & 0 & 0 \\ 0 & 0 & 0 & 2 & 8 & 0 & 0 & 0 \\ 0 & 0 & 1 & 2 & 7 & 0 & 0 & 0 \\ 0 & 0 & 2 & 4 & 4 & 0 & 0 & 0 \\ 2 & 1 & 1 & 2 & 1 & 1 & 1 & 1 \\ 0 & 1 & 0 & 4 & 0 & 0 & 0 & 5 \\ 1 & 1 & 0 & 7 & 0 & 1 & 0 & 0 \end{pmatrix}.$$

Part II

Rapidly mixing chains

In part I we have developed the basic theory of finite Markov chains. There the usual approach was to start with the state space S and the transition matrix P and then to calculate the relevant probabilities, expected running times and so on.

In the applications we have in mind, however, one is mainly interested in a fixed probability distribution π on a finite set S, and Markov chains come into play in order to simulate it.

More precisely: suppose that there are prescribed positive real numbers π_i for i in a finite – but generally huge – set S such that $\sum_i \pi_i = 1$. Assume that one needs a random generator which produces elements of S in such a way that each particular i occurs with probability π_i. In many cases π will be the uniform distribution, i.e., $\pi_i = 1/\mathrm{card}(S)$ for every i.

As an example fix an integer k_0 and two functions $\phi, \psi : \{0, \ldots, k_0\} \to \mathbb{Z}$ with $\phi \leq \psi$ and $\phi(0) = \psi(0)$.

Let us say that a mapping $\omega : \{0, \ldots, k_0\} \to \mathbb{Z}$ is a ϕ-ψ-path if $\phi \leq \omega \leq \psi$ and $|\omega(k) - \omega(k + 1)| = 1$ for $k = 0, \ldots, k_0 - 1$. The problem is to find a "typical" ϕ-ψ-path. (You can think of a game where one starts to play with $\omega(0)$ \$/DM/$\ldots$ and where one loses or gains 1 \$/DM/$\ldots$ at times $k = 1, \ldots, k_0$; $\phi(k)$ and $\psi(k)$ are certain maximal losses or gains which might depend on k. Another translation of this setting can be found in [32] where it is shown that the set S of all ϕ-ψ-paths corresponds to the collection of certain total orders which extend a given order.)

Clearly S is a finite set, it might be incredibly large, and in general the precise number of elements will be difficult to determine. Nevertheless it is comparitively simple to generate (approximately) uniformly distributed samples. One only has to define a Markov chain on S such that it is irreducible and the equilibrium is the uniform distribution, as samples one can use the position of the chain after "sufficiently many" steps. A natural candidate for such a chain declares as admissible the transitions $\omega \to \omega'$, where $\omega(k)$ and $\omega'(k)$ are different for at most one k. For a suitable choice of the transition probabilities from ω to the (at most $k_0 + 1$) ω' the transition matrix is doubly stochastic, and thus the equilibrium is the uniform distribution as desired.

The technique sketched in the preceding example is generalized to a **strategy to solve this problem with Markov chains** as follows.

- Find a transition matrix P such that the associated chain is irreducible and aperiodic and has $(\pi_i)_{i \in S}$ as its uniquely determined equilibrium distribution.

- Fix a "large" number k. We then know from theorem 7.4 that the probabilities $p_{ij}^{(k)}$ are "close to" π_j for all i, or – to phrase it otherwise – the algorithm

Start the chain at any i;

run it for k steps;

use the final position j as the output for a π–simulation

really produces the elements of S with (nearly) the correct probabilities.

One might suspect that this procedure is unnecessarily complicated: would it not be much simpler rather than to design a special P to work with a matrix where all rows are identical with entries (π_i)? Mathematically this is correct, one could even choose $k = 1$ in this case. For practical purposes, however, this idea is rather useless since in order to simulate π (*this* has to be achieved) one must be able to do just this if one wants to run the chain.

Therefore it is necessary to supplement the above description in that one wants to find (better: one must find) P such that in addition to the above requirements a random walk subject to P is easy to simulate.

There are some *obvious questions* in connection with this simulation procedure:

- What is meant precisely if one states that it is *simpler to run the chain* than to simulate π? Unfortunately I cannot provide a satisfactory answer. The meaning of "simple" depends on the power of the available computers and the built-in random generators. For the examples we have in mind even the fastest computers are unable to simulate the desired π *without* using Markov chains.

- *How* does one find P? Again there is no simple answer. We will study many examples, but there is no general rule for an appropriate choice of P for a new particular situation.

- *How large* must k (the number of steps) be in order to guarantee the desired precision? This is just the problem of the rate of convergence in theorem 7.4(i): how fast does P^k tend to W?

  Part II is devoted mainly to the investigation of techniques which might be used to deal with this aspect of the problem.

- *Why* should it be interesting to produce random elements in a finite set subject to prescribed probabilities, i.e., what are the *applications*?

  The answer is postponed to part III where we will study a number of examples.

Some of the methods we will study in part II have been known for a long time (like the estimations which use the second-largest eigenvalue of the transition matrix), others are more recent. Not all results apply to arbitrary Markov chains, for some it is assumed that they are defined by graphs, others can only be used for Markov chains on groups. (These latter chains will be discussed rather extensively in the chapters 15 and 16. What is proved there can also be considered as an introduction to harmonic analysis on finite groups.)

For more details on what will be done see the introductions to the various chapters or the extended table of contents at the beginning of this book. Some supplementary information can be found in the Notes-and-Remarks chapter 22.

9 Perron-Frobenius theory

In this chapter we aim at understanding why the *eigenvalues* of the transition matrix P of a chain play a fundamental role for the rate of convergence to the equilibrium distribution. Also we will discuss some other far-reaching algebraic consequences of the fact that a square matrix is stochastic[1].

We begin with a lemma concerning the eigenvalues of general stochastic matrices:

Lemma 9.1 *Let $P = (p_{ij})_{i,j=1,\ldots,N}$ be a stochastic matrix. Then 1 is an eigenvalue, and every eigenvalue λ satisfies $|\lambda| \leq 1$.*

Proof. Since P is a stochastic matrix we know that

$$P(1,1,\ldots,1)^\top = (1,1,\ldots,1)^\top$$

so that 1 is an eigenvalue.

Now let $\lambda \in \mathbb{C}$ be given such that there is a nontrivial $x = (x_1,\ldots,x_N)^\top$ with

$$P(x_1,\ldots,x_N)^\top = \lambda(x_1,\ldots,x_N)^\top.$$

Choose an index i_0 such that $|x_{i_0}| = \max_j |x_j|$. This means that all x_j lie in the disk with radius $|x_{i_0}|$ and center in the origin. The eigenvalue equation implies that λx_{i_0} is a convex combination of the x_j, and therefore it is obvious that necessarily $|\lambda| \leq 1$ holds. Here is a more formal proof:

$$
\begin{aligned}
|\lambda||x_{i_0}| &= \left| \sum_j p_{i_0 j} x_j \right| \\
&\leq \sum_j p_{i_0 j} |x_j| \\
&\leq \sum_j p_{i_0 j} |x_{i_0}| \\
&= |x_{i_0}|,
\end{aligned}
$$

hence $|\lambda| \leq 1$. $\qquad\square$

Simple examples show that nonreal eigenvalues of modulus one are possible, even for P which correspond to irreducible chains: fix an N'th root of unity, say w, and consider the $N \times N$-matrix

$$
P := \begin{pmatrix}
0 & 1 & 0 & 0 & \cdots & 0 \\
0 & 0 & 1 & 0 & \cdots & 0 \\
\vdots & \vdots & \vdots & \vdots & & \vdots \\
1 & 0 & 0 & 0 & \cdots & 0
\end{pmatrix},
$$

a deterministic cyclic walk; we have

---

[1] The algebraic theory of such matrices is usually called *Perron-Frobenius theory*. The present book contains only the very beginnings of this theory, for a more extensive study see [69].

$$P(1, w, \ldots, w^{N-1})^\top = w(1, w, \ldots, w^{N-1})^\top$$

so that w is an eigenvalue of P.

Let us now turn to properties of stochastic matrices P with strictly positive P^k for suitable k, i.e., to matrices which correspond to aperiodic and irreducible chains (see lemma 7.3). *Such* chains will be of importance in the applications we have in mind.

Proposition 9.2 *Let P^k have strictly positive entries for a suitable k.*
  (i) *$\lambda = 1$ is the only eigenvalue of P with modulus one[2].*
  (ii) *The geometric multiplicity of $\lambda = 1$ is one. The eigenspace associated with $\lambda = 1$ is spanned by $(1, \ldots, 1)^\top$.*
  (iii) *If a nontrivial vector π satisfies $\pi^\top P = \pi^\top$, then there are a $\mu \in \mathbb{C}$ with $|\mu| = 1$ and strictly positive $a_1, \ldots, a_N$ such that $\pi_1 = \mu a_1, \ldots, \pi_N = \mu a_N$.*
  (iv) *The algebraic multiplicity of $\lambda = 1$ is also one: 1 is a simple root of the characteristic equation $\det(\lambda Id - P) = 0$.*

Proof. (i) Suppose first that not only P^k but P itself has strictly positive components. As in the preceding proof we will use a convexity argument. This time it will be important that disks in the plane are not only convex, but *strictly convex:*

> Whenever $p_1, \ldots, p_N > 0$ with $\sum p_i = 1$ and $z_1, \ldots, z_N \in \mathbb{C}$ with $|z_1|, \ldots, |z_N| \leq r$ and $|\sum_i p_i z_i| = r$ are given, then $z_1 = \cdots = z_N$.

A proof of this obvious fact is simple. Without loss of generality suppose that $r = 1 = \sum p_i z_i$. If, e.g., we had $z_1 \neq 1$, then the real part of z_1 would be at most $1 - \epsilon$ for a strictly positive ϵ. But then the real part of $\sum p_i z_i$ could be estimated by $p_1(1 - \epsilon) + (1 - p_1) < 1$, a contradiction.

Now let λ with $|\lambda| = 1$ be such that there is a nontrivial vector x with $Px = \lambda x$. With the notation of the proof of the preceding lemma we know that λx_{i_0} is a convex combination of $x_1, \ldots, x_N$ with strictly positive weights. Since λx_{i_0} lies on the boundary of the disk under consideration it follows that $x_1 = \cdots = x_N$. Hence also $\lambda x_{i_0} = \sum p_i x_i = x_{i_0}$, and therefore – since $x_{i_0} \neq 0$ – we have shown that $\lambda = 1$.

The general case can be reduced to what has already been shown. $Px = \lambda x$ implies that $P^k x = \lambda^k x$, and consequently – since P^k has strictly positive entries by assumption – all x_i coincide and $\lambda^k = 1$. And with P^k also P^{k+1} is strictly positive so that $\lambda^{k+1} = 1$ as well. This proves that $\lambda = 1$.

(ii) This result has been shown as a by-product in the preceding proof.

(iii) Since $\pi^\top P = \pi^\top$ implies that $\pi^\top P^k = \pi^\top$ we may assume that P itself is strictly positive. Now recall that in the Cauchy-Schwarz inequality one has equality precisely if the vectors under consideration point into the same direction. For the case of complex numbers this means that $|z_1 + \cdots + z_N| = |z_1| + \cdots + |z_N|$ iff $z_i = a_i \mu$ for a suitable $\mu \in \mathbb{C}$ and $a_1, \ldots, a_N \geq 0$.

This is applied here as follows:

$$\sum_i |\pi_i| = \sum_i |\sum_j p_{ji} \pi_j|$$

---

[2] This result should be compared with the preceding example: this chain has a strictly positive period, and only for this reason eigenvalues λ with $\lambda \neq 1$, $|\lambda| = 1$ are possible.

$$\leq \sum_i \sum_j p_{ji} |\pi_j|$$

$$= \sum_j |\pi_j| \sum_i p_{ji}$$

$$= \sum_j |\pi_j|.$$

Consequently the inequality is in fact an equality, and we have $|\sum_j p_{ji}\pi_j| = \sum_j |p_{ji}\pi_j|$ for every i. It follows that the $p_{ji}\pi_j$ and thus also the π_j point into the same direction.

As a consequence every π_i, being of the form $\sum_j p_{ji}\pi_j$, will be different from zero if at least one π_j is nontrivial, and this completes the proof of (iii).
(*Note:* The result also follows from a combination of lemma 7.3 and theorem 7.4. We have preferred, however, to provide an independent algebraic proof.)

(iv) There seems to be no way to avoid some technicalities in order to prove this fact.[3] We start with

Claim 1: Let $Q = (q_{ij})$ be a matrix such that $0 \leq q_{ij} \leq p_{ij}$ for all i,j and $q_{ij} < p_{ij}$ at least once (as a short-hand notation we will express this by writing $0 \leq Q \leq P$, $Q \neq P$). Then $|\lambda| < 1$ for all eigenvalues λ of Q.

Proof of claim 1: Let y be a nontrivial vector and assume that $y^\top Q = \lambda y^\top$; here we use the fact that we have the choice to deal with left or right eigenvectors. Denote by z the vector $(|y_1|,\ldots,|y_N|)^\top$. We have $|\lambda| z^\top \leq z^\top Q \leq z^\top P$ (where "$\leq$" stands for the coordinate-wise order), and with

$$|\lambda| \sum_i z_i \leq \sum_i \sum_j p_{ji} z_j = \sum_j z_j \left(\sum_i p_{ji}\right) = \sum_j z_j \tag{9.1}$$

we arrive at $|\lambda| \leq 1$.

But we need more, the claim is "$<$" and not only "$\leq$". To this end we will show that $|\lambda| = 1$ leads to a contradiction.
Suppose that $|\lambda| = 1$. The first consequence is that then $z^\top \leq z^\top P$, but in this componentwise inequality there can't be any "$<$" since otherwise (9.1) would produce $\sum_i z_i < \sum_j z_j$.

Knowing that $z^\top = z^\top P$ we can conclude from (iii) that all components of z are strictly positive, and this finally leads to the desired contradiction: it implies that $z^\top Q$ is strictly smaller than $z^\top P$ at least at one component in contrast to $z^\top \leq z^\top Q \leq z^\top P$ and $z^\top = z^\top P$.

Claim 2: Let $\tilde{P}$ be the matrix $(p_{ij})_{i,j=1,\ldots,N-1}$ (just forget the last row and the last column in P). Then the modulus of all eigenvalues of $\tilde{P}$ is strictly less than one.

Proof of claim 2: Complete $\tilde{P}$ with zero entries to obtain an $N \times N$-matrix and call the resulting matrix Q. Then Q and $\tilde{P}$ have the same (nonzero) eigenvalues, $0 \leq Q \leq P$ holds, and $Q \neq P$ is surely also satisfied since Q is not a stochastic matrix. Hence the assertion is a consequence of claim 1.

---

[3] If P is self-adjoint or at least similar to a self-adjoint matrix then the result is covered by part (ii) since then the geometric equals the algebraic multiplicity. The general result is contained here only for the sake of a complete description of the spectral behaviour of P. It will not be needed in the sequel.

We now turn to the proof of (iv). Let $\phi(\lambda) := \det(\lambda Id - P)$ denote the characteristic polynomial of P and consider any λ where $\phi(\lambda)$ does *not* vanish. Then $\lambda Id - P$ is invertible, but more is true: the inverse can explicitly be described as the product of $1/\phi(\lambda)$ with a matrix $A(\lambda)$ where each coefficient is a polynomial in λ (usually $A(\lambda)$ is called the *adjugate* associated with $\lambda Id - P$, the possibility of this easy description is an immediate consequence of Cramer's rule). Hence for all λ with $\phi(\lambda) \neq 0$ we have

$$(\lambda Id - P)A(\lambda) = A(\lambda)(\lambda Id - P) = \phi(\lambda)Id. \tag{9.2}$$

But all entries of the matrices involved in these equations are polynomials, and since there are only finitely many zeroes of ϕ it follows that (9.2) holds for *all* λ. In particular equality obtains for $\lambda = 1$ and thus, if we put $A := A(1)$, it follows from $\phi(1) = 0$ that $A = PA = AP$. This has the remarkable consequence that all rows of A are in the left eigenspace and all columns are in the right eigenspace of the eigenvalue 1. By (ii) and (iii) this yields that either A is identically zero or nonzero at *every* component. We claim that the second alternative necessarily holds: the element a at the end of the last row of A is just the determinant of that matrix which arises from $Id - P$ after cancelling the last row and the last column; and this determinant is nonzero by our claim 2 since 1 is not an eigenvalue of the truncated P.

We need, however, a little bit more. Call – with the notation of claim 2 – $\tilde{\phi}(\lambda)$ the characteristic polynomial of $\tilde{P}$. The zeroes of $\tilde{\phi}$ lie in $\{|\lambda| < 1\}$, and $\tilde{\phi}(\lambda)$ is strictly positive for large positive real λ. Hence, by continuity, $\tilde{\phi}$ is strictly positive on $[1, \infty[$, and this implies in particular that $a > 0$. Therefore, by (ii) and (iii), all components of A are nonzero and positive.

To finish the proof we differentiate the matrix equation (9.2). With $B(\lambda) :=$ "the coordinate-wise derivative of $A(\lambda)$" we get

$$A(\lambda) + (\lambda Id - P)B(\lambda) = \phi'(\lambda)Id,$$

so that in particular $A + (Id - P)B(1) = \phi'(1)Id$ holds. We multiply this matrix equation from the left by any strictly positive $\pi^\top$ for which $\pi^\top = \pi^\top P$; such a $\pi^\top$ exists by lemma 9.1 and (iii). And now the vector equation $\pi^\top A = \phi'(1)\pi^\top$ proves that $\phi'(1) \neq 0$ as claimed. $\qquad\qquad\qquad\qquad\qquad\qquad\qquad\qquad\qquad\qquad\qquad\qquad\qquad\qquad\qquad\quad\square$

Combining the preceding results with standard matrix theory we arrive at

Theorem 9.3 *Let P be a stochastic $N \times N$-matrix such that P^k is strictly positive for a suitable k. Then there is an invertible matrix S such that $S^{-1}PS$ can be written as*

$$\begin{pmatrix} 1 & 0 & 0 & \cdots & 0 \\ 0 & J(\lambda_2, n_2) & 0 & \cdots & 0 \\ 0 & 0 & J(\lambda_3, n_3) & \cdots & 0 \\ \vdots & \vdots & \vdots & & \vdots \\ 0 & 0 & 0 & \cdots & J(\lambda_r, n_r) \end{pmatrix}.$$

Here $J(\lambda, n)$ stands for a typical Jordan block, i.e., for the $n \times n$-matrix

$$\begin{pmatrix} \lambda & 1 & 0 & \cdots & 0 & 0 \\ 0 & \lambda & 1 & \cdots & 0 & 0 \\ \vdots & \vdots & \vdots & & \vdots & \vdots \\ 0 & 0 & 0 & \cdots & \lambda & 1 \\ 0 & 0 & 0 & \cdots & 0 & \lambda \end{pmatrix},$$

and the symbol "0" denotes matrices of suitable dimension with zero entries. All $|\lambda_2|, \ldots,$
$|\lambda_r|$ *are strictly less than one.*

*If P is similar to a self-adjoint matrix, then the eigenvalues λ are real, the dimensions
n_ρ of the Jordan blocks are one and the λ may be enumerated such that $1 > \lambda_2 \geq \lambda_3 \geq
\cdots \geq \lambda_N > -1.$*

This theorem will be essential to derive mixing rates which are optimal, at least theo-
retically (see the next chapter). In these investigations the limit behaviour of P^k will be
of interest. That theorem 9.3 can be used when *explicit* formulas are needed is illustrated
by the following

Example: Let a and b be numbers with $0 < a, b < 1$. We define P by $\begin{pmatrix} 1-a & a \\ b & 1-b \end{pmatrix}$.
As eigenvalues we obtain 1 and $1 - (a+b)$, hence P will be similar to the diagonal ma-
trix $\tilde{P} := \begin{pmatrix} 1 & 0 \\ 0 & 1-(a+b) \end{pmatrix}$. A transformation matrix S can be calculated by finding
a nontrivial solution of $PS = S\tilde{P}$; this set of four linear equations gives (up to a constant)

$$S = \begin{pmatrix} 1 & 1 \\ 1 & -b/a \end{pmatrix}. \text{ Hence } S^{-1} = \begin{pmatrix} b/(a+b) & a/(a+b) \\ a/(a+b) & -a/(a+b) \end{pmatrix}.$$

Finally, from $P = S\tilde{P}S^{-1}$, we get $P^k = S\tilde{P}^k S^{-1}$ and thus the explicit formula

$$P^k = \frac{1}{a+b} \begin{pmatrix} b + a[1-(a+b)]^k & a - a[1-(a+b)]^k \\ b - b[1-(a+b)]^k & a + b[1-(a+b)]^k \end{pmatrix}.$$

All qualitative and quantitative questions concerning the limit behaviour can now be
answered easily.

Exercises

9.1: Let λ and μ be complex numbers such that $|\lambda| \leq |\mu| < 1$. Does there exist an
irreducible stochastic matrix P such that λ and μ are eigenvalues of P? Is it possible
to find P such that μ is the eigenvalue with maximal modulus among all eigenvalues
different from 1?

9.2: Suppose that $P = (p_{ij})_{i,j=1,\ldots,N}$ is a stochastic $N \times N$-matrix such that, for a suitable
r, $\{1,\ldots,r\}$ and $\{r+1,\ldots,N\}$ are closed sets for the associated chain. How are the
eigenvalues of P related to the eigenvalues of $(p_{ij})_{i,j=1,\ldots,r}$ and $(p_{ij})_{i,j=r+1,\ldots,N}$?

9.3: Let an irreducible chain be given. Prove that λ^* (= the maximum of the eigenvalues
different from 1) does not necessarily coincide with the absolute value of the second-
largest eigenvalue.

9.4: Let $P = (p_{ij})_{i,j=1,\ldots,N}$ be a stochastic matrix. The states are labelled such that
the i in $\{1,\ldots,r\}$ belong to minimal invariant subsets and the i in $\{r+1,\ldots,N\}$ are
transient. Now let λ be an eigenvalue of P such that $|\lambda| = 1$. Then, for any $(\pi_1,\ldots,\pi_N)$
such that $(\pi_1,\ldots,\pi_N)P = \lambda(\pi_1,\ldots,\pi_N)$ it follows that $\pi_{r+1} = \cdots = \pi_N = 0$.

9.5: Let P and λ be as in the preceding exercise. Prove that λ is a d'th root of unity,
where $1 \leq d \leq N$. Conversely, if $1 \leq d \leq N$, every d'th root of unity can be an eigenvalue
of a suitable stochastic $N \times N$-matrix.

9.6: Let the stochastic matrix P be such that P^k is strictly positive for a suitable number k. Then all eigenvalues λ of P with $|\lambda| = 1$ satisfy $\lambda^k = 1$.

10   Rapid mixing: a first approach

Here we start our investigations of rapid mixing for chains which are aperiodic and irreducible: how fast do the P^k tend to W, the matrix which contains in each row the equilibrium distribution? The structure of P has been described in the last chapter, and from this description it follows that the rate of convergence to W will depend on the number

$$\lambda^* := \max\{|\lambda| \mid \lambda \text{ is an eigenvalue}, \lambda \neq 1\}$$

which is known to be smaller than one.

> The qualitative argument is as follows. For a suitable S the matrix $S^{-1}PS$ is built up from blocks on the diagonal, the first one contains only the number 1, the others are matrices $J(\lambda, n)$ with $|\lambda| < 1$. Therefore the k'th powers of $S^{-1}PS$ will tend to a matrix S' which has a "1" at the top-left position and for which all other entries vanish. It remains to remark that the rate of convergence is that of the "worst" $(J(\lambda, n))^k$, i.e., it is determined by λ^*. Also,
>
> $$\begin{aligned} \lim P^k &= SS^{-1}\lim P^k SS^{-1} \\ &= S(\lim S^{-1}PS)S^{-1} \\ &= SS'S^{-1}, \end{aligned}$$
>
> and this is a matrix with identical rows.

We will start our investigations in *the second section* below by making this precise. There we will restrict ourselves to the special case where P and the equilibrium distribution are in *detailed balance* – the definition will be introduced shortly –, which in particular implies that we have to deal with self-adjoint matrices only; in this way we avoid the notational difficulties when treating powers of Jordan blocks. Such P are the favourite candidates for our purposes, they cover many of the applications we have in mind. We provide the definition and collect some properties in *the first section*.

The *convergence theorems in section 2* are in a sense the best that can be done by using eigenvalues. There are, however, *two drawbacks*. The first one is that the results which describe the *precise* asymptotic order of convergence can be proved only in the case of detailed balance. This is less important since the more interesting upper bounds could also be obtained for more general P. But it is extremely unsatisfactory that everything depends on λ^*, a number which in most cases cannot be determined explicitly. Therefore the theorems in section 2 are not the end of the story, in the following chapters they will be complemented by

- results by which one gets at least reasonable *upper estimates* for λ^* and

- bounds which do not use eigenvalues.

In the present chapter – in *the third section* – we will only point out how an upper bound for the rate of convergence can be read off from the entries of P directly. This is universally applicable, the provable rate of convergence, however, is rather poor in the case of large state spaces. For reasonably large S, however, the direct approach has many advantages. We will present *a first application to examples from renewal theory* in *section 4*.

This chapter will be introductory. Nevertheless the results will suffice to illustrate the *typical difficulties* with which one is faced in this area. For example, an estimate is of little practical use if one cannot identify the relevant numbers which are involved (like λ^*). Also we will see that often there are several meaningful choices to define what "is close to" means. Here we are interested in how fast P^k approximates W: should all $|p_{ij}^{(k)} - \pi_j|$ be small (the *absolute* error), or is it more desirable to measure the approximation by the size of the numbers $|p_{ij}^{(k)} - \pi_j|/\pi_j$ (the *relative* error)? Or does it suffice to know this only for a particular i and all j (the error with respect to a fixed starting position)? Should one demand small $\sum_j |p_{ij}^{(k)} - \pi_j|?\dots$

Having settled this question one aims at proving theorems of the form: whenever $k \geq k_0$, then the approximation is better than ε. However, such a result is rather worthless if k_0 is gigantic.

> This, of course, often happens when one tries to provide rigorous results for algorithms in applied mathematics "which work somehow". Here is a quotation from [44], p. 7, a book which deals with the numerical treatment of differential equations:
>
> "The statement falls into the broad category of statements like 'the distance between London and New York is less than 47 light years' which, although manifestly true, fail to contribute significantly to the sum total of human knowledge."

Hence it will always be desirable to complement the upper by lower estimates. Only then one can be sure that – up to a constant which hopefully is of reasonable size – one has got the best possible result.

Detailed balance

Let P be any stochastic matrix which we will assume to be irreducible. Suppose that there is a probability vector $\lambda = (\lambda_1, \dots, \lambda_N)^{\top}$ such that

$$\lambda_i p_{ij} = \lambda_j p_{ji} \text{ for all } i, j. \tag{10.1}$$

This has some *remarkable consequences*. The first is that λ necessarily *coincides with the equilibrium distribution* π from theorem 7.5 since (10.1) simply implies that $\lambda^{\top} P = \lambda^{\top}$. Thus, to avoid confusion, we replace (10.1) with the condition

$$\pi_i p_{ij} = \pi_j p_{ji} \text{ for all } i, j. \tag{10.2}$$

Now imagine that the chain is in equilibrium, either by choosing π as the starting distribution or by starting arbitrarily and waiting for a very, very long time. For arbitrary irreducible and aperiodic chains this means that the probability to find the chain in state i is π_i, no matter at what step you decide to have a look at the chain. What about the *probability to observe a certain jump*, from i to j, say? This is just the probability to find the chain in i times the conditional probability to jump from i to j, that is $\pi_i p_{ij}$; *this* is the meaning of the numbers in (10.2). For general chains, however, $\pi_i p_{ij}$ will be different from $\pi_j p_{ji}$, nobody really expects that jumps from i to j are equally likely as those from j to i.

As an example consider a deterministic cyclic random walk on $\{0, \ldots, N-1\}$ with $N \geq 3$: $p_{i,i+1} = 1$. If the chain is in equilibrium, that is if all states are equally probable, we "expect" a jump from i to $i + 1 \bmod N$, jumps into the other direction are not possible. Of course this expectation can be verified numerically: $\pi_i p_{i,i+1}$ is $1/N$ whereas $\pi_{i+1} p_{i+1,i}$ is zero. (By a simple modification we also could have an example with an irreducible *and aperiodic* chain: pass from P to $(Id + P)/2$.)

Remarkably this *reversal of the order of jumps* can always be modelled. Start with an irreducible P and an equilibrium distribution π and try to find another stochastic matrix $\tilde{P}$ with the same equilibrium and the following property: the probability of jumps from i to j subject to P is precisely the probability to observe a transition from j to i if the chain is driven by $\tilde{P}$. By the above remarks this means that

$$\pi_i p_{ij} = \pi_j \tilde{p}_{ji} \text{ for all } i, j \tag{10.3}$$

has to hold, and – surprisingly – such a $\tilde{P}$ always exists: one simply has to take (10.3) as the definition of $\tilde{P}$. Then the properties of P and π easily imply that $\tilde{P}$ is a stochastic matrix with $\pi^\top \tilde{P} = \pi^\top$. Also this new matrix is irreducible (and even aperiodic if P is) by lemma 7.3 since its entries are positive multiples of the p_{ij}.

For obvious reasons the chain associated with $\tilde{P}$ is called the *time reversal* of the original chain.

It is time for a formal definition:

Definition 10.1 If an irreducible stochastic matrix P and a probability vector π satisfy $\pi_i p_{ij} = \pi_j p_{ji}$ for all i, j, then one says that P and π are *in detailed balance* (or that P is *reversible*).

For later use we collect some properties the easy proofs of which are left to the reader:

Proposition 10.2 *Let P be an irreducible matrix with equilibrium π. Denote by D the matrix for which $d_{ii} = \sqrt{\pi_i}$ and for which all other entries vanish.*

(i) *P and π are in detailed balance iff DPD^{-1} is a symmetric matrix. In particular it follows that there exists a basis of $\mathbb{R}^N$ which consist of eigenvectors of P if this condition is satisfied.*

(ii) *Denote by H_π the N-dimensional real vector space equipped with the scalar product*

$$\langle (x_1, \ldots, x_N)^\top, (y_1, \ldots, y_N)^\top \rangle_\pi := \sum \pi_i x_i y_i,$$

and associate with P the linear map $T_P : H_\pi \to H_\pi, (x_i)_i \mapsto Px = (\sum_j p_{ij} x_j)_i$. Then P and π are in detailed balance iff T_P satisfies $\langle T_P x, y \rangle_\pi = \langle x, T_P y \rangle_\pi$ for all x, y, i.e., iff T_P is a self-adjoint operator on the Hilbert space $(H_\pi, \langle \cdot, \cdot \rangle_\pi)$.

(iii) *If P and π are in detailed balance, then the matrix $(\pi_i p_{ij})_{ij}$ is symmetric. Conversely, let $(a_{ij})_{i,j=1,\ldots,N}$ be a symmetric matrix with nonnegative a_{ij} such that $\sum_{ij} a_{ij} = 1$ and $\pi_i := \sum_j a_{ij}$ is strictly positive for every i. Then, with $p_{ij} := a_{ij}/\pi_i$, the matrix $P := (p_{ij})$ is a stochastic matrix with equilibrium $\pi := (\pi_1, \ldots, \pi_N)^\top$, and P and π satisfy the detailed balance condition. If, for some k, A^k has strictly positive components, then so has P^k. Thus there are essentially as many P which are in detailed balance with their equilibria as there are (sufficiently nontrivial) symmetric nonnegative matrices.*

Remark: Let $\langle \cdot, \cdot \rangle$ be a fixed scalar product on $\mathbb{R}^N$ and suppose that the map T_P from the preceding proposition is self-adjoint with respect to $\langle \cdot, \cdot \rangle$. Then the eigenvalues of T_P are real and there is an orthogonal basis $x_1, \ldots, x_N$ of eigenvectors[1]. If the x_i were also orthogonal with respect to $\langle \cdot, \cdot \rangle_\pi$ then it would follow that T_P is $\langle \cdot, \cdot \rangle_\pi$–self-adjoint and thus P would be reversible. In general, however, much less can be shown: if (without loss of generality) x_1 is an eigenvector with associated eigenvalue 1, then $\langle x_1, x_i \rangle_\pi = 0$ for $i = 2, \ldots, N$.

> Denote by S the matrix for which the i'th column is just x_i. Then $S^{-1}PS$ is diagonal with the eigenvalues on the diagonal. Hence $(S^{-1}PS)^k = S^{-1}P^kS$ has the k'th power of the eigenvalues on the diagonal. If k tends to infinity this proves that $S^{-1}WS$ is the diagonal matrix with $1, 0, \ldots, 0$ on the diagonal. And this yields
>
> $$\langle x_1, x_i \rangle_\pi = \langle (c, c, \ldots, c)^\top, x_i \rangle_\pi = c\pi^\top x_i = 0$$
>
> for $i > 1$.

Here is a concrete example of a P with real eigenvalues which is not reversible. Define, for $a, b \geq 0$ with $b \leq a$, $2a + b = 1$,

$$P := \begin{pmatrix} a & a-b & 2b \\ a+b & b & a-b \\ 0 & a+b & a \end{pmatrix}.$$

P is doubly stochastic, hence the equilibrium is the uniform distribution, and we see that P is *not* reversible if $b > 0$. On the other hand, the eigenvalues of P are easily calculated if $b = 0$: we obtain $1, 1/2, -1/2$. And thus for small positive b we have a chain which is not reversible but nevertheless has three different real eigenvalues and thus a basis of eigenvectors.

An estimate using eigenvalues

Suppose that an aperiodic and irreducible P is such that for a suitable invertible matrix S the product $S^{-1}PS$ is diagonal. As diagonal entries of this matrix we find the eigenvalues of P, that is one entry is one and the others are less than one in absolute value. Therefore the rate of convergence of the powers $(S^{-1}PS)^k = S^{-1}P^kS$ is known, and this allows one – in principle – to derive estimates for the difference between P^k and W in this general situation. However, since we want more explicit bounds we confine ourselves to the restricted class of reversible chains.

Let P be irreducible, aperiodic and reversible with equilibrium $\pi^\top$. We define $\delta(k)$ to be the *maximal relative error* when approximating the components of π by the $p_{ij}^{(k)}$:

$$\delta(k) := \max_{ij} |p_{ij}^{(k)} - \pi_j|/\pi_j,$$

and we recall that λ^* stands for the maximum of the $|\lambda|$, where λ runs through all eigenvalues of P which are different from 1. We then claim:

---

[1] Note that the converse is also true: if T_P has real eigenvalues and if it is possible to find a basis $x_1, \ldots, x_N$ of eigenvectors, then there is a scalar product such that T_P is self-adjoint and the $x_1, \ldots, x_N$ are orthonormal.

Theorem 10.3

   (i) $\delta(k) \le (\lambda^*)^k / \min_i \pi_i$ *for all* k.

   (ii) $\delta(k) \ge (\lambda^*)^k$ *for all even* k; *if all eigenvalues are nonnegative then this inequality holds for all* k.

Proof. (i) With the notation of proposition 10.2 we denote by A the symmetric matrix DPD^{-1}. Choose an orthogonal matrix $S = (s_{ij})_{ij}$ such that $B := SAS^{-1}$ is diagonal. Note that this means that the rows $e_1^\top, \dots, e_N^\top$ as well as the columns $f_1, \dots, f_N$ of S are orthogonal with respect to the *ordinary scalar product* on $\mathbb{R}^N$: $\langle (a_i), (b_i) \rangle := \sum_i a_i b_i$. Also, $S^{-1} = S^\top$ holds so that, with Kronecker's δ-notation, $e_i^\top f_j = \delta_{ij}$. Without loss of generality we may assume that

$$
B = \begin{pmatrix}
1 & 0 & 0 & \cdots & 0 \\
0 & \lambda_2 & 0 & \cdots & 0 \\
\vdots & \vdots & \vdots & & \vdots \\
0 & 0 & 0 & \cdots & \lambda_N
\end{pmatrix},
$$

with $1 > |\lambda_2| \ge \cdots \ge |\lambda_N|$.

From $SA = BS$ it follows that the $e_i^\top$ are left eigenvectors of A with associated eigenvalue λ_i. In particular $e_1^\top A = e_1^\top$ and thus $(e_1^\top D)P = e_1^\top D$ hold, and consequently there is a number c such that $e_1^\top D = c\pi^\top$. Hence $e_1^\top = c\pi^\top D^{-1}$, and therefore, since $e_1^\top$ as well as $\pi^\top D^{-1}$ are normalized with respect to $\langle \cdot, \cdot \rangle$, we know that $c^2 = 1$. We may and will assume that $c = 1$, i.e., $e_1^\top = (\sqrt{\pi_1}, \dots, \sqrt{\pi_N})$.

The $e_i^\top$ are orthonormal, i.e., $e_i^\top e_j = \delta_{ij}$. Hence the $N \times N$-matrix $e_i e_i^\top$ has the property that $e_i e_i^\top e_j = \langle e_j, e_i \rangle e_i$. Thus, by linearity, $e_i e_i^\top x = \langle x, e_i \rangle e_i$ for *all* x, and we get (for arbitrary k)

$$
\begin{aligned}
A^k x &= A^k \Big(\sum_i \langle x, e_i \rangle e_i \Big) \\
&= \sum_i \langle x, e_i \rangle A^k e_i \\
&= \sum_i \lambda_i^k \langle x, e_i \rangle e_i \\
&= \sum_i \lambda_i^k e_i e_i^\top x.
\end{aligned}
$$

Therefore the matrices A^k and $\sum_i \lambda_i^k e_i e_i^\top$ coincide, and this provides a useful representation of P^k:

$$
P^k = (D^{-1}AD)^k = D^{-1}A^k D = \sum_i \lambda_i^k D^{-1} e_i e_i^\top D.
$$

The first summand is easy to evaluate since we know e_1; the matrix $D^{-1} e_1 e_1^\top D$ is just the matrix $\lim P^k = W$ each row of which coincides with $\pi^\top$. Written explicitly this means that

$$
p_{ij}^{(k)} = \pi_j + \sqrt{\pi_j / \pi_i} \sum_{l=2}^{N} \lambda_l^k s_{li} s_{lj},
$$

and it follows with the help of the Cauchy-Schwarz inequality that

$$\frac{|p_{ij}^{(k)} - \pi_j|}{\pi_j} \;=\; \frac{1}{\sqrt{\pi_i \pi_j}} \Big| \sum_{l=2}^{N} \lambda_l^k s_{li} s_{lj} \Big|$$

$$\leq\; \frac{(\lambda^*)^k}{\min_l \pi_l} \sum_{l=1}^{N} |s_{li}| |s_{lj}|$$

$$\leq\; \frac{(\lambda^*)^k}{\min_l \pi_l} \|f_i\| \, \|f_j\|$$

$$=\; \frac{(\lambda^*)^k}{\min_l \pi_l}.$$

This proves that $\delta(k) \leq (\lambda^*)^k / \min_l \pi_l$.

(ii) It follows from the above calculations that

$$\delta(k) \;\geq\; \max_j \frac{1}{\pi_j} |p_{jj}^{(k)} - \pi_j|$$

$$=\; \max_j \Big[\frac{1}{\pi_j} \Big| \sum_{l=2}^{N} \lambda_l^k s_{lj}^2 \Big| \Big].$$

Now suppose that k is even or that all λ_i are nonnegative. Then we can continue with

$$\geq\; \max_j \Big[\frac{1}{\pi_j} (\lambda^*)^k s_{2j}^2 \Big]$$

$$=\; (\lambda^*)^k \max_j \frac{1}{\pi_j} s_{2j}^2,$$

where we have made use of the fact that $\lambda^* = |\lambda_2| \geq |\lambda_3| \geq \cdots$.

We claim that $\max_j s_{2j}^2 / \pi_j \geq 1$ (which will finish the proof): otherwise $s_{2j}^2 < \pi_j$ would hold for all j in contrast to $1 = \sum_j s_{2j}^2 = \sum \pi_j$. $\qquad \square$

The proposition shows that, as it was to be expected, the *order* of convergence is proportional to $(\lambda^*)^k$. Note that the more interesting upper bound involves the equilibrium distribution, and therefore the result will be rather useless if there exist no estimates for the minimum of the π_i.

However, in many applications π *is* known explicitly, often it will be the uniform distribution. In this case the theorem gives the best possible bound, namely an estimate of order $N(\lambda^*)^k$. Even this looks not promising if the cardinality N of the state space is large, but one should have in mind that $(\lambda^*)^k$ decreases rapidly if λ^* is not too close to one.

In order to apply the proposition one needs to know λ^* or at least some reasonable estimates. We will see in the next chapter how this can be achieved by introducing the "conductance", a number associated with the chain which sometimes can be easier calculated than the eigenvalues but which nevertheless helps to bound λ^* away from one.

Here we only will indicate how the problem can be transformed into an *optimization problem*. Suppose that – as before – the chain is reversible and that we know for some reason that *all eigenvalues are nonnegative*.

In general this might be difficult to decide. However, if P is irreducible, aperiodic and reversible, then $\tilde{P} := (Id + P)/2$ has the same properties and even the same equilibrium distribution; this can be checked easily. The eigenvalues $\tilde{\lambda}$ of $\tilde{P}$ are just the numbers $(1 + \lambda)/2$, where λ runs through the eigenvalues of P. Consequently all of them are nonnegative.

Note that the passage from P to $\tilde{P}$ means a "slowing down" of the chain: the chain behaves essentially as before, but on the average there are only k real moves among $2k$ possible steps.

Then λ^* is the *second-largest eigenvalue*, and this number can be determined as follows. Let $\lambda_1 = 1 > \lambda_2 \geq \ldots \geq \lambda_N \geq 0$ be the eigenvalues in decreasing order, we are interested in λ_2. Choose an orthonormal basis $e_1, \ldots, e_N$ of eigenvectors of $A := DPD^{-1}$, where $\mathbb{R}^N$ is provided with the usual scalar product $\langle \cdot, \cdot \rangle$; we will assume that $Ae_i = \lambda_i e_i$ for $i = 1, \ldots, N$.

Denote by V the linear span of the vectors $e_2, \ldots, e_N$. V is just the orthogonal complement of $\mathbb{R}e_1$, a typical normalized element x of V has the form $x = \sum_{i=2}^{N} a_i e_i$, where the a_i are real with $\sum_i a_i^2 = 1$. Consequently $\langle x, Ax \rangle = \sum_{i=2}^{N} \lambda_i a_i^2$, and we see that this number assumes λ_2 as its maximal value.

A similar argument holds for arbitrary self-adjoint matrices, and we arrive at

Proposition 10.4 *Let $\langle \cdot, \cdot \rangle$ be any scalar-product on $\mathbb{R}^N$ and A a matrix which is self-adjoint with respect to $\langle \cdot, \cdot \rangle$. If the largest eigenvalue is simple with associated eigenvector e, then the second-largest eigenvalue λ_2 satisfies*

$$\lambda_2 = \max\left\{ \frac{\langle x, Ax \rangle}{\langle x, x \rangle} \;\middle|\; x \perp e, \; x \neq 0 \right\}.$$

Estimates which use the entries of P directly

We start with an arbitrary stochastic matrix P and consider, for $k = 1, \ldots$, the minimum and the maximum of the j'th column of P^k:

$$m_j^{(k)} := \min_i p_{ij}^{(k)}, \quad M_j^{(k)} := \max_i p_{ij}^{(k)}.$$

A convex combination of some numbers always dominates their minimum, and thus

$$
\begin{aligned}
m_j^{(k+1)} &= \min_i p_{ij}^{(k+1)} \\
&= \min_i \sum_l p_{il} p_{lj}^{(k)} \\
&\geq \min_i \sum_l p_{il} m_j^{(k)} \\
&= m_j^{(k)}.
\end{aligned}
$$

Similarly one gets $M_j^{(k+1)} \leq M_j^{(k)}$ so that

$$m_j^{(1)} \leq m_j^{(2)} \leq \cdots \leq M_j^{(2)} \leq M_j^{(1)}. \tag{10.4}$$

Now suppose that P is irreducible and aperiodic. Then the $p_{ij}^{(k)}$ will tend to π_j with $k \to \infty$. From (10.4) it follows immediately that with $p_{ij}^{(k)}$ also π_j lies in the interval $[m_j^{(k)}, M_j^{(k)}]$, and therefore $|\pi_j - p_{ij}^{(k)}|$ can be bounded by $M_j^{(k)} - m_j^{(k)}$. We will try to find estimates for this "variation in the j'th column of P^k".

Let δ be the minimum of the p_{ij}, $i,j = 1,\ldots,N$. From $\sum_j p_{ij} = 1$ it follows that in every row there is a j with $p_{ij} \leq 1/N$ and thus δ can be bounded from above by $1/N$. Hence $\tau := 1 - N\delta$ lies between 0 and 1, and we claim that

$$M_j^{(k+1)} - m_j^{(k+1)} \leq \tau(M_j^{(k)} - m_j^{(k)});$$

this implies that

$$M_j^{(k+1)} - m_j^{(k+1)} \leq \tau^k(M_j^{(1)} - m_j^{(1)}),$$

and we get geometrically fast convergence of $p_{ij}^{(k)}$ to π_j if $\tau < 1$.

To prove the claim we fix arbitrary i_0 and j_0. It will be convenient to denote by Δ' (resp. Δ'') the collection of those j in $\{1,\ldots,N\}$ where $p_{i_0 j} \geq p_{j_0 j}$ (resp. $p_{i_0 j} < p_{j_0 j}$) and to abbreviate the clumsy expressions $\sum_{j \in \Delta'} \cdots$ and $\sum_{j \in \Delta''} \cdots$ by $\sum' \cdots$ and $\sum'' \cdots$. First we observe that

$$\sum{}' (p_{i_0 j} - p_{j_0 j}) + \sum{}'' (p_{i_0 j} - p_{j_0 j}) = \sum_j p_{i_0 j} - p_{j_0 j} = 0.$$

Also, from $\sum'' p_{i_0 j} + \sum' p_{j_0 j} \geq N\delta$ and $\sum' p_{i_0 j} + \sum'' p_{i_0 j} = 1$, we get

$$\sum{}' (p_{i_0 j} - p_{j_0 j}) = 1 - \sum{}'' p_{i_0 j} - \sum{}' p_{j_0 j} \leq 1 - N\delta,$$

and these calculations imply that, for arbitrary s,

$$
\begin{aligned}
p_{i_0 s}^{(k+1)} - p_{j_0 s}^{(k+1)} &= \sum_j (p_{i_0 j} - p_{j_0 j}) p_{js}^{(k)} \\
&= \sum{}' (p_{i_0 j} - p_{j_0 j}) p_{js}^{(k)} + \sum{}'' (p_{i_0 j} - p_{j_0 j}) p_{js}^{(k)} \\
&\leq \sum{}' (p_{i_0 j} - p_{j_0 j}) M_s^{(k)} + \sum{}'' (p_{i_0 j} - p_{j_0 j}) m_s^{(k)} \\
&= \sum{}' (p_{i_0 j} - p_{j_0 j}) M_s^{(k)} - \sum{}' (p_{i_0 j} - p_{j_0 j}) m_s^{(k)} \\
&= \left[\sum{}' (p_{i_0 j} - p_{j_0 j}) \right] (M_s^{(k)} - m_s^{(k)}) \\
&\leq \tau(M_s^{(k)} - m_s^{(k)}).
\end{aligned}
$$

This holds for arbitrary i_0, j_0, and our claim follows.

With these preparations at hand we are ready to prove

Proposition 10.5 *Let P be an irreducible and aperiodic stochastic matrix, and let k_0 be such that P^{k_0} is strictly positive[2]. Denote by δ the minimum of the entries of P^{k_0} and put $\tau := 1 - N\delta$; note that $0 \leq \tau < 1$.*

(i) If $k_0 = 1$, then $|p_{ij}^{(k)} - \pi_j| \leq \tau^k$ for all states i,j and all k.

---

[2] Recall that such a k_0 exists by lemma 7.3.

(ii) *In the case $k_0 > 1$ this inequality has to be modified:*
$$|p_{ij}^{(k)} - \pi_j| \leq (1/\tau)\tau^{k/k_0}.$$
Therefore the $p_{ij}^{(k)}$ approach π_j for every i geometrically fast.

Proof. Part (i) is a direct consequence of the calculations preceding the proposition, one only has to note that $M_j^{(1)} - m_j^{(1)} \leq \tau$ (if $M_j^{(1)} = p_{i_0 j}$, then $1 = M_j^{(1)} + \sum_{s \neq j} p_{i_0 s} \geq M_j^{(1)} + (N-1)\delta$ so that $M_j^{(1)} - m_j^{(1)} \leq 1 - N\delta$).

To prove (ii), we need an elementary fact about decreasing sequences. Let (c_k) be a decreasing sequence of nonnegative numbers such that $c_{k'k_0} \leq \tau^{k'}$ for some fixed k_0, $\tau < 1$ and all k'. If then k is arbitrary we may write $k = k'k_0 + s$ with an $s < k_0$, and thus
$$c_k \leq c_{k'k_0} \leq \tau^{k'} = (\tau^{1/k_0})^k \tau^{-s/k_0} \leq (\tau^{1/k_0})^k/\tau.$$
Here, this has to be applied to the sequence
$$c_k := M_j^{(k)} - m_j^{(k)}.$$

Monotonicity has been shown above, and the estimate follows by passing from P to P^{k_0} and an application of (i). $\qquad\qquad\square$

A comparison of theorem 10.3 with the preceding proposition reveals that they deal with *different aspects of closeness*. Whereas the former bounds the *relative* error the latter makes assertions about the *absolute* difference between the $p_{ij}^{(k)}$ and the π_j. Another difference is that theorem 10.3 provides the correct order of convergence whereas proposition 10.5 might produce poor bounds[3].

To illustrate proposition 10.5 consider the simple example
$$P := \frac{1}{4}\begin{pmatrix} 2 & 2 \\ 1 & 3 \end{pmatrix}.$$

Here we may choose $k_0 = 1$ with $\delta = 1/4$ so that $\tau = 1 - 2/4 = 1/2$. The equilibrium is $\pi^{\mathsf{T}} = (1/3,\, 2/3)$, and thus proposition 10.5 predicts $|p_{i1}^{(k)} - 1/3|$, $|p_{i2}^{(k)} - 3/4| \leq 1/2^k$. An explicit calculation shows that
$$P^2 = \frac{1}{16}\begin{pmatrix} 6 & 10 \\ 5 & 11 \end{pmatrix}, \text{ and therefore } P^4 = \frac{1}{256}\begin{pmatrix} 86 & 170 \\ 85 & 171 \end{pmatrix}.$$

Thus, for example, $|p_{11}^{(4)} - \pi_1| = |86/256 - 1/3| = 1/384$, but from the proposition we only know that the error is $\leq 1/16$.

With some care it is sometimes possible to get better estimates. Denote by δ_{k_0} the minimum of the coefficients of P^{k_0} and put $\tau_{k_0} := 1 - N\delta_{k_0}$. Then the error in the k'th step can be bounded (essentially) by $(\tau_{k_0})^{k/k_0}$, and it might happen that – for two admissible k_0, k_1 – the numbers $(\tau_{k_0})^{k_0}, (\tau_{k_1})^{k_1}$ differ. Here is *an example:* for $P := \begin{pmatrix} 1/10 & 9/10 \\ 5/10 & 5/10 \end{pmatrix}$

we have $P^2 = \begin{pmatrix} 23/50 & 27/50 \\ 15/50 & 35/50 \end{pmatrix}$. Hence $\delta_1 = 1/10, \tau_1 = 4/5$ and $\delta_2 = 15/50, \tau_2 = 2/5$; and since $2/5 < (4/5)^2$ it is better to apply proposition 10.5 with $k_0 = 2$ rather than with $k_0 = 1$. Usually one will try to find a compromise between choosing k_0 large (= good bounds, but many calculations) or small.

---

[3] If, e.g., P is strictly positive with identical rows, then the $p_{ij}^{(k)}$ coincide with the π_j. But τ in proposition 10.5 is zero only if all rows of P correspond to the uniform distribution.

We are now going to present *a second approach without eigenvalues*, it is based on contracting maps and *Banach's fixed point theorem*.

Let (M, d) be a non-empty metric space and $T : M \to M$ a map such that $d(Tx, Ty) \le Ld(x, y)$ for some $L < 1$ and all x, y. If x_0 is a *fixed point* of T, i.e., if $Tx_0 = x_0$, then

$$d(T^k x, x_0) = d(T^k x, T^k x_0) \le L^k d(x, x_0) \qquad (10.5)$$

for arbitrary $x \in M$ and all k. Therefore the iterations $T^k x$ converge geometrically fast to x_0, and in particular it follows that there is at most one fixed point. Banach's fixed point theorem asserts that a fixed point in fact exists if (M, d) is assumed to be complete.

This will now be applied to investigate the convergence of a Markov chain to its equilibrium. We fix an arbitrary stochastic $N \times N$-matrix P. The complete metric space M which will be of importance here is the set

$$M := \{(x_1, \ldots, x_N) \mid x_i \ge 0, \sum x_i = 1\},$$

a subset of $\mathbb{R}^N$ provided with the l^1-norm[4]. The map $T : M \to M$ is defined by

$$T : (x_1, \ldots, x_N) \mapsto \Big(\sum_j x_j p_{j1}, \ldots, \sum_j x_j p_{jN}\Big).$$

Lemma 10.6 *Denote by R_i the i'th row of P, $i = 1, \ldots, N$, and put*

$$C_P := \frac{1}{2} \max_{i,j} \|R_i - R_j\|_1.$$

(i)   *The l^1-diameter of the convex hull K of the R_i is $2C_P$.*

(ii) $\|Tx - Ty\|_1 \le C_P \|x - y\|_1$ *for $x, y \in M$.*

(iii) *Let x_0 be a fixed point of T. Then*

$$\|T^k x - x_0\|_1 \le 2(C_P)^k$$

for all x and all k.

Proof. (i) Denote the diameter of K by C'. Then $2C_P \le C'$ holds by definition. Fix any R_{j_0}. All R_j lie in the (convex!) ball $B(R_{j_0}, 2C_P)$ with center R_{j_0} and radius $2C_P$ by the definition of C_P. Thus $K \subset B(R_{j_0}, 2C_P)$, or $\|x - R_{j_0}\|_1 \le 2C_P$ for all $x \in K$. Now fix any $x_0 \in K$. By the first part all R_j are contained in the ball $B(x_0, 2C_P)$ so that – again by convexity – $K \subset B(x_0, 2C_P)$, i.e., $\|x_0 - y\|_1 \le 2C_P$ for all $y \in K$.
(For those who have some background in convexity we also include a "one-line-proof": the assertion is a consequence of the fact that a convex function (here: $(x, y) \mapsto \|x - y\|_1$, from $K \times K$ to $\mathbb{R}$) assumes its maximum at an extreme point; note that the extreme points of $K \times K$ are of the form (R_i, R_j).)
(ii) Let $a_1, \ldots, a_N$ be real numbers such that $\sum a_i = 0$, we claim that

$$\|\sum_i a_i R_i\|_1 \le C_P \|(a_1, \ldots, a_N)\|_1.$$

---

[4] That is $\|(x_1, \ldots, x_N)\|_1 := |x_1| + \cdots + |x_N|$.

Without loss of generality we may assume that $\|(a_1,\ldots,a_N)\|_1 = 2$. Denote by $\sum'$ (resp. $\sum''$) summation over those indices j where $a_j > 0$ (resp. $a_j \leq 0$). Then $\sum' a_j = -\sum'' a_j = 1$ so that the vectors $x := \sum' a_j R_j$, $y := -\sum'' a_j R_j$, being convex combinations of the R_j, lie in K. Hence, by (i), $2C_P \geq \|x - y\|_1 = \|\sum a_j R_j\|_1$, and this proves the claim.

For the proof of (ii) it only remains to consider, for given $x, y \in M$, the vector $(a_1,\ldots,a_N) := x - y$.

(iii) This follows from (10.5) and the fact that the diameter of M is two. $\qquad\square$

Now there are *two possibilities* to proceed further. The first one is to take the existence of an equilibrium for granted (theorem 7.4):

Proposition 10.7 *Let P be irreducible and aperiodic with equilibrium π. Then $\sum_j |p_{ij}^{(k)} - \pi_j| \leq 2(C_P)^k$ for all i and all k.*

Proof. This follows from part (iii) of the preceding lemma, one has to apply this assertion to $x_0 = (\pi_1,\ldots,\pi_N)$ and $x = (0,\ldots,0,1,0,\ldots,0)$ with the "1" at the i'th position. $\quad\square$

Also we can apply Banach's theorem to get at the same time the existence of the equilibrium and geometrically fast convergence:

Proposition 10.8 *Let P be irreducible and aperiodic, choose k_0 such that P^{k_0} is strictly positive[5]. Denote by C the number $C_{P^{k_0}}$ ($= 0.5$ times the maximal l^1-distance between the rows of P^{k_0}). Then:*

(i) *$C < 1$; therefore, by lemma 10.6 (ii) and Banach's fixed-point theorem, there is a unique fixed point $\pi^\top = (\pi_1,\ldots,\pi_N)$ of T^{k_0} in M, where $T : (x_1,\ldots,x_N) \mapsto (x_1,\ldots,x_N)P$.*

(ii) *π also satisfies $(\pi_1,\ldots,\pi_N)P = (\pi_1,\ldots,\pi_N)$ so that it is the equilibrium distribution. One has*

$$\sum_j |p_{ij}^{(k)} - \pi_j| \leq \frac{2}{C} C^{k/k_0}$$

for all i and k.

Proof. (i) This is clear: if two vectors x, y have have strictly positive components, then $\|x - y\|_1 < \|x\|_1 + \|y\|_1$. For the case under consideration this means that each two rows have an l^1-distance which is strictly less than two.

(ii) From

$$T^{k_0}(T(\pi^\top)) = T^{k_0+1}(\pi^\top) = T(T^{k_0}(\pi^\top)) = T(\pi^\top)$$

it follows that both $\pi^\top$ and $T(\pi^\top)$ are fixed-points of T^{k_0}. Thus, by uniqueness, $T(\pi^\top) = \pi^\top$.

Now let x be the vector $(0,\ldots,0,1,0,\ldots,0)$ (with 1 at the i'th position) and k an arbitrary integer. We write $k = rk_0 + k'$ with $0 \leq k' < k_0$ and $r \in \mathbb{N}_0$. Then

$$\sum_j |p_{ij}^{(k)} - \pi_j| \;=\; \|xP^k - \pi^\top\|_1$$

$$\;=\; \|xP^k - \pi^\top P^k\|_1$$

---

[5] Cf. lemma 7.3.

$$
\begin{aligned}
&= \; \|(x - \pi^{\mathsf{T}})P^{rk_0 + k'}\|_1 \\
&= \; \|(x - \pi^{\mathsf{T}})P^{rk_0}\|_1 \\
&\leq \; C^r \|x - \pi^{\mathsf{T}}\|_1 \\
&\leq \; 2\, C^{(k-k')/k_0} \\
&\leq \; \frac{2}{C}\, C^{k/k_0},
\end{aligned}
$$

and the proof is complete. $\qquad\square$

An application: bounds in the renewal theorem

As an application of the preceding results we want to find reasonable bounds for the rate of convergence in the renewal theorem. Recall that we have introduced the basic definitions at the end of chapter 6:

- We assume that we are given nonnegative numbers $f_1, f_2, \ldots$ such that $\sum f_i = 1$. In order to be able to work with matrices and to arrive at an irreducible and aperiodic chain we will assume that there is an N such that $f_i = 0$ for $i > N$ and $f_i > 0$ for $i = 1, \ldots, N$. (It is not too hard to get rid of these restrictions, in the general situation, however, there arise some notational and technical complications.)

- The f_i are used to play the following "game": start at zero, and in every step move from n to $n + m$ with probability f_m.

- Denote by ρ_n the probability that a player meets position n. Then the ρ_n satisfy the recurrence relations on page 46, and they converge to $1/\mu := 1/(1f_1 + \cdots + Nf_N)$.

How fast do they converge? For example, how close to $1/7$ is the probability that a player in the two-dice-game described at the end of chapter 6, page 46, will meet the position 101, say?

To deal with this problem, we will transform it such that the $|\rho_k - 1/\mu|$ correspond to certain $|p_{ij}^{(k')} - \pi_j|$ for a Markov chain on a *finite* state space. The idea is to pass from an investigation of single ρ_k to a study of *blocks* $\rho_{k+1}, \ldots, \rho_{k+N}$:

Lemma 10.9 *Define a matrix P to be the N'th power of*

$$
Q := \begin{pmatrix}
0 & 1 & 0 & \cdots & 0 \\
0 & 0 & 1 & \cdots & 0 \\
\vdots & \vdots & \vdots & & \vdots \\
0 & 0 & 0 & \cdots & 1 \\
f_N & f_{N-1} & f_{N-2} & \cdots & f_1
\end{pmatrix}.
$$

Then P is an irreducible and aperiodic stochastic matrix such that

$$
\begin{pmatrix}
\rho_{k+N} \\
\vdots \\
\rho_{k+2N-1}
\end{pmatrix}
= P
\begin{pmatrix}
\rho_k \\
\vdots \\
\rho_{k+N-1}
\end{pmatrix}.
\tag{10.6}
$$

Proof. The recurrence relations for the ρ_k (page 46) may be rephrased as

$$\begin{pmatrix} \rho_{k+1} \\ \vdots \\ \rho_{k+N} \end{pmatrix} = Q \begin{pmatrix} \rho_k \\ \vdots \\ \rho_{k+N-1} \end{pmatrix},$$

and by applying this relation N times it follows that multiplication by $P = Q^N$ describes the transition from $(\rho_k, \ldots, \rho_{k+N-1})^\top$ to $(\rho_{k+N}, \ldots, \rho_{k+2N-1})^\top$. With Q also P is a stochastic matrix, and since all f_i are strictly positive the same is true for all entries of P. Hence this matrix is irreducible and aperiodic. $\qquad\square$

Let us put $\rho_0 := 1$ and $\rho_i := 0$ for negative i. Then (10.6) also holds for $k = -N + 1$, and thus the vector $(\rho_{rN+1}, \ldots, \rho_{(r+1)N})^\top$ is the last column of P^r (= the product from the right of P^r by $(\rho_{-N+1}, \ldots, \rho_0)^\top = (0, \ldots, 1)^\top$) for every r. And therefore the distance of the $\rho_{rN+1}, \ldots, \rho_{(r+1)N}$ to π_N (= the N'th component of the equilibrium) is precisely the distance of the $p_{1N}^{(r)}, \ldots, p_{NN}^{(r)}$ to that number. In this way we get bounds for the convergence of the renewal sequence by using the estimates developed in this chapter.

E.g., from proposition 10.5 we get the following qualitative version of the discrete renewal theorem:

Proposition 10.10 *Let P be the strictly positive matrix of the preceding lemma, by δ we denote the minimum of its entries. Then the distance of the $\rho_{rN+1}, \ldots, \rho_{(r+1)N}$ to $1/\mu$ can be bounded by $(1 - N\delta)^r$.*

Here is an **example**, we consider the case $f_1 = 1/10, f_2 = 3/10, f_3 = 6/10$. Then

$$Q = \frac{1}{10} \begin{pmatrix} 0 & 10 & 0 \\ 0 & 0 & 10 \\ 6 & 3 & 1 \end{pmatrix} \text{ and thus } P = \frac{1}{1000} \begin{pmatrix} 600 & 300 & 100 \\ 60 & 630 & 310 \\ 186 & 153 & 661 \end{pmatrix}.$$

We have $N\delta = 81/100$, and thus we get the information that after $3r$ steps the ρ_k are close to $1/(f_1 + 2f_2 + 3f_3) = 2/5$ with an error of $(81/100)^r$.

Similarly one can use proposition 10.8, by this result we can bound the distance by $2(549/1000)^r$.

Exercises

10.1: Prove what has been claimed about the existence of the time-reversal chain: $\tilde{P}$ exists for irreducible P, and P and $\tilde{P}$ have the same equilibrium distribution.

10.2: If P is a doubly stochastic (resp. a symmetric) irreducible matrix, then so is $\tilde{P}$.

10.3: Let P be an arbitrary stochastic 2×2-matrix. Under what conditions on the entries is P irreducible? Calculate for these P the unique equilibrium and provide an explicit form of the time-reversal chain in terms of the entries of P.

10.4: Why is it necessary to restrict the definition of the time-reversal chain to irreducible chains?

10.5: In proposition 10.2 we have identified P with an operator T_P on the Hilbert space H_π. What is the relation between the adjoint operator T_P^* and the time reversal of P?

10.6: Let an irreducible and aperiodic stochastic matrix be such that all rows are identical. Does it follow that P is reversible?

10.7: Identify the reversible chains among the examples of chapter 2.

10.8: Prove that a strictly positive doubly stochastic matrix is in detailed balance iff it is symmetric.

10.9: Prove the assertions of proposition 10.2.

10.10: Consider the irreducible stochastic matrix

$$P = \frac{1}{10} \begin{pmatrix} 1 & 1 & 8 \\ 3 & 3 & 4 \\ 5 & 5 & 0 \end{pmatrix} .$$

Use the results of this chapter to decide how fast the rows of P^k tend to the equilibrium.

10.11: Let a renewal process be defined by

$$f_1 = f_2 = 1/100, \quad f_3 = 98/100.$$

What can be said about the rate of convergence of the ρ_k?

11   Conductance

The origin of the technique we are going to describe now lies in graph theory ([7], [8]) where the relation between the second-largest eigenvalue and certain structural properties of the graph under consideration was investigated first. This was extended to the Markov chain setting in [2].

The idea is to associate a constant – the *conductance* – with a Markov chain which measures the "strength of mixing". The definition is rather natural, it is, however, surprisingly difficult to prove rigorously that and how the conductance is related to the convergence of the chain to its equilibrium distribution. It will be shown that the knowledge of the conductance is essentially as good as information about the *size of the second-largest eigenvalue*. In view of proposition 10.3 this is not precisely what is needed since the second-largest eigenvalue λ_2 will not necessarily coincide with λ^*, the maximal modulus of the eigenvalues which are different from 1. However, this difficulty is easy to remedy since $\lambda^* = \lambda_2$ will hold if all eigenvalues are positive, and this can be achieved by passing from P to $(Id + P)/2$.

But there remains another problem, namely to *determine the conductance for a given chain numerically*. A tremendously large number of calculations has to be done – about 2^N if the state space has N elements –, and therefore a naive approach will usually not be successful. But one may find at least reasonable *bounds for the conductance* by a technique which was discovered by *Sinclair* (see [70], chapter 3.1), the method of the *canonical paths*.

These introductory remarks are also thought of as a schedule for the present chapter, we start with

Definition 11.1 Let $P = (p_{ij})_{i,j=1,\ldots N}$ be an irreducible, aperiodic and reversible stochastic matrix with equilibrium $(\pi_1,\ldots,\pi_N)^\top$. For $T \subset S := \{1,\ldots,N\}$ we define

$$C_T := \sum_{i \in T} \pi_i \text{ and } F_T := \sum_{i \in T, j \notin T} \pi_i p_{ij};$$

these numbers are called the *capacity of T* and *the ergodic flow from T to $S \setminus T$*, respectively. Φ_T denotes the quotient F_T/C_T, and the *conductance* of P (or the associated chain) is

$$\Phi := \min_{T \subset S} \max\{\Phi_T, \Phi_{S\setminus T}\}.$$

In order to see why this quantity can serve as a measure of mixing it is helpful to understand the *probabilistic meaning* of C_T, F_T and Φ_T. Fix T, start the chain anywhere and wait for some time. Then the chain will be close to its equilibrium, that is the probability to find it in a certain state i is (approximately) π_i, and consequently C_T is the probability that the chain occupies some position in T. Since we have already convinced ourselves[1] that $\pi_i p_{ij}$ is the probability to observe a jump from i to j it is clear

---

[1] See the discussion following (10.2).

that with probability F_T one sees a transition from T to its complement. And therefore Φ_T is a *conditional probability*: how likely is it that the chain leaves T in the next step under the condition that it is in fact in T now? Hence a *small* Φ_T means that T is something like a *trap* for the chain, and therefore it is to be expected that a good mixing rate will be related with a big conductance.

Note, however, that both T and $S \setminus T$ are involved. Thus a high conductance Φ is compatible with a small Φ_{T_0} for some T_0 provided that this is balanced by a big $\Phi_{T \setminus T_0}$.

Since P is reversible, the numbers F_T and $F_{S \setminus T}$ coincide. Therefore Φ_T will be bigger than $\Phi_{S \setminus T}$ iff C_T is smaller than $C_{S \setminus T}$. But $C_T + C_{S \setminus T} = 1$, and therefore the conductance could have alternatively been defined as

$$\Phi := \min\{\Phi_T \mid T \subset S, 0 < C_T \le 1/2\}.$$

Let us consider some *examples*. Suppose first that P is such that all rows are identical, they then necessarily will coincide with the equilibrium $\pi^\top$. In this case we have $F_T = C_T C_{S \setminus T}$ so that Φ_T equals $C_{S \setminus T}$, and therefore the conductance Φ is given by $\min\{\sum_{i \notin T} \pi_i \mid T \subset S \text{ such that } 0 < \sum_{i \in T} \pi_i \le 1/2\}$. Note that even in this particularly simple situation an exact calculation of Φ can be cumbersome if S is large.

As a second example consider $P := \begin{pmatrix} 1-a & a \\ b & 1-b \end{pmatrix}$ for a, b strictly between 0 and 1.

The equilibrium is $(b/(a+b), a/(a+b))$, and it follows easily that the conductance is $\max\{a, b\}$.

And finally let P be the chain associated with the cyclic random walk on $\{1, \ldots, N\}$:

$$P := \begin{pmatrix} 1-2p & p & 0 & 0 & \cdots & 0 & p \\ p & 1-2p & p & 0 & \cdots & 0 & 0 \\ 0 & p & 1-2p & p & \cdots & 0 & 0 \\ \vdots & \vdots & \vdots & \vdots & & \vdots & \vdots \\ p & 0 & 0 & 0 & \cdots & p & 1-2p \end{pmatrix},$$

where $0 < p < 1/2$. The equilibrium is the uniform distribution, and thus we have to consider the T such that $r := \operatorname{card}(T)$ is at most $N/2$. For the T with fixed $r \le N/2$ the ergodic flow Φ_T will depend on how "scattered" T is, the minimum will obviously be attained precisely when T is of the form $\{k, k+1, \ldots, k+r-1\} \pmod{N}$. Since in this case $F_T = 2p/N$ and thus $\Phi_T = 2p/r$ hold it follows that $\Phi = 2p/r'$, where $r' := [N/2]$ denotes the largest integer less than or equal to $N/2$. Therefore the conductance is only of the order $1/N$.

For general Markov chains – even in the case of moderate N – it will be difficult to determine Φ exactly, often it will only be possible to derive estimates. A rather intricate method which provides such results which was designed for a special class of chains will be presented below (proposition 11.7). Here we will only show a much weaker inequality:

Lemma 11.2 *Let the chain be irreducible, aperiodic and reversible. If $\gamma > 0$ is a number such that $\pi_j \le \gamma p_{ij}$ holds for all i, j, then*

$$\Phi \ge \frac{1}{2\gamma}.$$

Proof. Let $T \subset S$ be a subset with $C_T \leq 1/2$. Then $C_{S \setminus T} \geq 1/2$, and it follows that

$$
\begin{aligned}
C_T/2 &\leq C_T C_{S \setminus T} \\
&= \Big(\sum_{i \in T} \pi_i \Big) \Big(\sum_{j \notin T} \pi_j \Big) \\
&= \sum_{i \in T, j \notin T} \pi_i \pi_j \\
&\leq \sum_{i \in T, j \notin T} \gamma \pi_i p_{ij} \\
&= \gamma F_T.
\end{aligned}
$$

Hence $F_T/C_T \geq 1/2\gamma$, and this proves that $\Phi \geq 1/2\gamma$. $\qquad\square$

Here is the *main result* which relates the conductance with the eigenvalues:

Theorem 11.3 *Let P be aperiodic, irreducible and reversible. Write the (necessarily real) eigenvalues of P as $\lambda_1 = 1 > \lambda_2 \geq \cdots \geq \lambda_N > -1$. Then*
  (i) $\lambda_2 \leq 1 - \Phi^2/2$, *and*
  (ii) $\lambda_2 \geq 1 - 2\Phi$.

Proof. (i) Let λ be any eigenvalue different from 1 and $x^{\mathsf{T}} = (x_1, \ldots, x_N)$ an associated left eigenvector with real components:

$$
(x_1, \ldots, x_N)P = \lambda(x_1, \ldots, x_N).
$$

We aim at proving that $1 - \lambda \geq \Phi^2/2$.

First we observe that $\sum x_j = 0$ which is immediate from $\lambda \sum x_j = \sum_{ij} p_{ij} x_i = \sum x_i$. Let $T \subset \{1, \ldots, N\}$ be the collection of indices where $x_i > 0$. Then, without loss of generality,

- C_T, the capacity of T, is $\leq 1/2$ (otherwise pass to $(-x_1, \ldots, -x_N)$);

- T is the set $\{1, \ldots, r\}$ for a suitable r, and the (nonnegative) numbers $x_1/\pi_1, \ldots, x_r/\pi_r$ are decreasing (obvious).

We now put $w_{ij} := \pi_i p_{ij}$,

$$
y_i := \left\{ \begin{array}{ll} x_i/\pi_i & : \quad i \in T \\ 0 & : \quad \text{otherwise} \end{array} \right. ,
$$

and we define D to be the number

$$
\sum_{i<j} w_{ij}(y_i{}^2 - y_j{}^2) / \sum_i \pi_i y_i{}^2.
$$

By definition, D is positive, we claim that

$$
\begin{aligned}
D &\geq \Phi, \text{ and} & (11.1) \\
1 - \lambda &\geq D^2/2. & (11.2)
\end{aligned}
$$

Both inequalities together imply the desired $1 - \lambda \geq \Phi^2/2$, we now turn to their proofs.

Proof of (11.1): Denote, for $k = 1, \ldots, r$, by T_k the set $\{1, \ldots, k\}$. From the definition of Φ it follows that $F_{T_k} \geq \Phi C_{T_k}$, and this will be used to prove that

$$\sum_{i<j} w_{ij}(y_i{}^2 - y_j{}^2) \geq \Phi \sum_i \pi_i y_i{}^2.$$

The idea is to use *summation by parts*, a discrete variant of the more common *integration by parts*:

$$\begin{aligned}
\sum_{i<j} w_{ij}(y_i^2 - y_j^2) &= \sum_{i<j} w_{ij} \sum_{i \leq k < j} (y_k^2 - y_{k+1}^2) \\
&= \sum_{k=1}^r (y_k^2 - y_{k+1}^2)\Big(\sum_{i \leq k < j} w_{ij}\Big) \\
&= \sum_{k=1}^r (y_k^2 - y_{k+1}^2) F_{T_k} && (11.3) \\
&\geq \Phi \sum_{k=1}^r (y_k^2 - y_{k+1}^2) C_{T_k} && (11.4) \\
&= \Phi \sum_{k=1}^r (y_k^2 - y_{k+1}^2) \sum_{1 \leq i \leq k} \pi_i \\
&= \Phi \sum_{i=1}^r \pi_i \sum_{k=i}^r (y_k^2 - y_{k+1}^2) \\
&= \Phi \sum_{i=1}^r \pi_i y_i^2;
\end{aligned}$$

in (11.3) the definition of the F_{T_k} was inserted, and in (11.4) it was important to know that the $y_k^2 - y_{k+1}^2$ are nonnegative.

Proof of (11.2): The proof of this inequality is even harder. We set

$$E := \sum_{i<j} w_{ij}(y_i - y_j)^2 / \sum_i \pi_i y_i^2,$$

and it will be shown that $E \geq D^2/2$ (= *claim 1*) as well as $1 - \lambda \geq E$ (= *claim 2*).

Proof of claim 1: From $(a+b)^2 \leq 2(a^2 + b^2)$ and $\sum_{j, j \neq i} w_{ij} \leq \pi_i$ for all i it follows that

$$\begin{aligned}
\sum_{i<j} w_{ij}(y_i + y_j)^2 &\leq 2 \sum_{i<j} w_{ij}(y_i^2 + y_j^2) \\
&\leq 2 \sum_i \pi_i y_i^2; && (11.5)
\end{aligned}$$

here we have used once more the fact the $w_{ij} = w_{ji}$.

This is combined with a rather tricky application of the Cauchy-Schwarz inequality: we put $\alpha_{ij} := \sqrt{w_{ij}}(y_i - y_j)$ and $\beta_{ij} := \sqrt{w_{ij}}(y_i + y_j)$ for $i < j$. Then $(\sum_{i<j} \alpha_{ij}\beta_{ij})^2 \leq \sum_{i<j} \alpha_{ij}^2 \sum_{i<j} \beta_{ij}^2$ leads to $D^2 \leq E \sum_{i<j} w_{ij}(y_i + y_j)^2 / \sum_i \pi_i y_i^2$, and with the help of

(11.5) this can further be estimated by $2E$.

Proof of claim 2: We evaluate $(x_1, \ldots, x_N)(Id - P)(y_1, \ldots, y_N)^\top$ in two ways. On the one hand this matrix product is

$$(1 - \lambda)(x_1, \ldots, x_N)(y_1, \ldots, y_N)^\top = (1 - \lambda) \sum_i x_i y_i = (1 - \lambda) \sum_i \pi_i y_i^2.$$

On the other hand this number equals $\sum_{ij}(Id - P)_{ji} x_j y_i$ which we may first simplify as $\sum_{i \in T, j}(Id - P)_{ji} x_j y_i$ since the y_i vanish outside T. Further, for i in T and j not in T the coefficient $(Id - P)_{ji}$ is ≤ 0 and also $x_j \leq 0$ holds, and thus $\sum_{ij}(Id - P)_{ji} x_j y_i$ is bounded from below by $\sum_{i,j \in T}(Id - P)_{ji} y_i x_j$. We also note that $(Id - P)_{ji}$ equals $-p_{ji}$ for $i \neq j$ and $1 - p_{ii} = \sum_{l, l \neq i} p_{il}$ if $i = j$, and therefore we may continue with

$$
\begin{aligned}
\sum_{i,j \in T}(Id - P)_{ji} y_i x_j &= \sum_{i,j \in T, i \neq j} -p_{ji} y_i x_j + \sum_{i \in T} \sum_{l \neq i} p_{il} y_i x_i \\
&= \sum_{i,j \in T, i \neq j} -w_{ji} y_i y_j + \sum_{i \in T} \sum_{l \neq i} w_{il} y_i^2 \\
&= \sum_{i,j \in T, i \neq j} -w_{ij} y_i y_j + \sum_{i \in T} \sum_{l \neq i} w_{il} y_i^2 \\
&\geq \sum_{i,j \in T, i \neq j} -w_{ij} y_i y_j + \sum_{i,l \in T, i \neq l} w_{il} y_i^2 \\
&= -2 \sum_{i<j} w_{ij} y_i y_j + \sum_{i<j} w_{ij}(y_i^2 + y_j^2) \qquad (11.6) \\
&= \sum_{i<j} w_{ij}(y_i - y_j)^2;
\end{aligned}
$$

in (11.6) we have used that $w_{ij} = w_{ji}$.

We thus have established that $(1 - \lambda) \sum \pi_i y_i^2 \geq \sum_{i<j} w_{ij}(y_i - y_j)^2$, i.e., $E \leq 1 - \lambda$.

(ii) We have to show that $1 - \lambda_2 \leq 2\Phi_T$ for all $T \subset S$ such that $C_T \leq 1/2$.

Let such a T be given. The idea is to combine proposition 10.2(ii) with proposition 10.4: $(1, 1, \ldots, 1)^\top$ is an eigenvector with associated (simple) eigenvalue 1 of P, and thus

$$\lambda_2 \langle x, x \rangle_\pi \geq \langle x, Px \rangle_\pi \qquad (11.7)$$

whenever $x = (x_i)$ is such that $\sum \pi_i x_i = 0$.

To make use of this fact we define the x_i by

$$
x_i = \begin{cases} 1/C_T & : \quad \text{for } i \in T \\ -1/C_{S \setminus T} & : \quad \text{for } i \in S \setminus T. \end{cases}
$$

Then, by the definition of C_T, $C_{S \setminus T}$, we have $\sum \pi_i x_i = 0$, and in order to proceed we will have to evaluate the scalar products in (11.7).

The left-hand side is simply $\lambda_2(1/C_T + 1/C_{S \setminus T})$, again by the definition of the capacities. For the investigation of the right-hand side we write P as $Id - (Id - P)$. First we note that the entries on the diagonal of $Id - P$ are the numbers $(\sum_{j, j \neq i} p_{ij})_i$ and that we may replace $\pi_i p_{ij}$ by $\pi_j p_{ji}$:

$$
\begin{aligned}
\langle x, (Id - P)x \rangle_\pi &= \sum_i \pi_i x_i \sum_j (Id - P)_{ij} x_j \\
&= -\sum_i \sum_{j, j \neq i} \pi_i x_i p_{ij} x_j + \sum_i \sum_{j, j \neq i} \pi_i x_i p_{ij} x_i \\
&= \sum_i \sum_{j, j \neq i} \pi_i p_{ij} (x_i^2 - x_i x_j) \\
&= \sum_{i < j} \pi_i p_{ij} (x_i - x_j)^2 .
\end{aligned}
$$

Since the x_i are constant on T and on $S \setminus T$, there are contributions to this sum only when i lies in T and j in $S \setminus T$ or vice versa, and therefore the value is

$$
\Big(\frac{1}{C_T} + \frac{1}{C_{S \setminus T}} \Big)^2 \sum_{i \in T, j \notin T} \pi_i p_{ij} = F_T \Big(\frac{1}{C_T} + \frac{1}{C_{S \setminus T}} \Big)^2 .
$$

To finish the proof we evaluate (11.7):

$$
\begin{aligned}
\lambda_2 \Big(\frac{1}{C_T} + \frac{1}{C_{S \setminus T}} \Big) &\geq \langle x, Px \rangle_\pi \\
&= \langle x, x \rangle_\pi - \langle x, (Id - P)x \rangle_\pi \\
&= \Big(\frac{1}{C_T} + \frac{1}{C_{S \setminus T}} \Big) - F_T \Big(\frac{1}{C_T} + \frac{1}{C_{S \setminus T}} \Big)^2 .
\end{aligned}
$$

This yields $1 - \lambda_2 \leq F_T(1/C_T + 1/C_{S \setminus T})$, and since $C_{S \setminus T} \geq C_T$, this can further be estimated by $\leq 2F_T/C_T = 2\Phi_T$. $\square$

It has already been emphasized that it might be a difficult task to calculate the conductance of a given chain. Therefore we want to complement our investigations by describing a technique which provides reasonable estimates for a special case. What we have in mind are *chains which are defined by graphs*. There is given a finite set V (*the vertices*) together with a subset $E \subset V \times V$ (*the edges*). We will assume that E is symmetric: (i, j) belongs to E iff (j, i) does. Then $G := (V, E)$ will be called *an undirected graph* (or simply *a graph* since we will not discuss directed graphs). Graphs can easily be visualized if we let the vertices correspond to points in the plane and if we draw a line segment between two of these points i, j iff $(i, j) \in E$.

Graphs give rise to Markov chains as follows:

Definition 11.4 Let $G = (V, E)$ be any graph. Denote by d the maximal number of edges connecting any vertex with the others:

$$
d := \max_{i \in V} \operatorname{card}\{ j \mid j \neq i, (i, j) \in E \};
$$

we will assume that $d > 0$.

Now let β be a number, $0 < \beta \leq 1$. We define a Markov chain by declaring V to be the set of states and by fixing the transition probabilities according to the following rule: if there are d_i edges from "state" i to other vertices (i.e., $d_i = \operatorname{card}\{ j \mid j \neq i, (i, j) \in E \}$), then pass from i to any of the j with $j \neq i$, $(i, j) \in E$, with equal probability β/d and stay at i with probability $1 - d_i \beta/d$.

It should be clear that the graphs

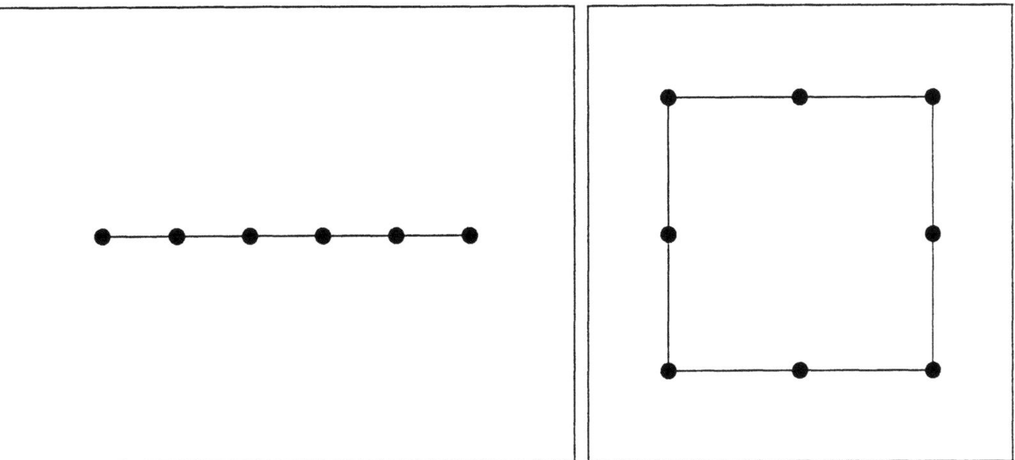

give rise to a reflecting and a cyclic random walk, respectively (cf. the pictures on page 14).

Some readers might wonder why we do not work with the more natural "pass to the next admissible vertex" (that is one for which there is a connecting edge) in such a way that – if there are d_i edges – each transition has probability $1/d_i$. The reason is that only definition 11.4 leads to "nice" chains:

Lemma 11.5 *The chain defined in 11.4 is symmetric so that the uniform distribution is an equilibrium distribution. If the graph is connected[2], then the chain is irreducible. If in addition β is strictly less than one, then we are dealing with an aperiodic and irreducible chain which is in detailed balance.*

(These facts are obvious, a proof is omitted.)

The following definition is the graph theoretical variant of "conductance".

Definition 11.6 Let (V, E) be a graph. Define μ to be the minimum of the numbers

$$\frac{\operatorname{card}(\{(i,j) \mid i \in T,\ j \notin T, (i,j) \in E\})}{\operatorname{card}(T)},$$

where T runs through all sets of vertices such that $\operatorname{card}(T) \leq \operatorname{card}(V)/2$. μ is called the *edge magnification* of the graph.

Thus μ counts the outbound edges in relation with the size of a collection of vertices, and therefore it is to be expected that in the following examples the first one has a small μ (because of the "bottleneck" in the middle) whereas the second one – a complete graph – leads to a maximal value among all graphs with four vertices:

---

[2] This means: for i, j there are $i_1, \ldots, i_n$ such that $i_1 = i$, $i_n = j$ and $(i_k, i_{k+1}) \in E$ for $k = 1, \ldots, n-1$.

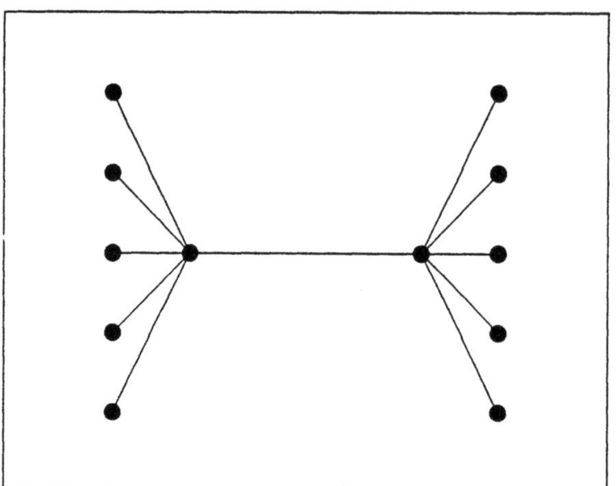 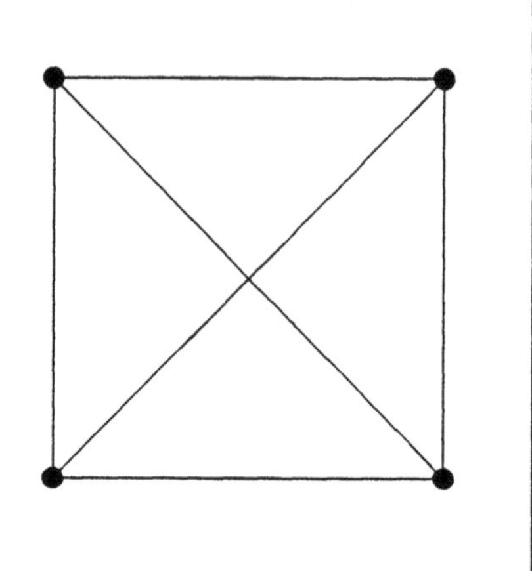

In fact, if we consider $T=$ "the six points on the left-hand side" in the first graph, then
the relevant quotient is $1/6$, and no other subset behaves worse. For the second graph we
have to investigate subsets with 1 and 2 elements. For such subsets the quotient is $3/1$
resp. $4/2$, and this means that $\mu = 2$.

Here is *another example*. Let N be an even number, we consider the *cyclic graph* with
vertices $\{0,\ldots,N-1\}$ for which there are edges precisely between i and $i+1$ (modulo
N). For a fixed k a subset T with k elements will have a minimum number of edges
joining it with the complement iff it is a cyclic segment. In this case there are 2 such
edges. It follows that the edge magnification is $2/[N/2] = 4/N$.

We note in passing that the edge magnification is related with the conductance of the
underlying Markov chain according to definition 11.4 in a simple way:

$$\Phi = \beta\mu/d.$$

This is obvious, more interesting is the question whether μ can be calculated more simply
than the conductance. An interesting technique which provides reasonable bounds is due
to Sinclair (see [70]). This *method of canonical paths* will be presented now.

Let (V, E) be a graph which we assume to be connected. For $i, j \in V$ a (directed) *path*
from i to j is nothing but a sequence $i_1,\ldots,i_n$ in V such that $i_1 = i$, $i_n = j$, and (i_k, i_{k+1})
is an edge for $k = 1,\ldots,n-1$. For $e \in E$ we say that *the path meets e* if there is a k
such that $(i_k, i_{k+1}) = e$.

Assume that, for every ordered pair of different vertices, a path from i to j is prescribed;
its length might depend on i and j.

> Of course a formal definition which uses "there is given a map such that ..." instead
> of "there is prescribed ..." is also possible. It would be rather clumsy, and therefore
> we prefer to give an illustration instead: imagine a city where there are certain
> points of interest (= the vertices) and certain streets joining them (= the edges).
> Then we assume that directions are given how to drive from i to j.

Usually there will be a gigantic number of such lists of "canonical paths". It is intuitively
clear that there should be a connection between the edge magnification μ and the "over-
lap" of the canonical paths: in the case of a small μ it is to be expected that there are
edges which are used by many paths. This is in fact true:

Proposition 11.7 *Let a family of canonical paths be prescribed and suppose that α is an integer such that the following holds: whenever e is an edge, then there are at most α pairs (i,j) of different vertices such that the canonical path $i_1, \ldots, i_n$ from i to j meets e. Then $\mu \geq \mathrm{card}(V)/2\alpha$.*

Proof. We will use the following simple argument: whenever $\phi : A \to B$ is a map between finite sets such that every preimage $\phi^{-1}(\{b\})$ contains at most α elements, then $\alpha \, \mathrm{card}(B) \geq \mathrm{card}(A)$.

Now let $T \subset V$ be arbitrary such that $\mathrm{card}(T) \leq \mathrm{card}(V)/2$. Denote by $e_1, \ldots, e_l$ the edges which join T to its complement. For $i \in T$ and $j \notin T$ the canonical path from i to j necessarily will meet some e_λ, and this fact can be used to define a map ϕ from $\{(i,j) \mid i \in T, \ j \notin T\}$ to $\{1, \ldots, l\}$: the path from i to j meets $\phi(i,j)$[3]. From our assumption we know that each preimage contains at most α elements, and in this way we arrive at the inequality

$$\alpha l \geq \mathrm{card}(T)(\mathrm{card}(V) - \mathrm{card}(T)).$$

But $\mathrm{card}(V) - \mathrm{card}(T) \geq \mathrm{card}(V)/2$ so that $l/\mathrm{card}(T) \geq \mathrm{card}(V)/2\alpha$. The proof is now complete since μ is the minimum of the numbers on the left-hand side of the inequality. $\square$

Here are two **examples** how to apply canonical paths:

1. Consider the cyclic graph with N vertices (see page 98). First we consider the following family of canonical paths:

> *Go from i to j in the clockwise direction, that is the canonical path is i, $i+1$, ..., $(i+k) \bmod N$, where $k \geq 0$ is the smallest number such that $i+k = j \bmod N$.*

(Thus, for example, one needs a path of length 10 to come from 11 to 9 in the case $N = 12$.)

Now fix any edge, say $e = \{0,1\}$. If $i = 0$, then there are $N-1$ vertices j (namely $j = 1, 2, \ldots, N-1$) where the canonical path from i to j meets e; for $i = 1$ (resp. 2, resp. 3, ...) there are 0 (resp. 1, resp. 2, ...) candidates, hence we get a total of $0 + 1 + 2 + \cdots + N-1 = N(N-1)/2$ paths which meet e. Therefore we might put $\alpha = N(N-1)/2$ in the preceding proposition, and we get $\mu \geq N/[N(N-1)] = 1/(N-1)$. Note that this is of the same order as the correct $\mu = 4/N$ and that, with the present value, we get the estimate $\Phi = \beta\mu/d \geq \beta/2(N-1)$ for the conductance of the associated chain.

As a variant we fix the canonical paths according to

> *Move from i to j on the shortest way, clockwise or counterclockwise. If both paths have the same length, i.e., if N is even and $j = i + N/2 \bmod N$, then choose the clockwise direction.*

To find the best α now we repeat the preceding analysis (let us assume for simplicity that N is even). For $e = \{0,1\}$, for example, this edge will be met by $N/2$ (resp. $N/2-1$, $N/2-1$, $N/2-2$, $N/2-2, \ldots$) paths if we start at 0 (resp. at $1, N-1, 2, N-2, \ldots$ and move according to the rules to the other points, and therefore the choice

---

[3] Note that ϕ is in general not uniquely determined, there might be several possibilities to associate a pair (i,j) with an edge.

$$\alpha = (N/2) + 2(1 + 2 + \cdots + (N/2-1)) = N^2/4$$

is admissible. Note that, expectedly, this is much better than before.

2. Let r be a fixed integer, and $V = \{0,1\}^r$. We define a graph by prescribing edges between points $(i_1,\ldots,i_r)$ and $(j_1,\ldots,j_r)$ for which for precisely one κ the coordinate i_κ is different from j_κ; this graph can be considered as an r-dimensional *hypercube* (cf. example 8 in chapter 2).

In order to estimate μ we use canonical paths: a canonical path from $u = (i_1,\ldots,i_r)$ to $v = (j_1,\ldots,j_r)$ is defined by flipping the coordinates which are different one after the other from left to right.

> As an example consider $r = 7$, $u = (0100101)$ and $v = (0010111)$. The canonical path is
> $$(0100101) \rightarrow (0000101) \rightarrow (0010101) \rightarrow (0010111).$$

Now let e be an arbitrary edge, from state $(i_1,\ldots,i_{s-1},0,i_{s+1},\ldots,i_r)$ to state $(i_1,\ldots,i_{s-1},1,i_{s+1},\ldots,i_r)$, say. By definition a canonical path from u to v will meet e precisely if u (resp. v) is of the form $(*,\cdots,*,0,i_{s+1},\ldots,i_r)$ (resp. $(i_1,\ldots,i_{s-1},1,*,\cdots,*)$) with arbitrary $*$ in $\{0,1\}$. This means that there are 2^{r-1} possibilities for such pairs (u,v) so that α in proposition 11.7 can be chosen as this number. We get

$$\mu \geq \operatorname{card}(V)/(2\alpha) = 2^r/(2 \cdot 2^{r-1}) = 1,$$

and the rate of rapid mixing could now be further discussed with the help of the estimate $\Phi = \beta\mu/d \geq \beta/r$.

How powerful is the method of canonical paths? There seem to be no general results which assert that one can always determine μ (at least up to a constant) by this technique. Thus we are in a situation which is different from the preceding ones. In theorem 10.3, propositions 10.5 and 10.8, and also in theorem 11.3 we had to calculate something and this gave rise to a convergence result. Now things have changed. We have to be *creative* to find an appropriate as possible definition of "canonical paths", a definition which provides the smallest possible α in order to find the best estimate for μ and thus for Φ. It should be clear that this will necessitate to take into account the particular structure of the graph and that one will have a chance to find satisfactory solutions only after some experience with this method.

> We have presented the method of canonical paths only for the case of chains which are induced by graphs. Here is a more general version, it can be considered as a refinement of lemma 11.2.
>
> Consider, as always in this chapter, an irreducible, aperiodic and reversible chain. By a prescription of *canonical paths* we mean a rule which for arbitrary different states s,t associates $(i_1,\ldots,i_n)$ with $s = i_1$ and $t = i_n$. Fix states i,j such that $i \neq j$. We say that a path $(i_1,\ldots,i_n)$ *meets* (i,j) if there is a k such that $i_k = i$ and $i_{k+1} = j$. Now let α be a number such that, for all i,j,
>
> $$\sum \pi_s \pi_t \leq \alpha \pi_i p_{ij},$$
>
> where the sum runs over all pairs (s,t) such that the canonical path from s to t meets (i,j). Then we have

Proposition 11.8 $\Phi \geq 1/2\alpha$.

The proof is left to the reader, it is canonical once one has understood the proofs of lemma 11.2 and proposition 11.7.

It should be emphasized that proposition 11.8 generalizes both results: lemma 11.2 follows if one chooses as a canonical path always the shortest one: $i_1 = s, i_2 = t$; and proposition 11.7 is contained in 11.8 if one considers chains which are determined by graphs as in definition 11.4.

Exercises

11.1: Calculate the conductance of the chain given by

$$P = \frac{1}{3} \begin{pmatrix} 1 & 1 & 1 \\ 1 & 0 & 2 \\ 1 & 2 & 0 \end{pmatrix}.$$

11.2: Consider the cyclic walk on $\{1, \ldots, N\}$, where, with the notation of example 1 of chapter 2, $a_i = c_i = \varepsilon$ and $b_i = 1 - 2\varepsilon$ for all i (with $0 < \varepsilon < 1/2$.) What is the conductance of this chain?

11.3: For what $\eta > 0$ does there exist a reversible chain such that the conductance is precisely η?

11.4: Determine the conductance of an arbitrary reversible chain provided that there are only two states.

11.5: What is the edge magnification μ of the complete graph with N vertices?

11.6: Calculate the edge magnification of a graph with vertices $a, b, a_1, \ldots, a_r, b_1, \ldots, b_s$, where edges are between a and b, between a and all a_i, and between b and all b_i (this graph is depicted in the case $r = s = 5$ as the left picture after definition 11.6).

11.7: Let N vertices be given. Try to find edges in such a way that the resulting graph is connected and the edge magnification is

a) as large as possible,

b) as small as possible.

11.8: Suppose that a graph has N vertices and r edges, that there are prescribed canonical paths and that it is known that the total length of all paths is L. What can be said about the edge magnification μ?

11.9: Let the vertices of a graph be the elements of the symmetric group S_r und suppose that there are edges between two permutations iff one can pass from one to the other by applying a transposition. Find a family of canonical paths for this graph and bound the edge magnification.

11.10: Prove lemma 11.5.

11.11: Prove proposition 11.8.

12   Stopping times and the strong Markov property

This chapter prepares our investigations of how *coupling methods* can be used to bound mixing rates, it also will be of importance in chapter 14.

To understand these approaches one needs to know what *stopping times* are. Whereas the underlying idea is simple, the formal definition is rather involved.

We will use the opportunity to complement our discussion of the Markov property from chapter 1: it will be shown that the Markov processes discussed in this book have in fact the *strong Markov property*[1].

Stopping times

Let us return to the situation from the very beginning: we are given a finite set S, probabilities $(p_i)_{i \in S}$ which determine where to start the walk and some stochastic rules to move in the k'th step from a state i to another position. You can think of a homogeneous Markov chain where the rules are encoded in a single stochastic matrix P, but this is not necessary here: the transition procedure can be prescribed as complicated as you wish[2].

> E.g.: Start at any point of $S := \{0, 1, 2, \ldots, 9\}$ with equal probability. To determine where to move in the k'th step, generate an integer j according to the Poisson distribution with parameter k (i.e., $\mathbb{P}(j) = k^j e^{-k}/j!$); if the present position is $i \in S$ and if this state has been occupied by the walk for an *odd* number of times, then move to $(i + j) \bmod 10$, otherwise go to $(i + k) \bmod 10$.

> Let, for example, start the walk at 7, and suppose that the Poisson random generator produces the numbers $0, 0, 1, 2, 5, 2, 3, \ldots$. Then the walk will begin with $7, 7, 9, 0, 2, 7, 9, 6, \ldots$

Now we want to define *a rule by which the walk can be stopped:* someone has to shout "STOP!" which will then result in terminating the walk immediately. Simple examples – they refer to the preceding walk – of such rules are:

1. stop after the 444'th step;

2. stop immediately after the starting position has been occupied;

3. stop as soon as state 2 has been occupied for the 5'th time;

4. stop after the first transition of the form $i \to i$;

---

[1] This second part of the present chapter is not essential for the investigations to come. Some readers, however, might consider it interesting to learn how carefully one has to deal with the notion "memoryless".

[2] This means that arbitrary S-valued stochastic processes are admitted.

5. have a careful look at the Poisson random generator and announce "STOP!" as soon as it has produced – for some k – in the k'th step the number 7 and in the $(k+1)$'th step the number 17.

We note in passing that it is also possible to prescribe "rules" which cannot reasonably be obeyed, like "stop three steps before 5 will be occupied for the second time".

> This is similar to the situation when you need help in a town where you are for the first time: rules like "turn to the left at the first intersection with a traffic light" are reasonable, but the advice "turn to the right three streets before you see the petrol station on the left-hand side" will be of limited use.

We are now going to be a little bit more formal. We start with an arbitrary stochastic process $X_0, X_1, \ldots : \Omega \to S$, where $(\Omega, \mathcal{A}, \mathbb{P})$ is any probability space. A "rule to stop" surely has to be formalized as a map $\mathbb{T} : \Omega \to \{0, 1, \ldots, \infty\}$. $\mathbb{T}(\omega)$ is the time when the process has to be stopped, where $\mathbb{T}(\omega) = \infty$ just means that it runs forever. But how is it possible to single out "reasonable" rules?

To this end, let us have another look at the above examples. If you investigate these rules more carefully, you will note a difference: whether or not they can reasonably be applied by an observer depends on the *information about the walk* he or she has. Rules 1 and 2 are deterministic and thus in a sense trivial: *no information* is needed. For rule 3 one needs at least *partial information*, it would suffice to observe state 2 and to notice how often it has been occupied. For rule 4 instead *one needs to know the whole walk* in order to be able to stop correctly, and rule 5 is the most demanding: one in fact has to have *full information*, not only about the positions of the walk but also how they have been produced, i.e., the values of the Poisson random generator.

Therefore we need something which formalizes "the information after the k'th step", then we will be able to say that a "reasonable" rule is one which uses only this information in order to decide whether to stop after this k'th step or not.

It was a remarkable idea in probability theory to *relate partial information with sub-σ-algebras*. It is now generally accepted that *partial information concerning a probability space $(\Omega, \mathcal{A}, \mathbb{P})$ is nothing but a σ-algebra $\mathcal{B}$ which is a subset of $\mathcal{A}$.*

> This looks strange when one is confronted with this fact for the first time, let's try to motivate it.
>
> Regard a probability space $(\Omega, \mathcal{A}, \mathbb{P})$ simply as some kind of machine which produces points ω in such a way that the results are "unforeseeable", but – on the long run – occur with a known frequency: one will find an ω in a prescribed set $A \in \mathcal{A}$ roughly in $\mathbb{P}(A)k$ of k random experiments if k is "sufficiently large". Then *partial information* about this probability space means that one knows something in advance, the most general variant seems to be the following: there is a collection $\mathcal{B}$ of measurable subsets such that, for any $B \in \mathcal{B}$, one knows in advance whether the ω which will be the result of the next experiment lies in B or not. Common examples are $\mathcal{B} = \{B\}$ (which gives rise to the notion of conditional probability $\mathbb{P}(A \mid B)$) or $\mathcal{B}$ = the σ-algebra generated by a fixed random variable.
>
> Now a moment's reflection shows that, if the above property holds for $\mathcal{B}$, it holds for the σ-algebra generated by $\mathcal{B}$ as well, and therefore one might assume from the beginning that $\mathcal{B}$ is already a σ-algebra.

Using σ-algebras we are now able to make precise what is meant by "the information of the process after the k'th step":

Definition 12.1 Let $X_0, X_1, \ldots$ be a stochastic process defined on some probability space $(\Omega, \mathcal{A}, \mathbb{P})$.

   (i) By a *filtration* we mean a sequence $(\mathcal{F}_k)_{k=0,1,\ldots}$ of σ-algebras such that $\mathcal{F}_0 \subset \mathcal{F}_1 \subset \cdots \subset \mathcal{A}$.

   (ii) A particularly important example is the *natural filtration associated with the process*: this filtration $(\mathcal{F}_k^{\mathrm{nat}})_{k=0,1,\ldots}$ is defined by $\mathcal{F}_k^{\mathrm{nat}} :=$ the smallest σ-algebra such that $X_0, \ldots, X_k$ are measurable.

   (iii) The process $X_0, X_1, \ldots$ will be called *adapted* to a given filtration $(\mathcal{F}_k)_{k=0,1,\ldots}$, if X_k is $\mathcal{F}_k$-measurable for every k.

This looks, admittedly, rather technical. It should be stressed, however, that the definition is natural once we have accepted to think of information as of sub-σ-algebras, here $\mathcal{F}_k$ is the information we have at our disposal after the k'th step. Then the first part of the definition only means that *we don't forget*, the information after $k + 1$ moves is not worse than that after k moves. The natural filtration, which is in a sense the smallest among the reasonable ones, contains the information given by the positions of the walk, and "the process is adapted" asserts nothing but the fact that the available information comprises the knowledge of all positions: *we are allowed to observe the walk*. In particular it is – trivially – true that the process is adapted to the natural filtration.

Now it is also clear how to explain the difference between the examples 1 through 4 and example 5: in the first four cases one only needs information which corresponds to *any* filtration such that the process is adapted, whereas in example 5 the filtration must contain information on the Poisson generator (cf. the discussion after the following definition).

Finally we can define *stopping times* as rules which use nothing but the information contained in a fixed (adapted) filtration. The definition is natural once we adopt the translation "having the information $\mathcal{B}$ (a subalgebra of $\mathcal{A}$) is nothing but the fact that all questions of the form 'does ω lie in B?' can be answered unambiguously for $B \in \mathcal{B}$".

Definition 12.2 Let $X_0, X_1, \ldots$ be as in the preceding definition, and $\mathcal{F}_0 \subset \mathcal{F}_1 \subset \cdots \subset \mathcal{A}$ a filtration such that $X_0, X_1, \ldots$ is adapted. A map $\mathbb{T} : \Omega \to \{0, 1, \ldots, \infty\}$ is called a *stopping time with respect to* $(\mathcal{F}_k)_k$ provided that $T^{-1}(\{k\})$ lies in $\mathcal{F}_k$ for every k.

Often it will suffice to replace the preceding formal definiton by the following **rule of thumb**: let a stopping rule be given which can be expressed as "Stop after step k provided that E_k holds", where E_k is an expression which contains (maybe) $X_0, \ldots, X_k$ but *not* $X_{k+1}, \ldots$; then this is a stopping time with respect to the natural filtration.

Here is a sketch what our examples 1 to 5 would look like if we were asked to treat them formally.

Consider any probability space $(\Omega, \mathcal{A}, \mathbb{P})$ such that it is possible to define independent random variables $U, Y_1, Y_2, \ldots$ such that $U : \Omega \to \{0, \ldots, 9\}$ is uniformly distributed and, for $k = 1, \ldots$, the random variable $Y_k : \Omega \to \{0, 1, 2, \ldots\}$ has a Poisson distribution with parameter k. Use these $U, Y_1, \ldots$ to define random variables $X_0, X_1, \ldots : \Omega \to \{0, \ldots, 9\}$ according to the above definition ($X_k :=$ the position after k steps, with $X_0 := U$ etc.; an explicit definition of the X_k in terms of the Y's surely would look rather ugly). Put $\mathcal{F}_k :=$ the σ-algebra generated by $U, Y_1, \ldots, Y_k$. Then it is easy to show that

   • the process $X_0, X_1, \ldots$ is adapted to the filtration $(\mathcal{F}_k)_{k=0,1,\ldots}$;

- the above rules 1 to 4, but not rule 5, are stopping times with respect to the natural filtration;
- rule 5 is a stopping time with respect to $(\mathcal{F}_k)_{k=0,\ldots}$.

The strong Markov property[3]

Now we will restrict ourselves again to the case of a homogeneous Markov chain on a finite set S which is defined by a stochastic matrix P and an initial distribution $(p_i)_{i \in S}$. We have seen already in the first chapter that the appropriate mathematical model is an S-valued Markov process, i.e., a sequence of random variables $X_0, X_1, \ldots : \Omega \to S$ defined on some probability space $(\Omega, \mathcal{A}, \mathbb{P})$ such that $\mathbb{P}(X_0 = i) = p_i$ and

$$\mathbb{P}(X_{k+1} = j \mid X_0 = i_0, X_1 = i_1, \ldots, X_{k-1} = i_{k-1}, X_k = i) = \mathbb{P}(X_{k+1} = j \mid X_k = i) = p_{ij}$$

for all i, j.

If you think of such a Markov chain as a collection of stochastic rules how to move on S, then the following holds:

> Suppose your walk is at "time" k_0 in state i_0. If you now set all counters to zero but continue your walk, then it will be impossible to distinguish what you see from a walk which starts deterministically at i_0.

This is more than obvious since we work with the same random generators as before, the fact has been used several times in part I of this book. It is, however, a little bit more complicated to check this fact in the mathematical model. There it reads as follows:

Lemma 12.3 *Let $X_0, X_1, \ldots : \Omega \to S$ be a homogeneous Markov process as above. Let $i_0 \in S$ and k_0 be given and suppose that $X_{k_0} = i_0$ happens with positive probability. Put $\Omega' := \{X_{k_0} = i_0\}$, provide this set with the restricted σ-algebra and the measure $\mathbb{P}' := \mathbb{P}/\mathbb{P}(\Omega')$, and define $Y_0, Y_1, \ldots : \Omega' \to S$ by $Y_k(\omega) := X_{k_0+k}(\omega)$. Then the $Y_0, Y_1, \ldots$ are a homogeneous Markov process which starts deterministically at i_0 and which has the same transition probabilities as the $X_0, X_1, \ldots$.*

Proof. The Markov property of the Y_k is reduced to that of the X_k once one knows that

$$\mathbb{P}(X_{k+1} = j_{k+1} \mid X_r = j_r, X_{r+1} = j_{r+1}, \ldots, X_k = j_k) = \mathbb{P}(X_{k+1} = j_{k+1} \mid X_k = j_k)$$

for all $0 \leq r < k$; the original Markov property from definition 1.3 corresponds to $r = 0$. But this equality follows immediately from

$$\mathbb{P}(X_r = j_r, X_{r+1} = j_{r+1}, \ldots, X_{k+1} = j_{k+1}) = \mathbb{P}(X_r = j_r) p_{j_r j_{r+1}} \cdots p_{j_k j_{k+1}},$$

an equality which follows easily from (1.2) in chapter 1. $\qquad\square$

It should be noted that, conversely, a process which has the property described in the lemma will be a homogeneous Markov process so that it can be used interchangeably with the original definition.

Let's now turn to a more involved construction, we start with the "obvious" part where the walk is defined by a collection of random generators:

---

[3] This subsection is also rather technical. What we present will not be a necessary prerequisite to understand couplings.

Let i_0 be fixed and suppose that a reasonable rule to stop the process is given; let it be stopped, e.g., three steps after a state j_0 (which might be different from i_0) has been occupied for the second time. Now start the process and suppose that it is stopped just at i_0. Then set all counters to zero and continue the walk. It is to be expected that one will observe something which looks like an ordinary walk which started at i_0.

That this "obvious" property is in fact true is the so-called **strong Markov property** which has been introduced by Hunt in [41]. For some time it has been tacitly assumed that it is shared by every homogeneous Markov process. The full truth, however, is more complicated: in our elementary situation (finite state space) the intuition is justified, but there are Markov processes which fail to have the strong Markov property (for an example cf. exercise 8.20 in [17]).

In order to work in the mathematical model we have to be more formal. In view of our discussion of stopping times we know what "reasonable stopping rules" are. It is also not too difficult to formalize what is meant by "start again after the process has been stopped":

Theorem 12.4 *Let $X_0, X_1, \ldots$ be a homogeneous S-valued Markov process. Further let $\mathbb{T} : \Omega \to \{0, 1, \ldots, \infty\}$ be a stopping time with respect to the natural filtration and suppose that $\mathbb{T}$ is finite almost surely; for simplicity we will even assume that $\{\mathbb{T} = \infty\} = \emptyset$. Then it is possible to define the stopped process $X_\mathbb{T} : \Omega \to S$ by $X_\mathbb{T}(\omega) := X_{\mathbb{T}(\omega)}(\omega)$. We claim that $X_\mathbb{T}$ is a random variable.*
Now let i_0 be fixed and suppose that $\Omega' := \{X_\mathbb{T} = i_0\}$ has positive probability. Ω' will be considerd as a probability space similarly to the preceding lemma, and a process $Y_0, Y_1, \ldots$ will be defined on this space by $Y_k(\omega) := X_{\mathbb{T}(\omega)+k}(\omega)$. Then the $Y_0, Y_1, \ldots$ form a homogeneous Markov process with the same transition probabilities as the $X_0, X_1, \ldots$, and the Y-process starts deterministically at i_0.

Proof. The measurability of $X_\mathbb{T}$ is easy to show: $\{X_\mathbb{T} = i\}$ coincides with

$$(\{\mathbb{T} = 0\} \cap \{X_0 = i\}) \cup (\{\mathbb{T} = 1\} \cap \{X_1 = i\}) \cup \cdots,$$

and this set obviously lies in $\mathcal{A}$.

Now let k be arbitrary, we consider the event $B := \{\mathbb{T} = k\}$. By assumption B lies in the σ-algebra generated by $X_0, \ldots, X_k$.

It can easily be described explicitly, it consists of all sets which are unions of sets of the form $C_{j_0, \ldots, j_k} = \{X_0 = j_0, \ldots, X_k = j_k\}$. (Proof: All $C_{j_0, \ldots, j_k}$ must be contained in any σ-algebra for which the $X_0, \ldots, X_k$ are measurable, and the union of the C's obviously *is* such a σ-algebra.)
For any fixed $C = C_{j_0, \ldots, j_k}$ one has

$$\mathbb{P}(C \cap \{X_{k+1} = i_1, \ldots, X_r = i_r\}) \;=\; \mathbb{P}(X_0 = j_0, \ldots, X_k = j_k, X_{k+1} = i_1, \ldots, X_{k+r} = i_r)$$
$$=\; p_{j_0} p_{j_0 j_1} \cdots p_{i_{r-1} i_r}.$$

If in particular it happens that $j_k = i_0$ then it follows that

$$\mathbb{P}(C \cap \{X_{k+1} = i_1, \ldots, X_{k+r} = i_r\}) = \mathbb{P}(C) p_{i_0 i_1} p_{i_1 i_2} \cdots p_{i_{r-1} i_r}.$$

Now observe that the set $B \cap \{X_k = i_0\}$ can be written as the *disjoint* union of such $C_{j_0, \ldots, j_k}$ with $j_k = i_0$, and in this way we arrive at

$$\mathbb{P}(B \cap \{X_k = i_0,\, X_{k+1} = i_1, \ldots, X_{k+r} = i_r\}) = \mathbb{P}(B \cap \{X_k = i_0\}) p_{i_0 i_1} p_{i_1 i_2} \cdots p_{i_{r-1} i_r}.$$

If we sum up these equations for $k = 0, 1, \ldots$ this gives

$$\mathbb{P}(X_{\mathbb{T}} = i_0,\, X_{\mathbb{T}+1} = i_1, \ldots, X_{\mathbb{T}+r} = i_r) = \mathbb{P}(X_{\mathbb{T}} = i_0) p_{i_0 i_1} p_{i_1 i_2} \cdots p_{i_{r-1} i_r},$$

and it follows immediately that the $Y_0, \ldots$ are a Markov process with (p_{ij}) as transition probabilities.

That the Y-process starts deterministically at i_0 is trivially true. $\qquad\square$

Remarks: 1. The result can be considered as a generalization of lemma 12.3 which corresponds to the case of a *deterministic stopping time* $\mathbb{T} = k_0$.

2. It was essential for the proof that we have dealt with a stopping time with respect to the *natural* filtration: the stopping rule must not depend on information which cannot be read off from the positions of the walk. The theorem does *not hold* in the case of arbitrary stopping times and adapted processes (see exercise 12.6 below).

Exercises

12.1: For $\Omega := \{0,1\}^{\mathbb{N}}$ and $i = 1, \ldots$ we denote by $X_i : \Omega \to \{0,1\}$ the natural i'th projection $(x_1, x_2, \ldots) \mapsto x_i$. Let $\mathcal{F}_N$ be the σ-algebra generated by $X_1, \ldots, X_N$. Which of the following subsets F of Ω are in $\mathcal{F}_{1000}$?

a) $F := \{(x_1, \ldots) \mid x_1 \neq x_{44} - x_{55} + 3x_{999}\}$.

b) $F := \{(x_1, \ldots) \mid x_{1001} \geq x_4 = x_1\}$.

c) $F := \{(x_1, \ldots) \mid (x_{1003} - 1)x_{1003} = 0\}$.

12.2: Suppose that $(\Omega, \mathcal{A}, \mathbb{P})$ is an arbitrary measure space and $T : \Omega \to \mathbb{N}_0$ a measurable map. Prove that there is a filtration $(\mathcal{F}_k)$ such that T is a stopping time with respect to this filtration. Is there a filtration with this property such that each $\mathcal{F}_k$ is as small as possible?

12.3: Fix a measure space $(\Omega, \mathcal{A}, \mathbb{P})$ together with a filtration $\mathcal{F} = (\mathcal{F}_k)$. $\mathbb{T}$, $\mathbb{T}_1$ and $\mathbb{T}_2$ are assumed to be stopping times with respect to $\mathcal{F}$.

a) What are the integers r such that $r\mathbb{T}$ is a stopping time?

b) Which of the random variables $\mathbb{T}_1 + \mathbb{T}_2$, $\mathbb{T}_1 \cdot \mathbb{T}_2$, $\max\{\mathbb{T}_1, \mathbb{T}_2\}$, $\min\{\mathbb{T}_1, \mathbb{T}_2\}$ is a stopping time?

12.4: Provide a probability space with $\Omega = \mathbb{N}$ together with random variables $X_0, X_1, \ldots$ such that

> "Stop at step $k_0 - 2$, where k_0 denotes the first time when the walk is in state 44."

is a stopping time with respect to the natural filtration.

12.5: To motivate stopping times we have discussed five examples at the beginning of this chapter. Find in all these cases a minimal filtration $\mathcal{F}$ such that the stopping procedure is a stopping time with respect to $\mathcal{F}$.

12.6: In the main theorem of this chapter, in theorem 12.4, it was important in the proof that we have dealt with stopping times *with respect to the natural filtration.*

a) Prove that there are cases where the theorem holds for stopping times with respect to a strictly larger filtration.

b) Give an example to show that one cannot replace the natural filtration by an arbitrary filtration in the theorem.

13   Coupling methods

Coupling methods have applications in many areas of probability theory. They were introduced by Doeblin ([29]) in the thirties, the reader will find a survey and a sketch of the history in [54]. Since the seventies they have been successfully used to estimate the mixing rate of Markov chains (see, e.g., [39] or [62]).

Stopping times are a necessary prerequisite to understand how one can bound mixing times by using coupled Markov chains, they were introduced in the last chapter. Our investigations of couplings begin with a motivation and the formal definition. Next we define the *total variation distance* between two probability measures, a distance which already was important in chapter 10. Then we show how the *coupling inequality* relates the variation distance with couplings, it will play an important role later.

Then we turn to *Markov chains*, we complement our study of the *how-fast-is-the-convergence-to-the-equilibrium*–problem by defining some appropriate quantities and by studying some of their properties. Next *coupled Markov chains* are defined. The main results can be found in theorem 13.9, they are applied in a number of illustrating *examples*. The final part of this chapter – which can be skipped at first reading – is addressed to the question: *how powerful are coupling methods?*

Couplings

The underlying *idea* of couplings is simple: you perform only one random experiment, and the result is used to determine the next step of more than one random walk. Consider for example the state space $S = \{0, \ldots, 9\}$, we want to perform two cyclic random walks by using a single fair die. Both walks start deterministically somewhere, for the choice of the respective next positions one throws the die.

Walk 1 now steps one unit (modulo 10) to the right resp. to the left depending on whether the die shows an even resp. an odd number. *Walk 2* instead moves one step to the right resp. to the left if the die shows $1, 2, 3$ resp. $4, 5, 6$. The remarkable fact is that both walks are perfect cyclic random walks, with equal probability they move one step clockwise or counter-clockwise. However, there is some dependency between the two walks: if walk 1 steps to the right (since the die shows 2, 4 or 6) it is more likely that walk 2 moves to the left than to the right (the probabilities are 2/3 and 1/3). This is different from a situation where they move independently, e.g., if one uses two fair coins.

> Consider as a variant the following rule for walk 2: move to the right resp. to the
> left in case of a result in $\{1, 3, 5\}$ resp. in $\{2, 4, 6\}$; this is precisely the opposite rule
> as for walk 1. Again both walks are perfect cyclic random walks, but now any step
> of walk 2 depends deterministically on what walk 1 does.

This will now be formalized:

Definition 13.1 Let μ and ν be two probability measures on $S := \{1, \ldots, N\}$.
By a *coupling* of μ and ν we mean any probability measure $\tilde{\mathbb{P}}$ on $S \times S$ with marginals μ and ν, that is $\mu(\Delta) = \tilde{\mathbb{P}}(\Delta \times S)$ and $\nu(\Delta) = \tilde{\mathbb{P}}(S \times \Delta)$ for $\Delta \subset S$.

It is sometimes convenient to think of μ and ν as of families of nonnegative numbers (μ_i) and (ν_i) with $\sum \mu_i = \sum \nu_i = 1$ and of $\tilde{\mathbb{P}}$ as an $N \times N$-matrix $[\rho_{ij}]$ with $\rho_{ij} \geq 0$, $\sum_{ij} \rho_{ij} = 1$. Then the coupling condition is (obviously) equivalent with $\sum_i \rho_{ij} = \mu_j$, $\sum_j \rho_{ij} = \nu_i$ for all i, j.

For example, let μ and ν be the measures on $\{1, 2, 3, 4\}$ which are defined by $(2/8, 2/8, 2/8, 2/8)$ and $(3/8, 3/8, 1/8, 1/8)$, respectively. Here are the ρ-matrices of two possible couplings[1]:

$$\frac{1}{8} \begin{bmatrix} 2 & 1 & 0 & 0 \\ 0 & 1 & 1 & 1 \\ 0 & 0 & 0 & 1 \\ 0 & 0 & 1 & 0 \end{bmatrix} ; \frac{1}{8} \begin{bmatrix} 2 & 0 & 1 & 0 \\ 0 & 2 & 0 & 1 \\ 0 & 0 & 1 & 0 \\ 0 & 0 & 0 & 1 \end{bmatrix} .$$

As another example, let $\mu = \nu$ be the uniform distribution on $\{L, R\}$. The four couplings

$$\frac{1}{2} \begin{bmatrix} 1 & 0 \\ 0 & 1 \end{bmatrix} , \frac{1}{2} \begin{bmatrix} 0 & 1 \\ 1 & 0 \end{bmatrix} , \frac{1}{4} \begin{bmatrix} 1 & 1 \\ 1 & 1 \end{bmatrix} , \frac{1}{6} \begin{bmatrix} 2 & 1 \\ 1 & 2 \end{bmatrix} .$$

could be used to control the next step of two random walks on $\{0, \ldots, 9\}$: μ corresponds to the first, ν to the second walk, where "L" and "R" yield "step to the left" resp. "step to the right". Do you recognize the above examples?

Even in the preceding very elementary case there are numerous couplings for two given measures μ, ν. The collection of all $\tilde{\mathbb{P}}$ is a compact convex set, it is a nontrivial task to describe its structure completely, e.g., by identifying the extreme points. However, this is not our concern here, we refer the reader to [61] and the literature cited there.

It is simple to convince oneself that for arbitrary μ, ν a *coupling exists*: put $\rho_{ij} := \mu_i \nu_j$, this coupling corresponds to the *product measure*, the associated random variables are *independent* in this case.

Couplings are contained in this book since we will use them to provide bounds for the mixing rate. The preceding μ, ν will correspond to the distributions of two random walks after a "large" number of steps, and couplings will come into play when we investigate the *distance* between μ and ν. The appropriate definition of "distance between two measures" here is

Definition 13.2 Let μ and ν be two probability measures on $S = \{1, \ldots, N\}$. Then the *total variation distance* between μ and ν is defined by

$$\|\mu - \nu\| := \sup_{\Delta \subset S} |\mu(\Delta) - \nu(\Delta)|.$$

This is very natural, the total variation distance is just the distance with respect to the sup-norm if we think of a measure on S as of a function from the power set to the reals. Surprisingly a simple description is possible, the reader cannot fail to be reminded of the l^1-norm difference we have met in proposition 10.7:

---

[1] We will use *square brackets* when dealing with coupling matrices in order not to confuse them with stochastic matrices.

Lemma 13.3 *Let μ and ν be identified with probability vectors (μ_i) and (ν_i). Then*

$$\|\mu - \nu\| = \frac{1}{2} \sum_i |\mu_i - \nu_i|.$$

Proof. Write S as the disjoint union of subsets S_1, S_2, where the elements of S_1 (resp. S_2) are precisely the i with $\mu_i > \nu_i$ (resp. $\mu_i \le \nu_i$). As a consequence of $\sum \mu_i = \sum \nu_i$ it follows that

$$\sum_{i \in S_1} |\mu_i - \nu_i| = \sum_{i \in S_2} |\mu_i - \nu_i| = \frac{1}{2} \sum_{i \in S} |\mu_i - \nu_i|.$$

Now let $\Delta \subset S$ be arbitrary, we suppose that, without loss of generality,

$$\sum_{i \in S_1 \cap \Delta} |\mu_i - \nu_i| \ge \sum_{i \in S_2 \cap \Delta} |\mu_i - \nu_i|.$$

Then we get

$$
\begin{aligned}
|\mu(\Delta) - \nu(\Delta)| &= \left| \sum_{i \in \Delta} \mu_i - \nu_i \right| \\
&= \sum_{i \in S_1 \cap \Delta} |\mu_i - \nu_i| - \sum_{i \in S_2 \cap \Delta} |\mu_i - \nu_i| \\
&\le \sum_{i \in S_1 \cap \Delta} |\mu_i - \nu_i| \\
&\le \sum_{i \in S_1} |\mu_i - \nu_i| \\
&= \frac{1}{2} \sum_{i \in S} |\mu_i - \nu_i|,
\end{aligned}
$$

and this proves that $\|\mu - \nu\| \le \sum_i |\mu_i - \nu_i|/2$.

The reverse inequality is easily proved: with $\Delta := S_1$ we have $\mu(\Delta) - \nu(\Delta) = \sum_i |\mu_i - \nu_i|/2$, and thus "$\ge$" has to hold in the lemma. $\square$

Remark: Note that, by the last step of the proof, it is not necessary to pass to absolute values in the definition of the total variation distance:

$$\|\mu - \nu\| = \sup_{\Delta \subset S} (\mu(\Delta) - \nu(\Delta)).$$

Here is the main result, the following *coupling inequality* relates couplings with the variation distance:

Proposition 13.4 *Let μ and ν be probability measures on $S = \{1, \dots, N\}$ and $\tilde{\mathbb{P}}$ a probability measure on $S \times S$ which is a coupling for μ, ν. Then*

$$\|\mu - \nu\| \le \tilde{\mathbb{P}}(\{(i,j) \mid i \ne j\}).$$

Also there exists a coupling $\tilde{\mathbb{P}}'$ such that in fact "=" holds in this inequality.

Proof. For any i, we may write $\mu_i = \sum_j \rho_{ji}$ and $\nu_i = \sum_j \rho_{ij}$. Now let $\Delta \subset S$ be given. If we express $\mu(\Delta) - \nu(\Delta)$ by the ρ_{ij}, then the summands which correspond to points of $\Delta \times \Delta$ will cancel, only the (i,j) in $\Delta' \times \Delta$ and in $\Delta \times \Delta'$ survive, where $\Delta' := S \setminus \Delta$:

$$\mu(\Delta) - \nu(\Delta) = \sum_{\Delta' \times \Delta} \rho_{ij} - \sum_{\Delta \times \Delta'} \rho_{ij}.$$

Since $\Delta' \times \Delta$ and $\Delta \times \Delta'$ are disjoint subsets of $\{(i,j) \mid i \neq j\}$ we can continue with the inequality $\leq \tilde{\mathbb{P}}(\{(i,j) \mid i \neq j\})$, and this – by the remark preceding the proposition – proves the first part.

For the proof of the second part put $\varepsilon_i := \min\{\mu_i, \nu_i\}$. The ε_i will serve as entries on the diagonal of a coupling $\tilde{\mathbb{P}}'$ which will provide "=" in the coupling inequality.

More precisely, $\tilde{\mathbb{P}}'$ will be defined by prescribing numbers ρ'_{ij}, and we start with the definition $\rho'_{ii} := \varepsilon_i$.

How can this be extended to give a coupling? Suppose, e.g., that $i \in S_2$ (the notation is as in the proof of the preceding lemma). Then $\varepsilon_i = \mu_i$, and it follows that all ρ'_{ji} for $j \neq i$ will have to vanish in order to achieve $\sum_j \rho'_{ji} = \mu_i$. Similarly the ρ'_{ij} have to be zero if $i \in S_1$ and $i \neq j$.

Thus it remains to fix the ρ'_{ij} with $i \in S_2, j \in S_1$ properly. We set $\tau := \sum_i(\mu_i - \varepsilon_i)$. Note that $\tau \geq 0$ and that $\tau = \sum_i(\nu_i - \varepsilon_i)$ (since $\sum_i \mu_i = \sum_i \nu_i$). Also, τ will vanish only if μ and ν coincide, a case in which the assertion is obvious: let all the ρ'_{ij} which remain to be found vanish. Therefore we may assume that τ is strictly positive. We define the remaining entries of $\tilde{\mathbb{P}}'$ as

$$\rho'_{ij} := (\mu_j - \varepsilon_j)(\nu_i - \varepsilon_i)/\tau \text{ for } i \in S_2,\ j \in S_1.$$

Then $\tilde{\mathbb{P}}'$ is a coupling for μ, ν, this follows at once from

$$\begin{aligned}
\tau &= \sum_i (\mu_i - \varepsilon_i) \\
&= \sum_{i \in S_1} (\mu_i - \varepsilon_i) \\
&= \sum_j (\nu_j - \varepsilon_j) \\
&= \sum_{j \in S_2} (\nu_j - \varepsilon_j).
\end{aligned}$$

Also it is clear from the construction that

$$\begin{aligned}
\tilde{\mathbb{P}}'(\{(i,j) \mid i \neq j\}) &= \sum_{i \in S_2, j \in S_1} \rho'_{ij} \\
&= \frac{1}{\tau} \sum_{i \in S_2, j \in S_1} (\mu_j - \varepsilon_j)(\nu_i - \varepsilon_i) \\
&= \sum_{j \in S_1} (\mu_j - \nu_j) \\
&= \|\mu - \nu\|,
\end{aligned}$$

and this completes the proof of our proposition. $\qquad\square$

Sometimes it will be convenient to use a slight *reformulation* of couplings and the coupling inequality. Let S be as before and $X, Y : \Omega \to S$ be two random variables defined on some probability space $(\Omega, \mathcal{A}, \mathbb{P})$; the measure $\mathbb{P}$ is then called a *coupling of X and Y*. If we pass to the induced measures – i.e., to $\mu_i := \mathbb{P}(X = i)$, $\nu_i := \mathbb{P}(Y = i)$ – and consider the coupling in the sense of definition 13.1 defined by $\tilde{\mathbb{P}}(i, j) := \mathbb{P}(X = i, Y = j)$ then the coupling inequality reads as follows:

Proposition 13.5 $\|\mu - \nu\| \leq \mathbb{P}(X \neq Y)$.

Coupled Markov chains and an estimate of the mixing rate

We now return to our main concern, the mixing rate of a chain. We will work with couplings, and – in view of the coupling inequality of proposition 13.4 – it will be necessary to start with some notions and facts concerning the variation distance of the measures we are interested in.

Definition 13.6 Let $P = (p_{ij})_{i,j=1,\ldots,N}$ be an irreducible and aperiodic stochastic matrix and $\pi^\top$ the associated equilibrium distribution.

(i)   For any $i \in S := \{1, \ldots, N\}$ and any integer $k \geq 0$ we denote by $d_i(k)$ the variation distance between the i'th row of P^k and $\pi^\top$:

$$d_i(k) := \frac{1}{2} \sum_j |p_{ij}^{(k)} - \pi_j|.$$

(ii) $d(k) := \max_i d_i(k)$.

(iii)   For $k \geq 0$ consider the N measures which correspond to the rows of P^k. By $\rho(k)$ we will denote the variation diameter of this collection:

$$\rho(k) := \frac{1}{2} \max_{i_1, i_2} \sum_j |p_{i_1 j}^{(k)} - p_{i_2 j}^{(k)}|.$$

(Note that we have already met the $d_i(k)$ in chapter 10; see proposition 10.7.)

Lemma 13.7

(i) $\rho(k) \geq \rho(k + 1)$,

(ii) $d(k) \geq d(k + 1)$,

(iii) $d(k) \leq \rho(k) \leq 2d(k)$,

(iv) $\rho(k + l) \leq \rho(k)\rho(l)$.

Proof. (i) Denote by Δ_k the convex hull of the N measures of part (iii) of definition 13.6. We have shown in lemma 10.6 that $\rho(k)$ is the diameter of this set. Also, since $P^{k+1} = PP^k$, the rows of P^{k+1} are convex combinations of the rows of P^k, and therefore the $(p_{ij}^{(k+1)})_{j=1,\ldots,N}$ lie in Δ_k. This proves that $\Delta_{k+1} \subset \Delta_k$, and in particular (i) follows. (ii) We start with the observation that $\pi^\top$ lies in all Δ_k: these sets are closed and decreasing, and if $\pi^\top$ failed to lie in some Δ_k we would obtain a contradiction to the fact that the rows of P^k converge to $\pi^\top$.

By definition, $d(k)$ is the radius of the smallest ball (with respect to the total variation distance) with center $\pi^\top$ which contains $(p_{ij}^{(k)})_{j=1,\ldots,N}$ for all i or, equivalently, which contains Δ_k. (ii) follows now from $\Delta_{k+1} \subset \Delta_k$.

(iii) The first inequality is an immediate consequence of what just has been shown, the second one follows from the triangle inequality.

(iv) In this proof we will use for the first time the *coupling inequality*.

Let k, l and i_1, i_2 be arbitrary and denote the i_1'th and i_2'th row of P^{k+l} by μ and ν, respectively. We have to show that $\|\mu - \nu\| \leq \rho(k)\rho(l)$, and the idea is to construct a coupling $\tilde{\mathbb{P}}$ on $S \times S$ of μ and ν such that $\tilde{\mathbb{P}}(\{(i,j) \mid i \neq j\}) \leq \rho(k)\rho(l)$; an application of (the easy part of) proposition 13.4 then will complete the proof of (iv).

$\tilde{\mathbb{P}}$ will be constructed with the help of the hard part of 13.4. First we consider the i_1'th resp. the i_2'th row in P^k which we will denote by α resp. β. By proposition 13.4, there is a coupling $\tilde{\tilde{\mathbb{P}}}$ for these two measures, i.e., we find nonnegative numbers a_{ij} such that $\sum_i a_{ij} = \alpha_j$, $\sum_j a_{ij} = \beta_i$ and $\sum_{i \neq j} a_{ij} = \|\alpha - \beta\|$.

Similarly we treat the l-step transitions. For – not necessarily distinct – j_1, j_2 we choose, again with the help of proposition 13.4, an optimal coupling for the measures which correspond to the j_1'th and j_2'th row of $P^l = (p_{ij}^{(l)})$. In this way we get $a_{ij}^{j_1 j_2} \geq 0$ such that

$$\sum_i a_{ij}^{j_1 j_2} = p_{j_1 j}^{(l)},$$

$$\sum_j a_{ij}^{j_1 j_2} = p_{j_2 i}^{(l)},$$

$$\sum_{i \neq j} a_{ij}^{j_1 j_2} = \frac{1}{2} \sum_j |p_{j_1 j}^{(l)} - p_{j_2 j}^{(l)}|.$$

(The case $j_1 = j_2$ can be treated directly: simply put $a_{ii}^{j_1 j_2} := p_{j_1 i}^{(l)}$ and $a_{ij}^{j_1 j_2} := 0$ for $i \neq j$.)

With these preparations at hand we now define the coupling $\tilde{\mathbb{P}} = (\rho_{ij})_{i,j \in S}$ by $\rho_{ij} := \sum_{s,t} a_{st} a_{ji}^{st}$. This is in fact a coupling for μ and ν:

$$\begin{aligned}
\sum_j \rho_{ij} &= \sum_j \sum_{s,t} a_{st} a_{ji}^{st} \\
&= \sum_{s,t} a_{st} \sum_j a_{ji}^{st} \\
&= \sum_{s,t} a_{st} p_{si}^{(l)} \\
&= \sum_s p_{si}^{(l)} \sum_t a_{st} \\
&= \sum_s p_{si}^{(l)} p_{i_2 s}^{(k)} \\
&= p_{i_2 i}^{(k+l)};
\end{aligned}$$

similarly it follows that $\sum_i \rho_{ij} = p_{i_1 j}^{(k+l)}$.

In order to estimate $\tilde{\mathbb{P}}(\{(i,j) \mid i \neq j\})$ we note that the sums $\sum_{i,j, i \neq j} a_{ij}^{j_1 j_2}$ vanish for $j_1 = j_2$ and can be estimated by $\rho(l)$ in general:

$$\begin{aligned}
\tilde{\mathbb{P}}(\{(i,j) \mid i \neq j\}) &= \sum_{i \neq j} \rho_{ij} \\
&= \sum_{i \neq j} \sum_{s,t} a_{st} a_{ji}^{st} \\
&= \sum_{s,t} \left(\sum_{i \neq j} a_{ji}^{st} \right) a_{st} \\
&\leq \rho(l) \sum_{s \neq t} a_{st} \\
&= \rho(l) \|\mu - \nu\| \\
&\leq \rho(k)\rho(l).
\end{aligned}$$

This completes the proof.   □

Now we are ready to put things together, we will introduce *coupled Markov chains*. Here is the *idea*, a formal definition will be given shortly.

As usual we start with a transition matrix P on a finite set S where we fix two arbitrary states i_0, j_0 with $i_0 \neq j_0$.

We want to let two walks start at the same time, one at i_0, the other at j_0. We will speak of a *coupled Markov chain* if the following two conditions are satisfied:

- If one observes only one of the two walks, then it is an ordinary random walk with transition probabilities given by the matrix P.

- Suppose that, for a certain k, both walks occupy the same state i in the k'th step. Then they stay together for all future steps.

Here are some **examples** to illustrate what is meant:

1. With $S = \{1,2\}$ let P be the matrix $\begin{pmatrix} 1/3 & 2/3 \\ 2/3 & 1/3 \end{pmatrix}$.
The walks start at 1 and 2, in each step they move according to the following rule:

> Throw a fair die. If it shows 1 or 2, both random walkers stay where they are, otherwise they exchange their positions.

Obviously both conditions are met, the second one in a trivial sense (since the walks will *never* be at the same state.)

2. With S and P as in the preceding example, we change the rule (the starting positions, however, are as before):

> Throw a fair die, let the result be d. If $d = 1$ or $d = 2$, then walk 1 stays where it is, otherwise it changes to the (unique) other state. Walk 2 instead holds the present position if $d = 5$ or $d = 6$ and moves if $d = 1, 2, 3, 4$.
> This rule applies until the walks have met for the first time. Then *both* continue to move according to the previous rule for walk 1.

Now *there is a chance that the walks meet*: if they have not occupied the same position until the k'th step they will do so in the $k + 1$'th step with probability $2/3$ (namely if $d = 1, 2, 5$ or 6). Or: only with probability $1/3$ – if $d = 3$ or $d = 4$ – they will exchange their positions and therefore again fail to meet. Thus the number of steps until the first coincidence is just the waiting time until the first "success" ($d = 1, 2, 5, 6$), and therefore the expected number of steps is $3/2$.

3. Here we define a coupled random walk for the following chain, a variant of example 8 in chapter 2: the states are the 0-1-sequences of length r, transitions are possible if they are of the form $i \to i$ or between states which differ at just one component; the former have probability $1/2$, the latter probability $1/2r$.

The chain pauses with a positive probability, and therefore it is aperiodic. It is also clear that every two states communicate, and hence the chain will converge to its equilibrium. Since P is doubly stochastic, the equilibrium is the uniform distribution: all states are (roughly) equally likely if the chain has run for a "long" time.

Let i_0, j_0 be fixed starting positions of two random walks. Here is the rule how to proceed if walk 1 resp. walk 2 occupies state $i = (\varepsilon_1, \ldots, \varepsilon_r)$ resp. $j = (\tau_1, \ldots, \tau_r)$:

> Choose a coordinate $\rho \in \{1, \ldots, r\}$ and a number $w \in \{0, 1\}$, both according to the equidistribution.
>
> *First case: i and j coincide at the ρ'th bit: $\varepsilon_\rho = \tau_\rho$.*
> Then, if $w = 0$, let the walks move to i' resp. j', where these states arise from i resp. j by switching the ρ'th bit; if $w = 1$, both walks stay where they are.
>
> *Second case: $\varepsilon_\rho \neq \tau_\rho$.*
> Now, if $w = 0$, switch the ρ'th bit of i; this gives rise to a state which will be the new position of the first walk. The second walk stays at j. If it happens that $w = 1$, then walk 1 stays at i and walk 2 switches its ρ'th bit.

A moment's reflection shows that both walks really move with the desired transition probabilities: the probability is $1/2$ for keeping the position and – for arbitrary ρ – $1/2r$ for switching the ρ'th bit. It is also clear that they will move forever together after they have met at the same position.

How long will it take them to meet? Suppose that i_0 and j_0 have different bits precisely at the components lying in $\Delta \subset \{1, \ldots, r\}$. By our rule, regardless of which ρ has been randomly chosen in the k'th step, the ρ'th bit of the positions in the $k+1$'th step will be the same for both walks. Therefore *the walks will meet precisely as soon as the random generator producing the ρ has provided all elements of Δ.*

It's time for a formal definition:

Definition 13.8 Let $P = (p_{ij})_{i,j=1,\ldots,N}$ be a stochastic matrix. We put $S := \{1, \ldots, N\}$, and we fix $i_0, j_0 \in S$ with $i_0 \neq j_0$. By a *coupled Markov chain* associated with P, i_0, j_0 we mean two Markov processes $X_0, X_1, \ldots : \Omega \to S$ and $Y_0, Y_1, \ldots : \Omega \to S$ defined on the same probability space $(\Omega, \mathcal{A}, \mathbb{P})$ such that

   (i)   both processes have transition probabilities according to P;
  (ii) $X_0 = i_0$, $Y_0 = j_0$;
 (iii) $X_k(\omega) = Y_k(\omega)$ implies that $X_{k+1}(\omega) = Y_{k+1}(\omega)$ (all k and ω).

With every coupled Markov chain there is associated a *stopping time* $\mathbb{T}$. It is defined by

$$\mathbb{T}(\omega) := \min\{k \mid X_k(\omega) = Y_k(\omega)\},$$

where we adopt the usual convention that the minimum of the empty set is ∞. This – obviously – is in fact a stopping time with respect to the filtration $(\mathcal{F}_k)$, where $\mathcal{F}_k$ is the σ-algebra generated by $X_0, \ldots, X_k, Y_0, \ldots, Y_k{}^2$.

---

[2] To phrase it less formally: it suffices to observe the walks in order to be able to stop correctly.

We have already started to investigate $\mathbb{T}$ in our previous *examples*:

1. Here $\mathbb{T}$ is the constant function ∞.

2. We have mentioned already that $\mathbb{P}(\mathbb{T} > k) = (1/3)^k$ and that the expectation $\mathbb{E}(\mathbb{T})$ of $\mathbb{T}$ is $3/2$.

3. In order to treat this example one has to recall some elementary facts from elementary probability.

Let $U_1, U_2, \ldots$ be independent, equidistributed $\{1, \ldots, r\}$-valued random variables (they generate our ρ). It has been noted above that $\mathbb{T}$ is the waiting time until the $U_1, \ldots$ have exhausted all ρ in Δ. This will depend on how large Δ is. In the worst case one has to wait until *all* $1, \ldots, r$ have been provided at least once[3]. It is known that the expected value of this time is $1 + r/(r-1) + r/(r-2) + \cdots + r/1$, a number which can be bounded from above by $r(1 + \log r)$.

Here is the main result, stopping times and couplings meet to make possible a new rapid mixing inequality:

Theorem 13.9 *Let $P = (p_{ij})_{i,j=1,\ldots,N}$ be (an arbitrary) stochastic matrix and $i_0, j_0 \in \{1, \ldots, N\}$. Further, let (X_k), (Y_k) be a coupled Markov chain associated with P, i_0, j_0 as in the preceding definition; by $\mathbb{T}$ we denote the waiting time until the walks meet. If k is arbitrary, then $\|\mu - \nu\| \leq \mathbb{P}(\mathbb{T} > k)$, where μ, ν stand for the measures associated with the i_0'th resp. the j_0'th row of P^k.*

Proof. μ and ν are just the image measures of X_k and Y_k, respectively. Hence, by the coupling inequality in proposition 13.5, $\|\mu - \nu\| \leq \mathbb{P}(X_k \neq Y_k)$. But $\{X_k \neq Y_k\}$ is a subset of $\{\mathbb{T} > k\}$ since coupled Markov chains walk together as soon as they have met, and thus the proof is already complete. $\square$

Here are **the most important consequences:**

- Suppose that you can treat all possible i_0, j_0 in a unified way. Then you obtain an estimate of the maximum over the possible $\|\mu - \nu\|$, i.e., of $\rho(k)$.

- Once you know that $\rho(k_0)$ is "small" for a suitable k_0 you can use lemma 13.7: $\rho(rk_0) \leq \rho(k_0)^r$. With this lemma one also gets bounds for $d(k)$, the maximal total variation distance of the $(p_{i_0 1}^{(k)}, \ldots, p_{i_0 N}^{(k)})$ to the equilibrium.

- If only the expectation of $\mathbb{T}$ is known but possibly not the explicit values of the $\mathbb{P}(\mathbb{T} > k)$, one can still apply the theorem:

  simply use the obvious inequality $(k+1)\mathbb{P}(\mathbb{T} > k) \leq \mathbb{E}(\mathbb{T})$.

- It is a priori by no means clear *how to choose an appropriate coupling* in order to get good estimates for the mixing rate with the help of the theorem. In this respect the situation is similar to that of chapter 11 where the strength of what can be shown by using conductance methods was limited by our ability to invent skillfully designed canonical paths.

  We will return to this question at the end of this chapter.

---

[3] Some readers will have recognized that the problem we are dealing with is the *coupon collector's problem* from elementary probability in disguise; in Feller's book [30] one finds a thorough discussion.

Typical applications of the theorem are illustrated by the following **examples:**

1. In the above example 1 (page 114) $\mathbb{T}$ is finite for no ω. Therefore our theorem only yields the poor estimate $\|\mu - \nu\| \leq 1$ (and thus $\rho(k) \leq 1$).

2. The coupling of example 2 for the same chain works better: we get

$$\rho(k) \leq \mathbb{P}(\mathbb{T} > k) = (1/3)^k.$$

If it were only known that $\mathbb{E}(\mathbb{T}) = 3/2$ we could only conclude that

$$(k+1)\mathbb{P}(\mathbb{T} > k) \leq 3/2,$$

that is,

$$\rho(k) \leq \frac{3}{2(k+1)}.$$

3. We already know the expected value of the stopping time associated with example 3, and this gives the estimate $\rho(k) \leq (r/r + r/(r-1) + r/(r-2) + \cdots + r/1)/(k+1)$, i.e., a convergence which is roughly of order $r \log r / k$.

To get a better result we recall that $\mathbb{T} > k$ just means that k independent and equidistributed choices out of $\{1, \ldots, r\}$ have *not* produced all elements. If $i_0 \in \{1, \ldots, r\}$ is fixed, then this state will *not* have been chosen in k trials with probability $(1 - 1/r)^k$. Consequently, if we sum over all i_0, we get $r(1 - 1/r)^k$ as a bound for the probability that *any* element is not present after k choices. Hence $\mathbb{P}(\mathbb{T} > k) \leq r(1 - 1/r)^k$, and we arrive at $\rho(k) \leq r(1 - 1/r)^k$.

To discuss this a little bit further we recall that $1 - x \leq e^{-x}$ so that $(1 - 1/r)^k \leq e^{-k/r}$. Consequently, if we want to have $\rho(k)$ bounded by e^{-c} for an arbitrary c, it suffices to take k of order $r \log r + cr$. It follows that – not surprisingly – a doubling of r essentially necessitates a doubling of the number of simulation steps to achieve a similarly small $\rho(k)$, i.e., a similar precision of approximation.

4. Let P be a strictly positive stochastic matrix, we will apply the preceding theorem to give another proof of the fact that (P^k) converges to a matrix with identical rows.

It will suffice to show that $(\rho(k))_k$ tends to zero, and this will be proved by considering suitable couplings. We fix i_0 and j_0 as the starting positions of two walks, and we prescribe transitions as follows:

> Suppose that the walks have not yet met. Then the next step is for both as prescribed by P, and the two new positions are generated *independently.*
> From the first meeting on they move together, *only one random choice* (according to the appropriate P-probabilities) is necessary.

It is plain that this rule meets the requirements of coupled Markov chains. To apply theorem 13.9 we have to analyse the associated stopping time $\mathbb{T}$.

Let p be the positive number $\min_{ij} p_{ij}$ and suppose that at step k the two walks still occupy different positions, i' and j' say. Since we choose the next position independently, we can assure that – for arbitrary i, j – they will be next at i (walk 1) resp. j (walk 2) with probability $p_{i'i} p_{j'j}$. Therefore the probability that they meet is $\sum_i p_{i'i} p_{j'i}$, a number which is bounded from below by Np^2. And thus only with probability $(1 - Np^2)^k$ a meeting will *not* have taken place in k steps:

$$\mathbb{P}(\mathbb{T} > k) \leq (1 - Np^2)^k.$$

This really implies that $\rho(k) \to 0$; the rate of convergence, however, is much worse than that we have obtained by other methods (see proposition 10.5).

5. We are now going to analyse the *random-to-top-shuffle* (cf. also example 5 in chapter 2). There is given a deck of r cards, they are – from top to bottom – labelled $1, \ldots, r$. Shuffling is defined by selecting a card at random and putting it to the top. How often will it be necessary to shuffle this way until the cards in the resulting deck lie such that all of the $r!$ permutations are (approximately) equally probable[4]?

In order to prescribe a coupling rule we start with two fixed permutations of the cards (that is, with two points of our state space), we think of them as two decks of r cards which are labelled as above and which are in some fixed order at the beginning. Here is the rule:

> Select $\rho \in \{1, \ldots, r\}$ uniformly at random. ρ is used to perform the transitions in the two decks: in deck 1, the ρ'th card – counted from above – is put on top. Let this card have the label l, say, then in deck 2 also the label-l-card will move to the top, regardless of where it is found[5].

By this rule, after any move the uppermost cards in both decks coincide. It is also clear that both decks "move" together as soon as their cards are in the same order and that they transform according to the correct transition probabilities: every individual card has the same chance to be the card on top next, the reason is that the uniform distribution is invariant with respect to permutations.

The problem to estimate $\mathbb{T}$ for this coupling leads to similar questions as in the above example 3. Both decks will coincide as soon as every number in $\{1, \ldots, r\}$ has been chosen as a ρ. We omit to repeat the above discussion.

How powerful is the coupling method?

Any coupling can be used to get bounds for the $\rho(k)$, but the discussion of the examples has shown that it is not a simple task to get "good" inequalities in this way. What are the theoretical limits of this method, is it always possible to get the best possible estimates? Does there exist, for arbitrary S, P, i_0, j_0, a coupling such that equality holds for every k in the inequality of theorem 13.9? The answer is yes, the proof is due to Griffeath ([39], [40]), for different approaches see [62], [38], and [71].

The construction of this optimal coupling, however, is extremely involved, it is far beyond the scope of this book. Let us try to understand the difficulties.
We consider $S = \{1, 2, 3, 4\}$ together with the stochastic matrix

$$\frac{1}{8}\begin{pmatrix} 2 & 2 & 2 & 2 \\ 3 & 3 & 1 & 1 \\ 2 & 2 & 2 & 2 \\ 3 & 3 & 1 & 1 \end{pmatrix},$$

and we want to study couplings for walks which start at $i_0 = 1$ and $j_0 = 2$. Our very modest task will be to find a coupled Markov chain for P, i_0, j_0 such that theorem 13.9 gives *the best possible result for the first two steps of the chain.*

---

[4] Note that also in this example the equilibrium is the uniform distribution, see exercise 7.1.

[5] If, e.g., $r = 6$ and the two decks are $D_1 = (124653)$, $D_2 = (653214)$, then the choice $\rho = 3$ would lead to $D_1 = (412653)$, $D_2 = (465321)$.

As a preparation we calculate P^2:

$$\frac{1}{16}\begin{pmatrix} 5 & 5 & 3 & 3 \\ 5 & 5 & 3 & 3 \\ 5 & 5 & 3 & 3 \\ 5 & 5 & 3 & 3 \end{pmatrix},$$

and we define measures

$$\mu = (2/8,\ 2/8,\ 2/8,\ 2/8),\quad \nu = (3/8,\ 3/8,\ 1/8,\ 1/8)$$

($=$ the first two rows of P) and

$$\mu' = \nu' = (5/16,\ 5/16,\ 3/16,\ 3/16)$$

($=$ the first two rows of P^2). We need a stochastic rule which controls the two walks such that:

1. After the first step the probability that they occupy different positions is $\|\mu - \nu\| = 1/4$; the various states of S have to be chosen in accordance with μ and ν for the two walks.

2. Also in the second step they move as prescribed by P (and they move in the same way if they have met in the first step). Also it is necessary that they now occupy the same position with probability one (in order to have $\|\mu' - \nu'\| = 0 = \mathbb{P}(X_2 \neq Y_2)$).

To satisfy "1." we need a coupling for μ and ν. Two of them have been presented after definition 13.1, but both fail to fulfill "1.". With a glance at the construction in proposition 13.4 we can be more successful, we see that every coupling of the form

$$\frac{1}{8}\begin{bmatrix} 2 & 0 & \rho_{13} & \rho_{14} \\ 0 & 2 & \rho_{23} & \rho_{24} \\ 0 & 0 & 1 & 0 \\ 0 & 0 & 0 & 1 \end{bmatrix}$$

– with a doubly stochastic matrix $\begin{bmatrix} \rho_{13} & \rho_{14} \\ \rho_{23} & \rho_{24} \end{bmatrix}$ – gives the desired result. To satisfy "2." one has to choose these ρ_{ij} carefully. Consider, e.g., $\rho_{23}/8$. This is the probability that after the first move walk 1 resp. walk 2 occupies state 2 resp. 3. The next step has to be according to P, that is as described by the measures which correspond to the rows 2 and 3 of this matrix. The variation distance of them is different from zero, and thus there is no hope to find a coupling with zero entries off the diagonal. Therefore we cannot succeed to fulfill requirement "2." if it happens that ρ_{23} is different from zero. By a similar reasoning we convince ourselves that ρ_{14} will have to vanish, and consequently the only coupling for the first step with which we might hope to be successful is

$$\frac{1}{8}\begin{bmatrix} 2 & 0 & 1 & 0 \\ 0 & 2 & 0 & 1 \\ 0 & 0 & 1 & 0 \\ 0 & 0 & 0 & 1 \end{bmatrix}.$$

In fact this works! If we continue to couple carefully we can achieve our goal that the walks move together after the second step. If it happens, for example, that they are at states 1 and 3 we let them *both* choose the next position according to the probability law $(1/4, 1/4, 1/4, 1/4)$. This is the first and the third row of P, thus they move in accordance with the prescribed transition probabilities, also they are at the same place after this move. Similarly one deals with the case "walk 1 at 2, walk 2 at 4", then for both the law $(3/8, 3/8, 1/8, 1/8)$ is relevant.

The moral of the story is that in choosing even the first coupling rule one has to take into account how the future moves are governed by P. This makes the construction extremely difficult, even an optimal rule for k_0 steps for moderate k_0 will be an extremely demanding task. And therefore the fact that coupling methods are able to provide the best possible bounds for the mixing rate is mainly of theoretical interest.

Exercises

13.1: Let $(\mu_1, \ldots, \mu_N)$ and $(\nu_1, \ldots, \nu_N)$ be fixed probability distributions and $C_{\mu,\nu}$ the collection of all couplings for μ and ν. Prove that $C_{\mu,\nu}$ is a nonvoid compact convex set (in $\mathbb{R}^{N^2}$, if we identify couplings with suitable matrices).

13.2: Let $\mu = (1/N, \ldots, 1/N)$ be the uniform distribution and $\nu = (1, 0, \ldots, 0)$ the point mass at 1. Determine all extreme points of $C_{\mu,\nu}$.

13.3: Couplings can be defined more generally. If, e.g., μ and ν are probability measures on $[0, 1]$, then a coupling of μ and ν is a measure $\tilde{\mathbb{P}}$ on the square with marginals μ and ν. Prove the easy part of the coupling inequality in this more general setting:

$$\sup |\mu(A) - \nu(A)| \leq \tilde{\mathbb{P}}(\{(x,y) \mid x \neq y\}),$$

where the supremum is taken over all measurable $A \subset [0, 1]$.

13.4: Denote by $C'_{\mu,\nu}$ the collection of maximal couplings for μ and ν (the notation is as in exercise 13.1; a coupling is called *maximal* if it provides "=" in the coupling inequality). Prove that this set is a nonvoid closed face in $C_{\mu,\nu}$ (cf. exercise 1.2).

13.5: Let μ be a fixed probability measure on $\{1, \ldots, N\}$. Characterize the probability measures ν such that there is precisely one maximal coupling for μ and ν.

13.6: Let μ be the uniform distribution on $\{1, \ldots, N\}$. For what probability measures ν is the total variation distance from μ to ν as large as possible?

13.7: What are the probability distributions $(\mu_1, \ldots, \mu_N)$ and $(\nu_1, \ldots, \nu_N)$ such that there exists a coupling with

$$\tilde{\mathbb{P}}(\{(x,y) \mid x \neq y\}) = 1?$$

13.8: Consider an arbitrary Markov chain on a finite state space. Does there exist, for different states i_0, j_0, a coupled Markov chain $(X_k), (Y_k)$ associated with P, i_0, j_0 such that the X_k *never* meet the Y_k?

13.9: Choose any stochastic 2×2-matrix P which is not the identity matrix. Find best possible couplings for the first two steps of the chain similarly to our example from the end of the chapter.

14   Strong uniform times

The technique we are going to describe now was introduced in the eighties by Aldous and Diaconis (see [24], [3], [4]). As in the previous chapter stopping times play an important role, the reader can find the necessary prerequisites in chapter 12.

The notation will be as before, in order to have a unique equilibrium π it is assumed that the matrix P under consideration is irreducible. Let $X_0, X_1, \ldots : \Omega \to S$ be any Markov process with transition probabilites prescribed by a stochastic matrix P, the starting distribution might be arbitrary. Sometimes it is possible to stop the process in such a way that the states where one decides to stop are distributed in accordance with π. To apply such a stopping rule it might be necessary to have more information than just the knowledge about the walk up to the present positions, and therefore we cannot restrict ourselves to stopping times with respect to the natural filtration.

The stopping times we have in mind must have the special property that *the necessary information to stop correctly does not spoil the Markov property*. To be more precise, denote by $\mathcal{F}_0^{\text{nat}} \subset \mathcal{F}_1^{\text{nat}} \subset \cdots$ the natural filtration[1]. In the proof of theorem 12.4 we have shown that

$$\mathbb{P}(B \cap \{X_k = i_0, X_{k+1} = i_1, \ldots, X_{k+r} = i_r\}) = \mathbb{P}(B \cap \{X_k = i_0\}) p_{i_0 i_1} p_{i_1 i_2} \cdots p_{i_{r-1} i_r}$$

for every $B \in \mathcal{F}_k^{\text{nat}}$. Summation over all $i_1, \ldots, i_{r-1}$ leads to

$$\begin{aligned} P(B \cap \{X_k = i_0, X_{k+r} = i_r\}) &= \mathbb{P}(B \cap \{X_k = i_0\}) p_{i_0 i_r}^{(r)} \\ &= \mathbb{P}(B \cap \{X_k = i_0\}) \mathbb{P}(X_k = i_0 \mid X_{k+r} = i_r), \end{aligned}$$

that is to $\mathbb{P}(X_{k+r} = i_r \mid X_k = i_0) = \mathbb{P}(X_{k+r} = i_r \mid X_k = i_0, B)$.

This can be rephrased by saying that the additional information given by such a B does not influence the transition probabilities. We are interested in filtrations which behave similarly:

Definition 14.1 Let $X_0, X_1, \ldots$ be as before and $\mathcal{F}_0 \subset \mathcal{F}_1 \subset \cdots$ be a filtration on $(\Omega, \mathcal{A}, \mathbb{P})$ such that $\mathcal{F}_k^{\text{nat}} \subset \mathcal{F}_k$ for all k (so that (X_k) is adapted). A stopping time $\mathbb{T} : \Omega \to \{0, 1, \ldots, \infty\}$ with respect to this filtration will be said to be *compatible with the Markov property of* (X_k) if

$$\mathbb{P}(X_{k+r} = i_r \mid X_k = i_0) = \mathbb{P}(X_{k+r} = i_r \mid X_k = i_0, B)$$

holds for every $B \in \mathcal{F}_k$ and all i_0, i_r in the state space[2].

We have just convinced ourselves that the natural filtration has this property, further examples will be given later.

Of particular importance will be stopping times $\mathbb{T}$ such that the stopped process is distributed like π:

---

[1] cf. definition 12.1.

[2] Note that this property only depends on the filtration and not on the stopping time. Therefore it is a slight abuse of language to speak of a *time* which is compatible with the Markov property.

Definition 14.2 Let $\mathbb{T}$ be a stopping time as in the preceding definition. We say that $\mathbb{T}$ is a *strong uniform time* if

(i) $\mathbb{T}$ is finite with probability one, and

(ii) $\mathbb{P}(X_k = i \mid \mathbb{T} = k) = \pi_i$ for every k and every i.

(This means that, regardless of *when* the chain is stopped, the states *where* this happens are distributed exactly in accordance with the equilibrium.)

As a *simple illustration* consider a chain where P is such that P^{k_0} has identical rows for a suitable k_0 (as it happened with $k_0 = 2$ and

$$\mathbb{P} = \frac{1}{8} \begin{pmatrix} 2 & 2 & 2 & 2 \\ 3 & 3 & 1 & 1 \\ 2 & 2 & 2 & 2 \\ 3 & 3 & 1 & 1 \end{pmatrix}$$

at the end of the previous chapter). Then all these rows are identical with $\pi^\top$, and consequently the deterministic time $\mathbb{T} = k_0$ will be a strong uniform time with respect to the natural filtration.

We aim to apply this new concept to estimate the mixing rate:

Proposition 14.3 *Let P be an irreducible stochastic matrix with equilibrium π and $(X_k)_{k=0,1,\ldots}$ a Markov process governed by P.*
If we denote, for fixed k, by ν the distribution of X_k (i.e., $\nu_i := \mathbb{P}(X_k = i)$), then the variation distance $\|\pi - \nu\|$ is bounded by $\mathbb{P}(\mathbb{T} > k)$ for any strong uniform time.

Proof. Fix $k \geq 0$, $i, j \in S$ and consider any $k' \leq k$. The set $\{\mathbb{T} = k'\}$ lies in $\mathcal{F}_{k'}$ so that

$$\mathbb{P}(X_k = i \mid X_{k'} = j, \mathbb{T} = k') = \mathbb{P}(X_k = i \mid X_{k'} = j).$$

This together with the strong uniform time property of $\mathbb{T}$ yields

$$
\begin{aligned}
\mathbb{P}(X_k = i, \mathbb{T} = k', X_{k'} = j) &= \mathbb{P}(X_k = i \mid \mathbb{T} = k', X_{k'} = j)\mathbb{P}(\mathbb{T} = k', X_{k'} = j) \\
&= \mathbb{P}(X_k = i \mid X_{k'} = j)\mathbb{P}(\mathbb{T} = k', X_{k'} = j) \\
&= \mathbb{P}(X_k = i \mid X_{k'} = j)\mathbb{P}(X_{k'} = j \mid \mathbb{T} = k')\mathbb{P}(\mathbb{T} = k') \\
&= p_{ji}^{(k-k')}\pi_j \mathbb{P}(\mathbb{T} = k').
\end{aligned}
$$

Now it is of importance that π satisfies $\pi^\top P = \pi^\top$ (and thus also $\pi^\top P^{k-k'} = \pi^\top$). It follows that the sum over the preceding equations with different j is just $\pi_i \mathbb{P}(\mathbb{T} = k')$. Hence

$$
\begin{aligned}
\mathbb{P}(X_k = i) &\geq \mathbb{P}(X_k = i, \mathbb{T} \leq k) \\
&= \sum_{k' \leq k, j \in S} \mathbb{P}(X_k = i, X_{k'} = j, \mathbb{T} = k') \\
&= \sum_{k' \leq k} \pi_i \mathbb{P}(\mathbb{T} = k') \\
&= \pi_i \mathbb{P}(\mathbb{T} \leq k) \\
&= \pi_i(1 - \mathbb{P}(\mathbb{T} > k)).
\end{aligned}
$$

Now let $A \subset S$ be arbitrary. By our inequalities we have

$$\begin{aligned}
\nu(A) &= \mathbb{P}(X_k \in A) \\
&= \sum_{i \in A} \mathbb{P}(X_k = i) \\
&\geq \sum_{i \in A} \pi_i (1 - \mathbb{P}(\mathbb{T} > k)) \\
&= \pi(A)(1 - \mathbb{P}(\mathbb{T} > k)),
\end{aligned}$$

and it follows that $\nu(A) - \pi(A) \leq \pi(A)\mathbb{P}(\mathbb{T} > k) \leq \mathbb{P}(\mathbb{T} > k)$. By the remark after the proof of lemma 13.3 this proves that $\|\nu - \pi\| \leq \mathbb{P}(\mathbb{T} > k)$. $\qquad\square$

Some **comments** are in order. First, the end of the proof (where we have estimated $\pi(A) \leq 1$) shows that we could have proved a sharper estimate. We refer the reader to [4] where a distance different from our variation distance is introduced to obtain the best possible bound by using $\mathbb{T}$. There it is also sketched how one can "construct" strong uniform times which provide sharp estimates for every k. (This, however, makes it necessary to extend the notion of stopping times, one has to deal with *randomized stopping times*. Also the construction necessitates to know explicitly all $p_{ij}^{(k)}$, and therefore it seems to be only of theoretical interest.) Also, all remarks corresponding to those after theorem 13.9 can also be made here, in particular an approach which is independent of the starting position will provide bounds for the $\rho(k)$.

We are now going to study some **examples**. Note that one is faced with *two problems*. The first is to invent a stopping rule $\mathbb{T}$ which uses "not too much" information and nevertheless provides the equilibrium distribution, the second is to estimate the numbers $\mathbb{P}(\mathbb{T} > k)$ in order to get bounds for the mixing time.

1. In the rather trivial example after definition 14.2 the theorem gives the obviously true estimate $\|\nu - \mu\| = 0 = \mathbb{P}(\mathbb{T} > k)$ whenever $k \geq k_0$.

2. Let, for the state space $\{1, 2, 3, 4\}$, the transition matrix be defined by

$$P = \begin{pmatrix} p_{11} & p_{12} & p_{13} & 1/4 \\ p_{21} & p_{22} & p_{23} & 1/4 \\ p_{31} & p_{32} & p_{33} & 1/4 \\ 1/4 & 1/4 & 1/4 & 1/4 \end{pmatrix},$$

where the p_{ij} are strictly positive such that $\sum_i p_{ij} = \sum_j p_{ij} = 3/4$ for all i, j. Since P is doubly stochastic, the equilibrium distribution is the uniform distribution.

We consider a random walk which starts at state $i_0 = 1$ (the cases $i_0 = 2, 3, 4$ can be treated similarly, for the case $i_0 = 4$ it is surely optimal to stop deterministically after the first step). We consider the following stopping rule which will be called $\mathbb{T}$:

Stop one step after the walk has been in position 4 for the first time.

One can stop in accordance with this rule by just observing the walk, and therefore $\mathbb{T}$ is a stopping time with respect to the natural filtration. $\mathbb{T}$ is in fact a strong uniform time: it is compatible with the Markov property of the process, it is almost surely finite (since the probability to jump from $\{1, 2, 3\}$ to 4 in the next step is $1/4$), and since the last row of P is the uniform distribution, all states are equally likely to be the position of the walk at time $\mathbb{T}$. The theorem asserts now that the variation distance between the first row of P^k and the equilibrium is at most $\mathbb{P}(\mathbb{T} > k) = (3/4)^k$.

3. We now investigate the *top-to-random-shuffle*. Our r cards are labelled $1,\ldots,r$, and at the beginning the deck is in canonical order with card number r at the bottom. How often is it necessary to "shuffle" in the top-to-random way to be able to guarantee that all possible permutations of the cards are (nearly) equally likely?

As in the case of the *random-to-top shuffle* (see example 5 after theorem 13.9) it is necessary to observe that the chain has the uniform distribution as its equilibrium; this is left to the reader. We will provide bounds of the mixing time by investigating suitable uniform times.

The stopping time we have in mind is defined as follows:

Stop one step after the card with label r has reached the top position.

Here is an example with $r = 4$, we start with the permutation (1234). Suppose that a random generator produces equidistributed elements in $\{1, 2, 3, 4\}$ as follows:

$$1, 2, 4, 4, 3, 4\ldots$$

Then the deck "walks" as follows (the starting position is included):

$$(1234) \rightarrow (1234) \rightarrow (2134) \rightarrow (1342) \rightarrow (3421) \rightarrow (4231) \rightarrow (2314),$$

and here we stop.

This is a stopping time with respect to the natural filtration, hence we only have to check whether it is almost surely finite and whether the stopped positions are equidistributed.

To this end, we analyse the behaviour of the last card, the label-r-card. It will move from its original position to the $r-1$'th position precisely when our random generator has provided the number r. This has a probability $1/r$ and therefore the expected waiting time for this to happen is r. The next move upwards of the r-card happens when an element of $\{r-1, r\}$ is chosen. This results in an expected waiting time of $r/2$ since the probability of "success" is $2/r$. Note that we find at the positions $r-1, r$ two cards c_1, c_2 from $\{1, \ldots, r-1\}$ where both relative orders have the same probability.

Continuing this way we observe sooner or later that our card arrives at the top. For the last step we only have had an expected waiting time of $r/(r-1)$, and in the positions 2 to r we find any of the permutations of $\{1, \ldots, r-1\}$, all being equally likely. Finally, the last step produces a perfectly random permutation of $\{1, \ldots, r\}$.

The expected waiting time to arrive at this point is the number $r + r/2 + \cdots + r/r$ which is $\leq (r+1) \log r$. Hence $\mathbb{T}$, having a finite expectation, must be finite almost surely. That $X_{\mathbb{T}}$ is equidistributed has also been shown, and thus $\mathbb{T}$ is a strong uniform time.

To get bounds for $\rho(k)$ we have two choices. The first one is to use the fact that always $\mathbb{P}(\mathbb{T} > k) \leq \mathbb{E}(\mathbb{T})/(k+1)$ which in our case provides $d(k) \leq (r+1)(1+\log r)/(k+1)$. Also one could try to bound $\{\mathbb{T} > k\}$ directly. Denote the waiting times we have considered in the above analysis by $\mathbb{T}_1, \mathbb{T}_2, \ldots, \mathbb{T}_r$: $\mathbb{T}_1$ (resp. $\mathbb{T}_2$ resp. $\ldots$) is the moment when the bottom card moves for the first (resp. second resp. $\ldots$) time. Then $\mathbb{T} = \mathbb{T}_1 + \cdots + \mathbb{T}_r$, and therefore $\mathbb{T}$ will be greater than k only if at least one of the $\mathbb{T}_\rho$ is greater than k/r. It follows that

$$
\begin{aligned}
\mathbb{P}(\mathbb{T} > k) \ &\leq\ \mathbb{P}(\mathbb{T}_1 > k/r) + \cdots + \mathbb{P}(\mathbb{T}_r > k/r)\\
&=\ (1 - 1/r)^{k/r} + (1 - 2/r)^{k/r} + \cdots + (1 - r/r)^{k/r}\\
&\leq\ r(1 - 1/r)^{k/r}\\
&\leq\ r \exp\left(-k/r^2\right).
\end{aligned}
$$

4. We treat once more the collection S of 0-1-sequences of length r. Transitions are possible if they are of the form $i \to i$ (probability $1/2$) or between states which are different at precisely one component (probability $1/2r$); see page 114. Note that, since the associated transition matrix is doubly stochastic, the equilibrium distribution is the uniform distribution.

A realization of this chain could be defined by the following rule (the chain is assumed to start, e.g., at $(0, \ldots, 0)$):

> Choose a random position ρ in $\{1, \ldots, r\}$ and throw a fair coin. If it shows head, then don't move, otherwise switch the ρ'th component.

This rule obviously produces the desired transition probabilities, a more formal realization can also easily be given.

> One only has to choose a probability space $(\Omega, \mathcal{A}, \mathbb{P})$ together with independent random variables $U_1, U_2, \ldots, V_1, V_2, \ldots$ such that the U_k (resp. the V_k) are identically distributed, and $\mathbb{P}(U_k = \rho) = 1/r$ for $\rho = 1, \ldots, r$ and $\mathbb{P}(V_k = \varepsilon) = 1/2$ for $\varepsilon = 0, 1$. Then the U_k, V_k give rise to an appropriate Markov process $X_0, X_1, \ldots : \Omega \to S$: $X_0 := (0, \ldots, 0)$, and the choice of X_{k+1} given X_k depends on the values of V_k, U_k according to the above rule.
>
> If, e.g., $r = 4$ and the first pairs (U_k, V_k) happen to be $(4, 0)$, $(3, 0)$, $(2, 1)$, $(3, 0)$, $\ldots$, then the $X_0, X_1, \ldots$ are $(0, 0, 0, 0) \to (0, 0, 0, 0) \to (0, 0, 0, 0) \to (0, 1, 0, 0) \to (0, 1, 0, 0)$.

Our candidate for a *strong uniform time* for this process is defined as follows:

Stop as soon as the $U_1, U_2, \ldots$ have covered all of $\{1, \ldots, r\}$, that is

$$\mathbb{T}(\omega) := \min\{k \mid \{U_{k'}(\omega) \mid 1 \le k' \le k\} = \{1, \ldots, r\}\}.$$

It has to be emphasized that this is *not* a stopping time with respect to the natural filtration, and therefore we have to check more carefully whether proposition 14.3 can be applied. It is clear that we will have to deal with the filtration $(\mathcal{F}_k)$ defined by $\mathcal{F}_k :=$ the σ-algebra generated by $U_1, \ldots, U_k, V_1, \ldots, V_k$. Surely this is an appropriate filtration to make $\mathbb{T}$ a stopping time, and the process will also be adapted.

Claim 1: $\mathbb{T}$ is almost surely finite.

This is clear, see the discussion of the examples at the end of chapter 13.

Claim 2: $\mathbb{T}$ is compatible with the Markov property.

Naively this is obvious, the next state only depends on the present position and not on the special way it has been produced by the first U's and V's. For the sake of easy reference we include a general argument in the next lemma which covers the case under consideration.

Claim 3: $\mathbb{T}$ is a strong uniform time. Let ρ be an element of $\{1, \ldots, r\}$. Then the ρ'th component of X_k is 0 or 1 *with equal prability*, regardless of for how many $k' \le k$ we have had $U_{k'} = \rho$ (provided there is at least one such k'); this follows from the independence of the $U_1, \ldots, V_1, \ldots$. Hence, if we stop after all ρ's have occurred – immediately or later – we necessarily are at a state for which at every component both values 0 and 1 are equally likely, and these values are independent for different ρ. To phrase it otherwise: the distribution of $X_{\mathbb{T}}$ is the r-fold product of the uniform distribution on $\{0, 1\}$, i.e., the equidistribution on S.

In the last example we have claimed that the way how we have generated the process and defined $\mathbb{T}$ has given rise to a stopping time which respects the Markov property. This is true in many similar situations, an appropriate generalization reads as follows:

Lemma 14.4 *Let S be a finite set and $(\Omega, \mathcal{A}, \mathbb{P})$ a probability space together with equidistributed independent random variables $W_1, W_2, \ldots$ which are defined on Ω and have values in an arbitrary measurable space $(\Omega', \mathcal{A}')$. Suppose that there is given a measurable function $f : S \times \Omega' \to S$ which we use to define a process $X_0, X_1, \ldots : \Omega \to S$ as follows: X_0 is the constant function i_0 (where i_0 is a fixed state), $X_k(\omega) := f(X_{k-1}(\omega), W_k(\omega))$ for $k \geq 1$. Then:*

    (i) *(X_k) is a Markov process with transition probabilities $p_{ij} = \mathbb{P}(f(i, W_1) = j)$.*

    (ii) *If $\mathcal{F}_0$ denotes the trivial σ-algebra and, for $k \geq 1$, $\mathcal{F}_k$ the σ-algebra generated by $W_1, \ldots, W_k$, then X_0, X_1 is adapted and every stopping time with respect to this filtration respects the Markov property.*

Proof. First one has to check that the X_k are measurable, but this follows easily by induction on k from the formula

$$\{X_k = i\} = \bigcup_j (\{X_{k-1} = j\} \cap \{f(j, W_k) = i\}).$$

That the (X_k) are a Markov process with the p_{ij} as transition properties is left to the reader (see exercise 14.6). Clearly the X_k are adapted to $(\mathcal{F}_k)$, it remains to show that

$$\mathbb{P}(X_{k+r} = j \mid X_k = i) = \mathbb{P}(X_{k+r} = j \mid X_k = i, B)$$

for B in $\mathcal{F}_k$.

We denote by $g_r : S \times (\Omega')^r \to S$ the r-fold "composition" of f with itself: $g_1 := f$, and

$$g_r(i, \omega_1', \ldots, \omega_r') := f(g_{r-1}(\omega_1', \ldots, \omega_{r-1}'), \omega_r').$$

Then, by definition, $X_{k+r} = g_r(X_k, (W_{k+1}, \ldots, W_{k+r}))$. Now consider the set $B' := \{g_r(i, (W_{k+1}, \ldots, W_{k+r})) = j\}$. This set is independent of the $B \in \mathcal{F}_k$, and therefore it follows that

$$\mathbb{P}(B', X_k = i, B) = \mathbb{P}(B')\mathbb{P}(X_k = i, B)$$

as well as

$$\mathbb{P}(B', X_k = i) = \mathbb{P}(B')\mathbb{P}(X_k = i).$$

But $\{X_{k+r} = j, X_k = i\} = B' \cap \{X_k = i\}$, and we conclude that

$$
\begin{aligned}
\mathbb{P}(X_{k+r} = j, X_k = i, B)\mathbb{P}(X_k = i) \ &= \ \mathbb{P}((B', X_k = i, B)\mathbb{P}(X_k = i) \\
&= \ \mathbb{P}(B')\mathbb{P}(X_k = i, B)\mathbb{P}(X_k = i) \\
&= \ \mathbb{P}(X_k = i, B)\mathbb{P}(B', X_k = i) \\
&= \ \mathbb{P}(X_k = i, B)\mathbb{P}(X_{k+r} = j, X_k = i).
\end{aligned}
$$

This is precisely the formula for the conditional probabilities we have to show, and hence the proof of the lemma is complete. $\qquad\square$

Remark: Did you recognize the construction of the previous example? There $\Omega' = \{1, \ldots, r\} \times \{0, 1\}$, W corresponds to (U, V), and f is the function

$$f((\varepsilon_1, \ldots, \varepsilon_r), \rho, \varepsilon) = (\varepsilon_1, \ldots, \varepsilon_{\rho-1}, \varepsilon_\rho + \varepsilon \bmod 2, \varepsilon_{\rho+1}, \ldots, \varepsilon_r).$$

Exercises

14.1: Characterize the stochastic matrices P such that the constant time $\mathbb{T} = k_0$ is a strong uniform time.

14.2: Let $\mathbb{T}$ be a strong uniform time. Is $\mathbb{T} + 1$ also a strong uniform time? Or $2\mathbb{T}$?

14.3: Is it possible to find a strong uniform time for an arbitrary homogeneous Markov chain?

14.4: Construct a filtration together with a stopping time $\mathbb{T}$ such that $\mathbb{T}$ is not compatible with the Markov property.

14.5: Prove that the time $\mathbb{T}$ of example 4 before lemma 14.4 is not a stopping time with respect to the natural filtration.

14.6: In lemma 14.4 we have defined a process (X_k) from random variables (W_k). Prove that (X_k) is a Markov process.

15 Markov chains on finite groups I (commutative groups)

We now turn to state spaces which have the additional structure of a group; implicitly we have already met them, e.g., when prescribing rules like "with equal probability go to $i + 1$ mod N or to $i - 1$ mod N" (on $\{0, \ldots, N-1\}$). In a group it is possible to move from a state i to a new position by composing i with the elements j of the group, where j is chosen in accordance with a certain probability law which is *independent of i* (in the preceding example $+1$ and -1 have been chosen each with probability $1/2$).

Let $(G, \circ)$ be any finite group, we will denote the elements by g, h, ... Every probability measure $\mathbb{P}_0$ on G gives rise to a Markov chain with state space G if we define the transition probabilities by

$$p_{g, h \circ g} := \mathbb{P}_0(\{h\}) \ (\text{or } p_{g, h} := \mathbb{P}_0(\{h \circ g^{-1}\})) \text{ for } g, h \in G.$$

To state it otherwise: if the chain is now in position g, we choose an h in accordance with the probability law $\mathbb{P}_0$; the next position will then be $h \circ g$.
(Note that we multiply the random element h *from the left*; multiplication from the right leads to similar results.)

Here are **two examples:**

1. Denote by G the group $\{0, 1\}^r$ (addition is component-wise modulo 2). A probability measure $\mathbb{P}_0$ on G is defined by

$$
\begin{aligned}
\mathbb{P}_0(0, \ldots, 0) &:= 1/2 \\
\mathbb{P}_0(1, 0, \ldots, 0) &:= 1/2r \\
\mathbb{P}_0(0, 1, 0, \ldots, 0) &:= 1/2r \\
&\ \ \vdots \quad \vdots \quad \vdots \\
\mathbb{P}_0(0, \ldots, 0, 1) &:= 1/2r.
\end{aligned}
$$

The associated chain is equivalent with the random walk on the hypercube which we have already met several times (cf. example 3 on page 114).

2. Let $G = S_r$ be the group of permutations of r elements. Denote, for $k = 1, \ldots, r$, by Π_k the following permutation:

$$
\Pi_k := \begin{pmatrix} 1 & 2 & \ldots & k-1 & k & k+1 & \ldots & r \\ k & 1 & \ldots & k-2 & k-1 & k+1 & \ldots & r \end{pmatrix}
$$

(Π_1 is the identical permutation). We define $\mathbb{P}_0$ such that all Π_k have the same probability $1/r$. Do you recognize the *top-to-random-shuffle chain* (chapter 14, page 124)?

Lemma 15.1 *With the preceding definition of the transition probabilities the following assertions are true:*

(i) *The $p_{g,h \circ g}$ are the entries of a stochastic matrix; in fact this matrix is doubly stochastic so that the uniform distribution is the equilibrium distribution.*

(ii) *Let H be the subgroup generated by* $\operatorname{supp} \mathbb{P}_0 := \{h \mid \mathbb{P}_0(h) > 0\}$, *the* support *of $\mathbb{P}_0$. The irreducible subsets of the chain are precisely the sets of the form $H \circ g$ with $g \in G$, that is, the left conjugacy classes. In particular the chain is irreducible iff* $\operatorname{supp} \mathbb{P}_0$ *generates G.*

(iii) *The chain is aperiodic and irreducible iff there is a k such that every element of G can be written as the product of k elements, each lying in* $\operatorname{supp} \mathbb{P}_0$.

Proof. (i) It should be obvious that the sum over each row of the transition matrix is one. For the second part fix any g_0, we have to calculate the sum over all probabilities to jump from an arbitrary g to g_0. Since there is precisely one h which gives rise to this transition, namely $g_0 g^{-1}$, we have to evaluate the sum $\sum_g \mathbb{P}_0(g_0 g^{-1})$. But the collection of all $g_0 g^{-1}$ coincides with G, hence this sum is the total mass of $\mathbb{P}_0$.

(ii) Let g_0 be arbitrary. If the chain starts there, one may arrive with a positive probability at all states of the form $g_r \circ \cdots \circ g_1 \circ g_0$ with arbitrary r and $g_1, \ldots, g_r$ in the support of $\mathbb{P}_0$. Therefore the claim is that H coincides with

$$ H' := \{ g_r \circ \cdots \circ g_1 \mid r \in \mathbb{N}, \ g_1, \ldots, g_r \in \operatorname{supp} \mathbb{P}_0 \}. $$

H' is clearly a subset of H, it remains to show that H' is a group. Let g be an arbitrary element of the support. Since the group is finite there is a k with $g^k = e \, (= \text{the neutral element})$, and therefore H' contains the inverse g^{k-1} of g. It follows that inverse elements of *arbitrary* elements of H' are also in this set: $(g_r \circ \cdots \circ g_1)^{-1} = g_1^{-1} \circ \cdots \circ g_r^{-1}$. That H' is closed with respect to multiplication is clear.

(iii) The property described in (iii) is nothing but the statement that the k'th power of the transition matrix is strictly positive. Therefore the assertion follows from part (ii) of lemma 7.3. $\qquad\qquad\square$

Given G and $\mathbb{P}_0$, what are the properties of the associated chain? In particular, what are the relevant objects to investigate rapid mixing? It will turn out that an answer can be given which depends on *harmonic analysis*.

In the present chapter we discuss the case of *commutative finite groups*, in this particularly simple setting it is easier to understand the underlying ideas: *characters, Fourier transform of functions and measures, convolutions and the role of the Plancherel theorem* to bound the variation distance between measures. The investigation of *arbitrary finite groups* is postponed to the next chapter. The strategy is essentially the same, the technicalities, however, are considerably more demanding.

In this chapter $(G, +)$ will be a *finite commutative group*, the "group multiplication" is written "$+$", the neutral element will be denoted by "0". Whereas it is known what such groups look like explicitly we prefer to regard $(G, +)$ as an abstract object. We are mainly interested in the following questions:

- How is it possible to relate the *abstract* group G with more concrete objects like numbers?

- If $\mathbb{P}_0$ is a probability measure on G, how can one describe explicitly the measures which correspond to the k-step transitions of the Markov chain associated with G and $\mathbb{P}_0$?

- What quantities have to be known in order to decide how fast this chains converges to its equilibrium?

> Note that the distance to the equilibrium π^T does not depend on the starting distribution since π^T is the uniform distribution: walks which start at a state g_0 are equivalent with a translation by g_0 of walks starting at 0; and the uniform distribution is invariant with respect to translation.

Characters

The crucial idea to tackle the first problem successfully is to consider appropriate maps from G to the complex numbers:

Definition 15.2 Denote by $(\Gamma, \cdot)$ the multiplicative group of all complex numbers of modulus one. Then a *character* on G is a group homomorphism χ from G to Γ:

$$\chi(g + h) = \chi(g)\chi(h)$$

for all $g, h \in G$.

It is not difficult to check that characters have the following properties:

> With χ, χ_1, χ_2 also $\overline{\chi}$ (= the map which assigns to g the complex conjugate $\overline{\chi(g)}$ of $\chi(g)$) and $\chi_1\chi_2$ are characters on G; $\overline{\chi}$ is the inverse $1/\chi$ of χ; the constant map $g \mapsto 1$ is a character (the *trivial character* χ_{triv} *on* G); the collection $\widehat{G}$ of all characters forms a commutative group with respect to pointwise multiplication ($\widehat{G}$ is called the *character group* of G); if G has N elements, then the range of any character on G is contained in the set of the N'th roots of unity, i.e., in
>
> $$\{\exp(2\pi i j/N) \mid j = 0, \ldots, N - 1\};$$
>
> in this and in the following chapter i will denote the complex number $\sqrt{-1}$.

(The proofs are simple; the last property, for example, is a consequence of the fact that $g + g + \cdots + g$, a sum with N summands, is the neutral element 0 and that 0 is mapped to 1 by every character.)

Here are some *examples of characters.*

1. Consider first $\mathbb{Z}_N = (\mathbb{Z}/N\mathbb{Z}, +)$, the group $\{0, 1, \ldots, N - 1\}$ of residues modulo N with addition modulo N. A moment's reflection shows that $\chi_j : \alpha \mapsto \exp(2\pi i j\alpha/N)$ defines a character for every j. Conversely, if χ is an arbitrary character of this group, then – since G has N elements – $\chi(1)$ must be of the form $\exp(2\pi i j/N)$ for a suitable j. This means that χ and χ_j coincide at 1, and since this element generates G and since both χ and χ_j are characters it follows that $\chi = \chi_j$.

Therefore $j \mapsto \chi_j$ is an onto mapping from G to $\widehat{G}$. Obviously it is also one-to-one and a group homomorphism, and thus $(G, +)$ and $(\widehat{G}, \cdot)$ are isomorphic.

2. Let $G = \{0,1\}^r$, with component-wise addition modulo 2. How is it possible to assign complex numbers to the $g \in G$ such that sums are mapped to products?

Since $g + g = 0$ for every g the range of any character will be in $\{-1,+1\}$. Also it is clear that a character is determined by its values on the "unit vectors" $e_1 := (1,0,\ldots,0)$, $e_2 := (0,1,0,\ldots,0)$, $\ldots$, $e_r := (0,\ldots,0,1)$ since these elements generate G. Therefore it is natural to try a definition which assigns an arbitrary number ε_k in $\{-1,+1\}$ to e_k:

$$\chi_{\varepsilon_1\ldots\varepsilon_r} : (i_1,\ldots,i_r) \mapsto \varepsilon_1^{i_1}\varepsilon_2^{i_2}\cdots\varepsilon_r^{i_r}$$

for $(\varepsilon_1,\ldots,\varepsilon_r) \in \{-1,+1\}^r$ and $(i_1,\ldots,i_r) \in G$. It is clear that all these maps are characters, that all of them are different and that every character has this form. Thus the map $(\varepsilon_1,\ldots,\varepsilon_r) \mapsto \chi_{\varepsilon_1\ldots\varepsilon_r}$ is a bijection between $\{-1,+1\}^r$ and $\widehat{G}$, and it is also a group isomorphism if $\{-1,+1\}^r$ is provided with component-wise multiplication. This group is isomorphic with G so that – as in in the preceding example – G and its character group are isomorphic.

This is a general fact, G and $\widehat{G}$ are always the same groups if G is finite and commutative:

Lemma 15.3 *Let $(G,+)$ be a commutative group with N elements. The N-dimensional vector space of all mappings from G to $\mathbb{C}$ will be denoted by X_G, and this space will be provided with the scalar product $\langle f_1, f_2 \rangle_G := \sum_g f_1(g)\overline{f_2(g)}/N$.*

   (i)   *Let χ be a character which is not the trivial character χ_{triv}. Then $\sum_g \chi(g) = 0$.*

  (ii)   *In the Hilbert space $(X_G, \langle \cdot,\cdot \rangle_G)$ the family of characters forms an orthonormal system.*

 (iii)   *Any collection of characters is linearly independent.*

 (iv)   *$\widehat{G}$ has at most N elements.*

  (v)   *In fact there exist N different characters so that $\widehat{G}$ is an orthonormal basis of X_G. Also $(G,+)$ is isomorphic with $(\widehat{G},\cdot)$.*

Proof. (i) Fix an arbitrary g_0 and calculate $\sum_g \chi(g_0 + g)$. On the one hand – since χ is multiplicative – this sum equals $\chi(g_0)\sum_g \chi(g)$. On the other hand the $g_0 + g$ run through every group element precisely once if g runs through G, and consequently we have

$$\chi(g_0)\Big(\sum_g \chi(g)\Big) = \sum_g \chi(g).$$

Thus $\sum_g \chi(g) = 0$ provided that we can find some g_0 with $\chi(g_0) \neq 1$.

(ii) That two different characters χ_1,χ_2 are *orthogonal* follows from (i) since the scalar product $\langle \chi_1, \chi_2 \rangle_G$ is (up to the factor $1/N$) the sum $\sum \chi(g)$ with the character $\chi = \chi_1\chi_2^{-1} = \chi_1\overline{\chi_2}$. They are also *normalized* due to the factor $1/N$ in the scalar product.

(iii), (iv) These assertions follow immediately from (ii) (note that X_G is N-dimensional).

(v) Whereas the preceding proofs have been self-contained this part needs the result that commutative finite groups are products of cyclic groups[1] (see [46], theorem 3.13). With the help of this fact the proof is simple, it even provides another verification of part (iv).

The assertion is true if G is a cyclic group $\mathbb{Z}_N$, this is just what we have shown in example 1 above. Also, the characters of a product group $G_1 \times G_2$ are precisely the maps $(g_1,g_2) \mapsto \chi_1(g_1)\chi_2(g_2)$ with $\chi_1 \in \widehat{G_1}$ and $\chi_2 \in \widehat{G_2}$ (this is an easy exercise), a fact which can be rephrased by saying that

---

[1] We will give a self-contained (and rather lengthy) proof of the corresponding statement for arbitrary finite groups in the next chapter.

$$(\chi_1, \chi_2) \mapsto ((g_1, g_2) \mapsto \chi_1(g_1)\chi_2(g_2))$$

is a bijection from $\widehat{G_1} \times \widehat{G_2}$ to $\widehat{G_1 \times G_2}$. This map is even a group isomorphism, and thus (v) is true for a product if it holds for the factors.

Since G can be built up this way from cyclic groups the result follows. $\square$

By the lemma the set of characters $\widehat{G}$ is an orthonormal basis of the Hilbert space $(X_G, \langle \cdot, \cdot \rangle_G)$. This has some remarkable consequences:

Corollary 15.4

    (i)   *Let f be any element of X_G. Then f can be written as a linear combination of the $\chi \in \widehat{G}$ as follows:*

$$f = \sum_{\chi} \langle f, \chi \rangle_G \chi.$$

    (ii) *For different $g, h \in G$ there is a character χ such that $\chi(g) \neq \chi(h)$.*

Proof. (i) is an explicit restatement of the fact that the χ form an orthonormal basis, and (ii) follows easily from (i): it suffices to consider an f with $f(g) \neq f(h)$. $\square$

Fourier transform of functions

Now we introduce the *Fourier transform of functions on G*; this is a preparation to treat *the Fourier transform of probability measures* which will be of crucial importance in the sequel.

Definition 15.5 Let $f : G \to \mathbb{C}$ be any function. We define the *Fourier transform $\hat{f}$ of f* by

$$\hat{f} : \widehat{G} \to \mathbb{C}, \quad \chi \mapsto \frac{1}{N} \sum_g f(g)\chi(g).$$

The following properties are easy consequences of the definition or of lemma 15.3:

- $f \mapsto \hat{f}$ is a linear map from the complex-valued functions on G to the complex-valued functions on $\widehat{G}$.

- The Fourier transform of a character χ is the indicator function of the set $\{\overline{\chi}\}$: $\hat{\chi}(\chi')$ vanishes for $\chi' \neq \overline{\chi}$ and is one at $\chi' = \overline{\chi}$.

- There are N different $\overline{\chi}$ and therefore the range of the Fourier transform is N-dimensional. Consequently – as an onto linear map between N-dimensional vector spaces – it is also one-to-one.

An explicit description of the inverse map is easy:

Lemma 15.6 (Inverse Fourier transform)
Any $f \in X_G$ can be reconstructed from $\hat{f}$ by the formula

$$f(g) = \sum_{\chi} \hat{f}(\chi)\overline{\chi(g)} \quad \text{for } g \in G.$$

Proof. We already know that $f \mapsto \hat{f}$ is a bijection, and therefore it suffices to check the formula for the elements of $\widehat{G}$, they are a basis of X_G. But for $f = \chi_0 \in \widehat{G}$ the Fourier transform is zero resp. one for $\chi \neq \overline{\chi_0}$ resp. $\chi = \overline{\chi_0}$ so that the sum reduces to $\overline{\chi_0(g)} = \chi_0(g)$. $\qquad\square$

The Fourier transform is not only a bijection between the spaces of complex-valued functions on G and $\widehat{G}$, respectively. It even is an isometry if we measure the size of a function by its (suitably normalized) L^2-norm:

Proposition 15.7 (The Plancherel formula)
Let f_1, f_2 be functions on G. Then

$$\langle f_1, f_2 \rangle_G = \sum_\chi \hat{f}_1(\chi)\overline{\hat{f}_2(\chi)}.$$

In particular,

$$\frac{1}{N}\sum_g |f(g)|^2 = \sum_\chi |\hat{f}(\chi)|^2$$

holds for all f.

Proof. Suppose first that $f_1 = \chi_1$ and $f_2 = \chi_2$ are characters. Then, by lemma 15.3, the left-hand side is zero resp. one depending on whether χ_1, χ_2 are different or equal. The Fourier transforms are the indicator functions of the sets $\{\overline{\chi_1}\}$ and $\{\overline{\chi_2}\}$, and therefore the sum on the right-hand side also is zero for different χ_1, χ_2 and one otherwise. This proves the assertion for this special situation.

The general case follows since arbitrary f_1, f_2 can be written as linear combinations of characters. $\qquad\square$

Fourier transform of measures

As we have already noted we will need in particular the *Fourier transform of measures* $\mathbb{P}_0$, it is defined by

$$\widehat{\mathbb{P}_0} : \widehat{G} \to \mathbb{C}, \quad \chi \mapsto \sum_g \chi(g)\mathbb{P}_0(\{g\});$$

we note in passing that this is just the integral of χ with respect ot the measure space $(G, \mathbb{P}_0)$.

Note that $\widehat{\mathbb{P}_0}$ is *not* the Fourier transform of the function $g \mapsto \mathbb{P}_0(\{g\})$, for functions one has to multiply the sum by $1/N$.

The reason for the different treatment of functions and measures lies in the role of the uniform distribution on G, it is the only probability measure $\mathbb{P}_0$ which respects the group structure in that it is *translation invariant*[2]: $\mathbb{P}_0(A + g) = \mathbb{P}_0(A)$ for every $A \subset G$. And every $\mathbb{P}_0$ has a *density* $f_{\mathbb{P}_0}$ with respect to the uniform distribution, namely $f_{\mathbb{P}_0} : g \mapsto N\mathbb{P}_0(\{g\})$. Therefore the Fourier transform of the measure $\mathbb{P}_0$ is nothing but the (ordinary) Fourier transform of this density.

---

[2] In harmonic analysis it is called the *Haar measure* of G.

By lemma 15.6 all information contained in $\mathbb{P}_0$ is contained in its Fourier transform. The reason why this "translation" is important will become clear immediately, first let's calculate some **examples:**

1. We start with the Bernoulli probability space: here G is $\mathbb{Z}_2 = \{0,1\}$, and the measure $\mathbb{P}_0$ assigns to 1 resp. to 0 the probability p resp. $1-p$. The character group of G consists of the trivial character and of $\chi : g \mapsto (-1)^g$. Then $\widehat{\mathbb{P}_0}$ maps χ_{triv} to $1 \cdot (1-p) + 1 \cdot p = 1$ and χ to $1 \cdot (1-p) - 1 \cdot p = 1 - 2p$.

2. More generally, let $p_0, \ldots, p_{N-1}$ be nonnegative numbers with $\sum p_j = 1$. They give rise to a probability measure $\mathbb{P}_0$ on $\mathbb{Z}_N$, and we will calculate its Fourier transform. The typical character is $\chi_j : \alpha \mapsto \exp\left(2\pi i j \alpha / N\right)$, it is mapped by $\widehat{\mathbb{P}_0}$ to

$$\sum_{\alpha=0}^{N-1} p_\alpha \exp\left(2\pi i j \alpha / N\right),$$

that is to a certain convex combination of N'th roots of unity.

3. Now we investigate the very first example where we have introduced a measure $\mathbb{P}_0$ on the hypercube. Let $\chi_{\varepsilon_1 \ldots \varepsilon_r}$ be any character (see page 131). Its image under $\widehat{\mathbb{P}_0}$ is

$$\frac{1}{2}(\varepsilon_1^0 \cdots \varepsilon_r^0) + \frac{1}{2r} \sum_{\rho=1}^{r} \varepsilon_1^0 \cdots \varepsilon_{\rho-1}^0 \varepsilon_\rho^1 \varepsilon_{\rho+1}^0 \cdots \varepsilon_r^0 = \frac{1}{2} + \frac{1}{2r}(\varepsilon_1 + \cdots + \varepsilon_r).$$

Consider in particular in the preceding example 2 the special case when $\mathbb{P}_0$ is the uniform distribution. Then $\widehat{\mathbb{P}_0}$ at χ_j is $\sum_\alpha \exp\left(2\pi i \alpha j / N\right)/N$, and this number is 1 for $j = 0$ and 0 for $j = 2, \ldots, N-1$.

Proof: With $\xi := \exp\left(2\pi i j / N\right)$ this sum is $(1 + \xi + \cdots + \xi^{N-1})/N$, and for $j = 1, \ldots, N-1$ we have $\xi \neq 1$ so that it can be evaluated as $(1 - \xi^N)/[N(1-\xi)] = 0$.

This is a special case of part (i) of the following

Lemma 15.8 *Let $\mathbb{P}_0$, $\mathbb{P}_1$, $\mathbb{P}_2$ be probability measures on the finite commutative group G; by U we denote the uniform distribution.*

 (i) $\mathbb{P}_0 = U$ *iff* $\widehat{\mathbb{P}_0}(\chi)$ *is one for the trivial character and zero for the other* χ.

 (ii) *The variation distance* $\|\mathbb{P}_1 - \mathbb{P}_2\|$ *can be estimated by* $(\sum_\chi |\widehat{\mathbb{P}_1}(\chi) - \widehat{\mathbb{P}_2}(\chi)|^2)^{1/2}/2$; *in particular* $\|\mathbb{P}_1 - U\|$ *is less than or equal to* $(\sum_{\chi \neq \chi_{\mathrm{triv}}} |\widehat{\mathbb{P}_1}(\chi)|^2)^{1/2}/2$, *where the summation runs over all* nontrivial *characters* χ.

 (iii) *Conversely, the distance of* $\widehat{\mathbb{P}_1}$ *and* $\widehat{\mathbb{P}_2}$ *with respect to the maximum norm is bounded by* $2\|\mathbb{P}_1 - \mathbb{P}_2\|$.

Proof. (i) $\widehat{U}(\chi) = \sum_g U(\{g\})\chi(g) = \sum_g \chi(g)/N$, and this sum is one resp. zero if $\chi = \chi_{\mathrm{triv}}$ resp. $\chi \neq \chi_{\mathrm{triv}}$ by lemma 15.3. Since $\mathbb{P}_0 \mapsto \widehat{\mathbb{P}_0}$ is one-to-one this happens only for the uniform distribution.

(ii) For measures $\mathbb{P}_1$, $\mathbb{P}_2$ the Plancherel formula has the special form

$$\frac{1}{N} \sum_g N^2 |\mu_g - \nu_g|^2 = \sum_\chi |\widehat{\mathbb{P}_1}(\chi) - \widehat{\mathbb{P}_1}(\chi)|^2,$$

where $\mu_g := \mathbb{P}_1(\{g\}), \nu_g := \mathbb{P}_2(\{g\})$. Thus it is only necessary to relate the L^1-norm with the L^2-norm:

$$
\begin{aligned}
4\|\mathbb{P}_1 - \mathbb{P}_2\|^2 &= \left(\sum_g |\mu_g - \nu_g|\right)^2 \\
&\leq N \sum_g |\mu_g - \nu_g|^2 \\
&= \sum_\chi |\widehat{\mathbb{P}_1}(\chi) - \widehat{\mathbb{P}_2}(\chi)|^2;
\end{aligned}
$$

here we have used the inequality $(\sum_{j=1}^N a_j)^2 \leq N \sum a_j^2$ for real a_j, it is just the Cauchy-Schwarz inequality for the Hilbert space $\mathbb{R}^N$ applied to the two vectors $(1,\ldots,1)$ and $(a_1,\ldots,a_N)$.

The second part of (ii) follows from the observation that the Fourier transform of any probability measure at the trivial character is 1 (so that the corresponding term in the sum is zero) and that $\widehat{U}$ vanishes at the nontrivial characters.

(iii) Let χ be arbitrary. Then

$$
\begin{aligned}
|\widehat{\mathbb{P}_1}(\chi) - \widehat{\mathbb{P}_2}(\chi)| &= \left|\sum_g \chi(g)(\mu_g - \nu_g)\right| \\
&\leq \sum_g |\mu_g - \nu_g| \\
&= 2\|\mathbb{P}_1 - \mathbb{P}_2\|.
\end{aligned}
$$

$\square$

This lemma allows one to transform the question "how close is a distribution $\mathbb{P}_0$ to the uniform distribution?" to the investigation of "how small are the $\widehat{\mathbb{P}_0}(\chi)$ for the nontrivial characters χ?". In order to apply it to the present situation it remains to have a closer look at the measures which correspond to the k-step transitions.

Convolutions

Our starting point was a probability $\mathbb{P}_0$ on G which was used to define the one-step transitions: if we start at an arbitrary g_0 we will be next at $g_0 + h$ with probability $\mathbb{P}_0(\{h\})$. From $g_0 + h$ we continue to go to $(g_0 + h) + h'$ where h' again is chosen in accordance with $\mathbb{P}_0$. Also – this was tacitly assumed throughout – the choices of h and h' (and also the choices for the moves to come) are *independent*. Hence the position which is occupied after the second step will be a certain $g_0 + h_0$, where h_0 has the form $h + h'$. Therefore the probability of the transition $g_0 \to g_0 + h_0$ is $\sum_{h+h'=h_0} \mathbb{P}_0(\{h\})\mathbb{P}_0(\{h'\})$.

> To argue a little bit more formally start with a sequence $W_1, \ldots$ of G-valued random variables which are independent and have distribution $\mathbb{P}_0$. Then the positions of the walk are $g_0, g_0 + W_1, g_0 + W_1 + W_2, \ldots$. We are interested in the distribution of $W_1 + W_2$. This can easily be calculated if we condition on W_1 and use the fact that W_1, W_2 are independent:

$$
\begin{aligned}
\mathbb{P}(W_1 + W_2 = h_0) &= \sum_h \mathbb{P}(W_1 + W_2 = h_0,\, W_1 = h) \\
&= \sum_h \mathbb{P}(W_1 = h,\, W_2 = h_0 - h) \\
&= \sum_h \mathbb{P}(W_1 = h)\mathbb{P}(W_2 = h_0 - h) \\
&= \sum_{h + h' = h_0} \mathbb{P}(W_1 = h)\mathbb{P}(W_2 = h') \\
&= \sum_{h + h' = h_0} \mathbb{P}_0(\{h\})\mathbb{P}_0(\{h'\}).
\end{aligned}
$$

Some readers will be reminded of a result from elementary probability where one derives a similar formula for the distribution of the sum of two independent $\mathbb{Z}$- or $\mathbb{R}$-valued random variables. The present investigations are the appropriate version for arbitrary commutative (finite) groups, later we will also discuss the non-commutative case.

We can summarize the preceding discussion by saying that the two-step transitions $g_0 \mapsto g_0 + h_0$ of our chain are governed by a probability measure which assigns to h_0 the number $\sum_h \mathbb{P}_0(\{h\})\mathbb{P}_0(\{h_0 - h\})$. This motivates

Definition 15.9 Let $\mathbb{P}_1$, $\mathbb{P}_2$ be probability measures on G.

(i) We define the *convolution* $\mathbb{P}_2 * \mathbb{P}_1$ of $\mathbb{P}_1$, $\mathbb{P}_2$ by

$$
(\mathbb{P}_2 * \mathbb{P}_1)(\{h_0\}) := \sum_h \mathbb{P}_1(\{h\})\mathbb{P}_2(\{h_0 - h\}).
$$

(ii) In the special case $\mathbb{P}_1 = \mathbb{P}_2 = \mathbb{P}_0$ we put $\mathbb{P}_0^{(2*)} := \mathbb{P}_0 * \mathbb{P}_0$. This is extended to a definition for arbitrary integer exponents by $\mathbb{P}_0^{((k+1)*)} := \mathbb{P}_0^{(k*)} * \mathbb{P}_0$.

(It is left to the reader to show that the convolution is again a probability measure.) With this definition we know:

If the one-step transitions are governed by $\mathbb{P}_0$, then one will observe k-step transitions of the form $g_0 \mapsto g_0 + h_0$ with probability $\mathbb{P}_0^{(k*)}(\{h_0\})$. Consequently the problem of how fast the chain converges to its equilibrium is equivalent with the question of how fast the $\mathbb{P}_0^{(k*)}$ tend to the uniform distribution.

Since we have proved that the variation distance can be calculated with the help of the Fourier transform it will be necessary to relate the Fourier transform of a convolution with the Fourier transforms of the factors. There is a surprisingly simple connection, a fact which makes the Fourier transform an extremely useful tool:

Proposition 15.10 *For probability measures $\mathbb{P}_1, \mathbb{P}_2$ on $(G, +)$ the Fourier transform of $\mathbb{P}_2 * \mathbb{P}_1$ is just the (pointwise) product of the functions $\widehat{\mathbb{P}_1}$ and $\widehat{\mathbb{P}_2}$. In particular it follows that, for any probability $\mathbb{P}_0$, the Fourier transform of $\mathbb{P}_0^{(k*)}$ is the k'th power of the Fourier transform of $\mathbb{P}_0$.*

Proof. Let χ be arbitrary. Then

$$
\begin{aligned}
\widehat{\mathbb{P}_2 * \mathbb{P}_1}(\chi) &= \sum_{g_0} \chi(g_0)(\mathbb{P}_2 * \mathbb{P}_1)(\{g_0\}) \\
&= \sum_{g_0} \chi(g_0) \sum_{g} \mathbb{P}_1(\{g\})\mathbb{P}_2(\{g_0 - g\}) \\
&= \sum_{g_0,g} \chi(g + g_0 - g)\mathbb{P}_1(\{g\})\mathbb{P}_2(\{g_0 - g\}) \\
&= \sum_{g_0,g} \chi(g)\chi(g_0 - g)\mathbb{P}_1(\{g\})\mathbb{P}_2(\{g_0 - g\}) \\
&= \sum_{g',g} \chi(g)\chi(g')\mathbb{P}_1(\{g\})\mathbb{P}_2(\{g'\}) \\
&= \left(\sum_{g} \chi(g)\mathbb{P}_1(\{g\})\right)\left(\sum_{g'} \chi(g')\mathbb{P}_2(\{g'\})\right) \\
&= \widehat{\mathbb{P}_2}(\chi)\widehat{\mathbb{P}_1}(\chi).
\end{aligned}
$$

$\square$

Therefore we arrive at the remarkable result that for a chain on G which is defined by a probability $\mathbb{P}_0$ the rate of convergence to the equilibrium solely depends on the size of the numbers $\widehat{\mathbb{P}_0}(\chi)$ for the nontrivial characters χ. They are always convex combinations of the $\chi(g)$, that is of complex numbers of modulus one. Therefore they lie in the unit disk, but only if they are "not too close" to the boundary their powers will converge sufficiently fast to zero; this convergence will be studied in more detail later.

Before we turn to applications to the mixing rate we prove an interesting consequence of the preceding proposition, it is an extension of lemma 15.1:

Proposition 15.11 *Let $\mathbb{P}_0$ be a probability measure on a finite commutative group $(G, +)$, its support will be denoted by Δ. Then the following assertions are equivalent:*

  (i)   *The associated chain is irreducible and aperiodic.*
  (ii) $\Delta - \Delta \ (:= \{g - h \mid g, h \in \Delta\})$ *generates G.*
  (iii)   *There are no proper subgroup H of G and $g_0 \in G$ such that Δ lies in $g_0 + H$.*
  (iv)   *There is a k such that every $g \in G$ can be written as a sum $g = g_1 + \cdots + g_k$ with $g_1, \ldots, g_k \in \Delta$.*
  (v) $|\widehat{\mathbb{P}_0}(\chi)| < 1$ *for every nontrivial character χ.*
  (vi)   *The measures $\mathbb{P}_0^{(k*)}$ converge to the uniform distribution on G with respect to the total variation norm.*

Proof. By lemma 15.1, (i) and (iv) are equivalent. Hence, under the assumption of (iv), the k-step transitions converge in variation norm to the equilibrium (lemma 7.4) which is the uniform distribution (lemma 15.1(i)), and this establishes "(iv)$\Rightarrow$(vi)". From (vi) the assertion (v) follows easily with the help of lemma 15.8 and proposition 15.10: the $\mathbb{P}_0^{(k*)}$ tend to U iff the numbers $(\widehat{\mathbb{P}_0}(\chi))^k$ tend to zero for all nontrivial χ.

(v)$\Rightarrow$(ii): Let $H_{\mathbb{P}_0}$ be the subgroup generated by $\Delta - \Delta$, suppose that $H_{\mathbb{P}_0}$ is a proper subgroup of G. Then the quotient $G/H_{\mathbb{P}_0}$ contains a nontrivial element so that there is a nontrivial character ψ on $G/H_{\mathbb{P}_0}$ (recall that commutative groups always have "many" characters: for any $g \neq 0$ there is – by corollary 15.4 – a character which maps g *not* to 1). Then $\chi :=$ "the natural quotient map composed with ψ" is a nontrivial character which is identically 1 on $H_{\mathbb{P}_0}$. In particular, $\chi(g_0 - g) = 1$ holds for all $g_0, g \in \Delta$.

Therefore, if we fix g_0, it follows that $\widehat{\mathbb{P}_0}(\chi)$ is the convex combination of certain $\chi(g)$ all of which are identical with $\chi(g_0)$. Thus $\widehat{\mathbb{P}_0}(\chi) = \chi(g_0)$ which is of modulus one. This proves that (v) implies (ii).

(ii)$\Rightarrow$(iii): Suppose that (iii) does not hold. Then, with a proper subgoup H and a suitable g_0, we have $\Delta \subset g_0 + H$, i.e., $\Delta - \Delta \subset g_0 - g_0 + H - H = H$. Consequently $H_{\mathbb{P}_0}$, being a subset of H, would be properly contained in G.

(iii)$\Rightarrow$(ii): Fix any $g_0 \in \Delta$. Then $\Delta - g_0 \subset H_{\mathbb{P}_0}$ holds so that $\Delta \subset g_0 + H_{\mathbb{P}_0}$. By assumption this implies $H_{\mathbb{P}_0} = G$.

(ii)$\Rightarrow$(iv): Let $g \in G$ be arbitrary. We already noted in the proof of lemma 15.1 that the subgroup which is generated by a certain subset is just the collection of finite sums from elements of this subset. Thus, in our case, there are $g_1, \ldots, g_r, h_1, \ldots, h_r \in \Delta$ such that $g = g_1 - h_1 + \cdots + g_r - h_r$. Now we observe that $(N-1)h(= h + \cdots + h$ with $N-1$ summands) equals $-h$ if N denotes the cardinality of G; this follows from $Nh = 0$, the order of an element divides the order of the group. Hence g can be written as a sum of $k_g := r + (N-1)r$ elements of Δ. This is also true for 0:

$$0 = h_1 + (N-1)h_1 + \cdots + h_r + (N-1)h_r.$$

And therefore (iv) holds with $k := \sum_g k_g$. $\square$

In order to check whether the conditions of the preceding proposition are satisfied we have to calculate the $|\widehat{\mathbb{P}_0}(\chi)|$. Let's review the **examples** from page 134:

1. In the case of the Bernoulli probability space there is only one nontrivial character. The associated $\widehat{\mathbb{P}_0}$-value is $1 - 2p$, and this number has modulus less than one precisely if p lies properly between 0 and 1.

2. For an arbitrary probability on $\mathbb{Z}_N$ we have to consider the characters with labels $j = 1, \ldots, N-1$. A convex combination of the numbers $\exp(2\pi i j\alpha/N)$ can lie on the boundary of the unit circle only if all weights are concentrated on the same number. And this happens only if the support of the measure is contained in a set of the form $\{a + kb \mid k = 0, \ldots, N-1\}$ with $\gcd\{b, N\} > 1$.

3. In the hypercube example the nontrivial characters correspond to the $\chi_{\varepsilon_1 \ldots \varepsilon_r}$ with $(\varepsilon_1 \ldots \varepsilon_r) \neq (1, \ldots, 1)$. The associated value of $\widehat{\mathbb{P}_0}$ is $1/2 + (\varepsilon_1 + \cdots + \varepsilon_r)/2r$, and this number is smaller than one since at least one of the ε_ρ is -1.

Rapid mixing

To derive results on rapid mixing we only have to combine the preceding results. If an arbitrary $\mathbb{P}_0$ on the commutative group G is given, then the distribution after k steps of a walk which starts at 0 corresponds to $\mathbb{P}_0^{(k*)}$. (A starting position at any other g_0 only means a translation of the walks; this is unimportant if we are interested in the distance to the uniform distribution, see the note on page 130.) A combination of lemma 15.8 with proposition 15.10 leads to

Proposition 15.12

$$\|\mathbb{P}_0^{(k*)} - U\|^2 \leq \frac{1}{4} \sum_{\chi \neq \chi_{\mathrm{triv}}} |\widehat{\mathbb{P}_0}(\chi)|^{2k}.$$

Here are some **examples** to illustrate the result:

1. Consider on $(\mathbb{Z}_N, +)$ a $\mathbb{P}_0$ which is supported by $\{0, 1\}$. What is the optimal choice of $p := \mathbb{P}_0(\{1\})$ to have a mixing rate as fast as possible? It is to be expected that very small or large p are not favourable: in the former case the chain stays too long close to its starting position, and in the latter it behaves nearly like a deterministic chain.

With the preceding theorem one can analyse the problem as follows. We have to discuss the numbers $\widehat{\mathbb{P}_0}(\chi_j) = p \exp\left(2\pi i j / N\right) + (1 - p)$ for $j = 1, \ldots, N - 1$ which are certain points on the line segment between one and $\exp\left(2\pi i j / N\right)$. Their absolute values decrease for p between 0 and $1/2$ and increase for p between $1/2$ and 1. Therefore the minimum value is – regardless of j – attained at $p = 1/2$: this is the optimal choice.

2. This time we consider on $(\mathbb{Z}_N, +)$ the measure $\mathbb{P}_0$ which is defined by $\mathbb{P}_0(\{-1\}) = \mathbb{P}_0(\{+1\}) = 1/2$; note that $\mathbb{P}_0$ gives rise to the cyclic random walk on $\{0, \ldots, N - 1\}$.

To avoid periodicity we assume that N is an odd number. The computation of the $\widehat{\mathbb{P}_0}(\chi_j)$ is easy:

$$\widehat{\mathbb{P}_0}(\chi_j) = \frac{1}{2}\left(\exp\left(2\pi i j / N\right) + \exp\left(-2\pi i j / N\right)\right) = \cos(2\pi j / N).$$

Therefore, by our proposition, it follows that

$$\|\mathbb{P}_0^{(k*)} - U\|^2 \leq \frac{1}{4} \sum_{j=1}^{N-1} \left(\cos(2\pi j / N)\right)^{2k}.$$

In the present form this estimate is of little use, we will try to simplify it. First we recall that $\cos(-x) = \cos(x)$ and $\cos(\pi - x) = -\cos x$ for all x. Thus the numbers

$$\cos(2\pi j / N), \; j = 1, \ldots, N - 1$$

are identical with the two times repeated sequences

$$\cos(\pi j / N), \; j = 2, 4, \ldots, K$$

and

$$-\cos(\pi j / N), \; j = 1, 3, \ldots, K',$$

where K (resp. K') denotes the maximum of the even (resp. odd) numbers which are $\leq (N - 1)/2$. Therefore – since only even powers occur – we may rewrite the above sum as

$$\frac{1}{2} \sum_{j=1}^{(N-1)/2} \left(\cos(\pi j / N)\right)^{2k}.$$

We continue by applying the inequality $\cos x \leq e^{-x^2/2}$ for $0 \leq x \leq \pi/2$.

One observes first that $x \cos x \leq \sin x$ for these x: the inequality holds at 0 and the derivative of the left-hand side is not greater than that of the right-hand side.

With this preparation at hand one considers $h(x) := \log(e^{x^2/2} \cos x)$. h vanishes at zero, and its derivative $(= x - \tan x)$ is not greater than zero by the first step. This proves that $h \leq 0$, hence the result.

It follows that

$$\|\mathbb{P}_0^{(k*)} - U\|^2 \;\leq\; \frac{1}{2} \sum_{j=1}^{(N-1)/2} \exp\left(-\pi^2 j^2 k/N^2\right)$$

$$\leq\; \frac{1}{2} \exp\left(-\pi^2 k/N^2\right) \sum_{j=1}^{\infty} \exp\left(-\pi^2(j^2-1)k/N^2\right)$$

$$\leq\; \frac{1}{2} \exp\left(-\pi^2 k/N^2\right) \sum_{j=0}^{\infty} \exp\left(-3\pi^2 j k/N^2\right)$$

$$=\; \frac{1}{2} \frac{\exp(-\pi^2 k/N^2)}{1 - \exp(-3\pi^2 k/N^2)};$$

in the last inequality we have used the fact that $j^2 - 1 \geq 3(j-1)$, the final expression has resulted from the formula for the geometric series.

The denominator is $\geq 2(1 - \exp\left(-3\pi^2\right)) \geq 1$ for $k \geq N^2$, and thus we finally arrive at the bound

$$\|\mathbb{P}_0^{(k*)} - U\|^2 \leq \exp\left(-\pi^2 k/N^2\right).$$

This means that one has to run the chain $k = O(N^2)$ steps in order to be sure that the distribution is close to the uniform distribution.

3. We modify the hypercube example from the beginning of this chapter: G is $\{0,1\}^r$ with component-wise addition modulo 2, and $\mathbb{P}_0$ assigns equal mass $1/(r+1)$ to the points

$$(0,\ldots,0),\ (1,0,\ldots,0),\ (0,1,0,\ldots,0),\ldots,\ (0,\ldots,0,1).$$

The characters have been identified on page 131, the associated values of $\widehat{\mathbb{P}_0}$ are

$$\widehat{\mathbb{P}_0}\left(\chi_{\varepsilon_1\ldots\varepsilon_r}\right) \;=\; \frac{1}{r+1} \sum_{i_1+\cdots+i_r \leq 1} \varepsilon_1^{i_1}\cdots\varepsilon_r^{i_r}$$

$$=\; \frac{1}{r+1}(1 + \varepsilon_1 + \cdots + \varepsilon_r)$$

$$=\; 1 - 2\omega(\varepsilon_1,\ldots,\varepsilon_r)/(r+1),$$

where $\omega(\varepsilon_1,\ldots,\varepsilon_r)$ denotes the number of -1's in the sequence $\varepsilon_1,\ldots,\varepsilon_r$.

We have to take into account all $(\varepsilon_1,\ldots,\varepsilon_r) \neq (1,\ldots,1)$. For $s = 1,\ldots,r$ there are precisely $\binom{r}{s}$ such vectors with $\omega(\varepsilon_1,\ldots,\varepsilon_r) = s$, and therefore the estimate from proposition 15.12 implies that

$$\|\mathbb{P}_0^{(k*)} - U\|^2 \leq \frac{1}{4} \sum_{s=1}^{r} \binom{r}{s}\left(1 - \frac{2s}{r+1}\right)^{2k}.$$

The summand with $s = 1$ is the relevant one if convergence for $k \to \infty$ is studied. It follows that

$$\|\mathbb{P}_0^{(k*)} - U\| \leq C\left(1 - \frac{2}{r+1}\right)^k,$$

with C depending on r but not on k.

Exercises

15.1: Which measure on the symmetric group S_r corresponds to the random-to-random shuffle, which one to the random-to-bottom shuffle?

15.2: In example 2 of chapter 2 we have introduced certain random walks. Which of these can be thought of as a random walk induced by a measure on the cyclic group $\mathbb{Z}_N$? (Your answer might depend on certain properties of the numbers a_i, b_i, c_i by which the walks are defined.)

15.3: Consider the collection of all probability measures $\mathbb{P}_0$ on a (not necessarily commutative) finite group $(G, \circ)$ which give rise to an irreducible chain. Is this set convex, is it open as a subset of all probability measures on G?

15.4: A chain is called *deterministic* if for every i there is a j with $p_{ij} = 1$. What precisely are the measures $\mathbb{P}_0$ on a finite group which give rise to deterministic chains?

(If not stated otherwise, $(G, +)$ denotes a commutative finite group in the sequel.)

15.5: Let $\mathbb{P}_0$ be a probability measure on G. If $\mathbb{P}_0(\{0\}) > 0$, then the associated chain is aperiodic. Is the reverse implication also true?

15.6: Let U be a subgroup of G. Prove that there is a probability measure $\mathbb{P}_0$ on G such that the closed subsets of the associated Markov chain are precisely the residue classes $\{u + g \mid u \in U\}$, where g runs through G.

15.7: Let $0 \le a < b < N$, we consider the measure $\mathbb{P}_0$ on the cyclic group $\mathbb{Z}_N$ which has mass $1/2$ on a and b. Characterize the numbers a, b, N such that the associated chain is irreducible (resp. irreducible and aperiodic).

15.8: Prove that, for commutative groups G_1 and G_2, the map

$$(\chi_1, \chi_2) \mapsto ((g, h) \mapsto \chi_1(g)\chi_2(h))$$

is a bijection between $\widehat{G_1} \times \widehat{G_2}$ and $\widehat{G_1 \times G_2}$.

15.9: Let $\mathbb{P}_0$ be a probability measure on G such that, for some k, the convolution product $\mathbb{P}_0^{(k*)}$ is the uniform distribution U. Then also $\mathbb{P}_0 = U$ holds.

15.10: For a probability measure $\mathbb{P}_0$ on G, prove that $\mathbb{P}_0(\{g\}) = \mathbb{P}_0(\{-g\})$ implies that all $\widehat{\mathbb{P}_0}(\chi)$ are real. Is the converse also true?

15.11: All characters on G are real valued iff $g + g = 0$ for all g.

15.12: $f_1 * f_2 = f_2 * f_1$ holds for arbitrary complex-valued functions on G.

15.13: More generally than in the present context one can define characters on arbitrary commutative groups $(G, +)$ with a topology: a *character* χ is a continuous group homomorphism from G to Γ. (Clearly the collection $\widehat{G}$ of all characters is a group with respect to pointwise multiplication also in this more general setting.)
Identify $\widehat{G}$ for the groups $(\mathbb{Z}, +)$, $(\mathbb{R}, +)$, and $(\Gamma, \cdot)$, where each of these spaces is provided with its natural topology.

16   Markov chains on finite groups II (arbitrary groups)

In this chapter we are going to generalize the preceding considerations to the case of an *arbitrary finite group* $(G, \circ)$, the approach will be similar: relate the abstract group with something more concrete and solve problems for G by transforming them into problems concerning numbers. However, it is not to be expected that the notions introduced up to now will suffice, the reason is simple:

> Whenever a group homomorphism, say ϕ, on G has a commutative range, then
>
> $$\phi(g \circ h) = \phi(g)\phi(h) = \phi(h)\phi(g) = \phi(h \circ g)$$
>
> for all g, h. Therefore there is no hope that such ϕ distinguish between different elements of a non-commutative group. In particular, homomorphisms from G to Γ will not suffice.

The *idea* is to pass *from characters to representations, i.e., to certain matrix-valued maps*. They will be introduced and studied in the first section, of particular importance will be the "essential" (the irreducible) representations. The further structure is similar to what has been done in the last chapter: *Fourier transform of functions, Fourier transform of measures, the Plancherel theorem, convolutions, connections with rapid mixing*. At the end of the chapter we will discuss *some examples*.

Representations

Before we introduce the relevant definition we recall that, for a complex $d \times d$-matrix $M = (a_{jk})_{j,k=1,\ldots,d}$, one defines the *adjoint matrix* M^* by $M^* = (\overline{a_{kj}})_{j,k=1,\ldots,d}$ (note that one not only passes to the complex conjugates of the entries, but that also a reflection at the main diagonal is necessary). M is called *unitary* provided that MM^* is the identity matrix Id; then $M^*M = Id$ also holds.

It is easy to see that the collection $\mathcal{U}_d$ of unitary $d \times d$-matrices is a group with respect to matrix multiplication, *this* is our candidate to serve as an appropriate range space.

A *d-dimensional representation*[1] ρ of $(G, \circ)$ is nothing but a group homomorphism from G to $\mathcal{U}_d$.

Examples

1. In the commutative case every character is a (one-dimensional) representation if we identify numbers with 1×1-matrices. Similarly, every map $\chi : G \to \Gamma$ with $\chi(g_1 \circ g_2) = \chi(g_1)\chi(g_2)$ on an *arbitrary* $(G, \circ)$ is a one-dimensional representation. In particular, every group admits the *trivial representation* ρ_{triv} which corresponds to the constant map $g \mapsto 1$.

---

[1] What we introduce are in fact *unitary* representations. Since these are the only representations we will consider in this book no confusion should arise.

But often there exist other one-dimensional ρ. A simple example is the map which assigns the *sign* to a permutation, that is $+1$ (resp. -1) if the permutation can be written as an even (resp. odd) number of transpositions. This induces a one-dimensional representation on every symmetric group $S_n (=$ the permutations of $\{1, \ldots, n\})$.

2. It is easy to get new representations from known ones. A first technique is to start with a d-dimensional ρ and to fix an M in $\mathcal{U}_d$. It is plain that then $\rho_M : g \mapsto M\rho(g)M^{-1}$ is a (d-dimensional) representation as well[2].

Also, if ρ_1, ρ_2 are representations which are d_1- and d_2-dimensional, respectively, then

$$g \mapsto \begin{pmatrix} \rho_1(g) & 0 \\ 0 & \rho_2(g) \end{pmatrix}$$

obviously defines a $(d_1 + d_2)$-dimensional representation. It is called *the product of* ρ_1, ρ_2 and written $\rho_1 \oplus \rho_2$.

3. We define the Hilbert space X_G as in the commutative case as the space of $\mathbb{C}$-valued mappings on G together with the scalar product $\langle f_1, f_2 \rangle_G := \sum_g f_1(g)\overline{f_2(g)}/N$; as before, N denotes the cardinality of G.

Every $g \in G$ induces a map $T_g : X_G \to X_G$ by

$$f \mapsto (g' \mapsto f(g^{-1} \circ g')).$$

The T_g are obviously linear, and they satisfy $T_{g_1 \circ g_2} = T_{g_1} \circ T_{g_2}$ (this is due to the rather artificial definition). They are also isometric:

$$\begin{aligned} \langle T_g(f_1), T_g(f_2) \rangle_G &= \frac{1}{N} \sum_{g'} f_1(g^{-1} \circ g')\overline{f_2(g^{-1} \circ g')} \\ &= \frac{1}{N} \sum_{g''} f_1(g'')\overline{f_2(g'')} \\ &= \langle f_1, f_2 \rangle_G. \end{aligned}$$

Thus they are *unitary linear operators* on the Hilbert space X_G so that, if we fix an orthonormal basis, every T_g correponds to a unitary matrix M_g, and we end up with an N-dimensional representation $g \mapsto M_g$.

> If $T : X_G \to X_G$ is any linear map and $f_1, \ldots, f_N$ any orthonormal basis, then – as is well-known – T is described by the matrix $A = (a_{kj})$, where $a_{kj} = \langle Tf_j, f_k \rangle_G$ (*not* in the reverse order). With this definition T corresponds to the map $(a_1, \ldots, a_N)^{\top} \mapsto A(a_1, \ldots, a_N)^{\top}$ from $\mathbb{C}^N$ to $\mathbb{C}^N$ if this space is identified with X_G via the map $(a_1, \ldots, a_N)^{\top} \mapsto \sum a_j f_j$.

It is called *the left-regular representation of* G, we will denote it by ρ_{regular}.

$\boxed{\textit{Irreducible representations}}$

Since we will use representations to describe G it is surely desirable to try to identify the "essential" ones:

---

[2] If the $\rho(g)$ are regarded as unitary operators on the d-dimensional Hilbert space, then the transition from ρ to ρ_M corresponds to the transition to a new orthonormal base.

Can one find a class $\mathcal{C}$ of representations

- which is sufficiently rich to reconstruct the structure of G
- and which is at the same time as small as possible among classes $\mathcal{C}$ of representations with the first property?

We will see that such a class in fact always does exist. To motivate the approach we first observe that a possible candidate for $\mathcal{C}$ must not contain representations which are of the form $\rho = \rho_1 \oplus \rho_2$ since such ρ can be built up as soon as one has access to the "atoms" ρ_1 and ρ_2.

Also, since equivalent representations contain essentially the same information, $\mathcal{C}$ should contain at most one representative from each equivalence class.

Therefore it is natural to start with

Definition 16.1 Let ρ, ρ_1 and ρ_2 be representations of $(G, \circ)$.
- (i) ρ_1 and ρ_2 are said to be *equivalent* provided that they have the same dimension (say d) and there exists an $M \in \mathcal{U}_d$ such that $\rho_2(g) = M\rho_1(g)M^{-1}$ for all g.
- (ii) ρ is called *irreducible* if it is *not* equivalent with a representation of the form $\rho_1 \oplus \rho_2$.

Examples will be given later, here we only note that one-dimensional representations trivially are irreducible. Also let's remark that possible candidates for the above class $\mathcal{C}$ necessarily are collections of irreducible representations two of which are not equivalent. That every such candidate behaves as desired will be the main content of the Peter-Weyl theorem below. It needs some preparations.

$\boxed{\textit{Schur's lemma and some consequences}}$

Lemma 16.2 (Schur's lemma)
Let ρ be a d-dimensional representation of $(G, \circ)$. Then the following conditions are equivalent:
- (i) *ρ is irreducible.*
- (ii) *If A is any $d \times d$-matrix such that $A\rho(g) = \rho(g)A$ for all g, then A is a multiple of the identity matrix.*
- (iii) *Let V be a subspace of $\mathbb{C}^d$ such that $\rho(g)x$ lies in V whenever $g \in G$ and $x \in V$ (V is said to* reduce ρ *if this is the case[3]). Then $V = \{0\}$ or $V = \mathbb{C}^d$.*

Proof. (i)$\Rightarrow$(ii): For the proof we remind the reader of some facts from linear algebra:

- A matrix A is called *self-adjoint* if $A = A^*$. If this is the case, then there is a unitary matrix M such that MAM^{-1} is a diagonal matrix.

- If $B = (b_{jk})$ is a diagonal $d \times d$-matrix, $Q(x) = a_0 + a_1 x + \cdots + a_r x^r$ is a polynomial and $C = Q(B) := a_0 Id + a_1 B + a_2 B^2 + \cdots + a_r B^r$, then C is also diagonal and the entries on the diagonal of C are the Q-images of the corresponding B-entries. In particular, if $b_{11} = b_{22} = \cdots = b_{d'd'}$ and $b_{11} \neq b_{jj}$ for some $1 \leq d' < d$ and all $j = d' + 1, \ldots, d$, then one may choose Q in such a way that the associated $C = (c_{jk})$ satisfies $c_{11} = \cdots = c_{d'd'} = 1$ and $c_{jk} = 0$ for the other j, k.

---

[3] Sometimes reducing subspaces are also called subspaces which are *invariant* with respect to the $\rho(g)$.

Now let A be a a $d \times d$-matrix which is *not* in $\mathbb{C}\,Id$ and which commutes with all $\rho(g)$. We will show that ρ is reducible.

As a first step we show that A may be chosen such that it is *self-adjoint*. To this end we observe that with A also A^* commutes with the $\rho(g)$. This is a consequence of the fact that $M^* = M^{-1}$ for unitary matrices M so that $(\rho(g))^* = \rho(g^{-1})$:

$$(\rho(g)A^*)^* = A(\rho(g))^* = A\rho(g^{-1}) = \rho(g^{-1})A = (\rho(g))^*A = (A^*\rho(g))^*,$$

hence $\rho(g)A^* = A^*\rho(g)$.

Therefore the $\rho(g)$ commute with the two selfadjoint matrices

$$A_1 = \frac{A + A^*}{2} \text{ and } A_2 = \frac{A - A^*}{2i},$$

and since $A = A_1 + iA_2$, one of these will *not* be in $\mathbb{C}\,Id$.

Thus, let A be self-adjoint. Choose an $M \in \mathcal{U}_d$ such that $B := MAM^{-1}$ is diagonal; we may assume that the entries $\lambda_1, \ldots$ on the diagonal are such that $\lambda_1 = \cdots = \lambda_{d'}$ and $\lambda_j \neq \lambda_1$ for a suitable $d' < d$ and all $j > d'$.

B obviously commutes with all $M\rho(g)M^{-1}$ $(= \rho_M(g))$, and so does every polynomial $Q(B)$ of B. Thus, with a proper choice of Q, we find a matrix $C = Q(B) = (c_{jk})$ with

- $c_{11} = \cdots = c_{d'd'} = 1$ and $c_{jk} = 0$ for the other entries;

- $C\rho_M(g) = \rho_M(g)C$ for all g.

But this means that $\rho_M(g)$ has a nonzero entry at position j, k only if $1 \leq j, k \leq d'$ or if $d' + 1 \leq j, k \leq d$, and thus ρ_M is of the form $\rho_1 \oplus \rho_2$ with two representations ρ_1 and ρ_2 (of dimension d' and $d - d'$, respectively). It follows that $\rho = (\rho_1 \oplus \rho_2)_{M^{-1}}$ so that ρ is reducible.

(ii)$\Rightarrow$(iii): We will prove that the existence of a nontrivial reducing subspace enables us to find a matrix not lying in $\mathbb{C}\,Id$ which commutes with all $\rho(g)$.

Let V be a reducing subspace, we assume for the moment that V is of the form

$$V_{d'} := \{(x_1, \ldots, x_{d'}, 0, \ldots, 0)^\top \mid x_1, \ldots, x_{d'} \in \mathbb{C}\}$$

with $1 \leq d' < d$. Then, since V reduces ρ, the entries of the $\rho(g)$ at the positions j, k with $j = d' + 1, \ldots, d$, $k = 1, \ldots, d'$ must vanish. Since $\rho(g) = (\rho(g^{-1}))^*$, this holds also true with the roles of j and k reversed. This implies that the $\rho(g)$ commute with the matrix C from the preceding part of the proof.

The general case can be reduced to the preceding argument: for any V which is neither $\{0\}$ or $\mathbb{C}^d$ there is a unitary matrix M such that $x \in V$ is equivalent with $Mx \in V_{d'}$ for a suitably chosen d', $1 \leq d' < d$. The result follows if we now argue with the representation ρ_M instead of ρ.

(iii)$\Rightarrow$(i): If $\rho = (\rho_1 \oplus \rho_2)_M$ holds, then $V := \{x \mid Mx \in V_{d'}\}$ defines a nontrivial reducing subspace; here d' stands for the dimension of ρ_1. $\square$

Schur's lemma has a number of interesting consequences:

Corollary 16.3 *If G is a commutative group, then every irreducible representation is one-dimensional. Therefore the collection of irreducible representations can be identified with the character group $\widehat{G}$ in this case.*

Proof. Let ρ be irreducible. Under the assumption of commutativity every fixed $\rho(g_0)$ commutes with all $\rho(g)$ and thus is of the form $a_{g_0} Id$. If the dimension were larger than one there would exist an abundance of matrices A *not* lying in $\mathbb{C}\,Id$ commuting with the $\rho(g)$ (or an abundance of nontrivial reducing subspaces), a contradiction. $\qquad\square$

Corollary 16.4 *Let ρ_1 and ρ_2 be irreducible representations of G with dimension d_1 and d_2, respectively. We will say that a $d_2 \times d_1$-matrix A connects ρ_1 and ρ_2 if*

$$A\rho_1(g) = \rho_2(g)A$$

for every g (note that this definition is not symmetric in ρ_1, ρ_2).
- (i) *If A connects ρ_1 and ρ_2, then either $A = 0$ or $d_1 = d_2$ in which case A is a nonnegative multiple of a unitary matrix.*
- (ii) *ρ_1 is equivalent with ρ_2 iff there is a nonzero A which connects ρ_1 and ρ_2.*

Proof. (i) Let $V \subset \mathbb{C}^{d_1}$ be the collection of the vectors x such that $Ax = 0$. From $A\rho_1(g) = \rho_2(g)A$ it follows that $\rho_1(g)x \in V$ provided that $x \in V$. Consequently V is either the zero space or all of $\mathbb{C}^{d_1}$. If we argue similarly with $W = \{Ax \mid x \in \mathbb{C}^{d_1}\}$ and ρ_2 it follows that W is $\{0\}$ or $\mathbb{C}^{d_2}$. Therefore, if A is not the zero matrix, it has independent rows, and the column rank is d_2. This is possible only if $d_1 = d_2$ and if A is invertible.

The equation $A\rho_1(g^{-1}) = \rho_2(g^{-1})A$ (for all g) implies that $\rho_1(g)A^* = A^*\rho_2(g)$ holds; this follows as usual by taking adjoints. Therefore A^*A commutes with all $\rho_1(g)$, and Schur's lemma provides an $a \in \mathbb{C}$ with $A^*A = aId$. This number a is necessarily real and strictly positive since we have, in the Hilbert space $\mathbb{C}^{d_1}$,

$$0 \leq \langle Ax, Ax \rangle = \langle A^*Ax, x \rangle = a\langle x, x \rangle,$$

and at least for some x strict positivity obtains.
It remains to note that $M := A/\sqrt{a}$ satisfies $M^*M = Id$ so that it is unitary.
(ii) One implication is trivial: in the case of equivalence we have $M\rho_1(g)M^{-1} = \rho_2(g)$ with a suitable unitary M so that M connects ρ_1 with ρ_2. The converse follows from the first part of the proof. (Note that with A also $A/\sqrt{a}$ connects ρ_1 with ρ_2 and that A^*A is nonzero for nonzero A; this follows easily from the equation $\langle A^*Ax, x \rangle = \langle Ax, Ax \rangle = \|Ax\|^2$.) $\qquad\square$

In order to apply the preceding facts it is useful to know how one can find connecting matrices:

Lemma 16.5 *Let ρ_1 and ρ_2 be irreducible representations of G with dimensions d_1 and d_2. Then, for any $d_2 \times d_1$-matrix A, the matrix*

$$\tilde{A} := \frac{1}{N} \sum_g \rho_2(g^{-1})A\rho_1(g)$$

connects ρ_1 and ρ_2.
Consequently $\tilde{A}$ is zero if ρ_1 and ρ_2 are not equivalent, and $\tilde{A}$ is a positive multiple of a unitary operator otherwise. In the case $\rho_1 = \rho_2$ the matrix $\tilde{A}$ is a multiple of the identity matrix. The constant is the trace[4] $\operatorname{tr}(A)$ of A, divided by the dimension of ρ_1.

---

[4] The *trace* of a square matrix is the sum over the diagonal elements.

Proof. Fix any g_0. Then

$$\tilde{A}\rho_1(g_0) = \rho_2(g_0)\tilde{A}$$

is equivalent with

$$\sum_g \rho_2(g^{-1})A\rho_1(g \circ g_0) = \sum_g \rho_2(g_0 \circ g^{-1})A\rho_1(g).$$

That this equation holds can easily be seen by a change of summation: with $g' := g \circ g_0$ the first sum is $\sum_g \rho_2(g_0 \circ (g')^{-1})A\rho_1(g')$.

Now suppose that $\rho := \rho_1 = \rho_2$. Since $\tilde{A}$ commutes with all $\rho(g)$, Schur's lemma provides an a such that $\tilde{A} = a\,Id$. Since the trace of AB is the same as the trace of BA for arbitrary square matrices it follows that the trace of A equals the trace of all $\rho_1(g^{-1})A\rho_1(g)$ and thus the trace of $\tilde{A}$. This completes the proof. $\qquad\square$

$\boxed{\textit{Duals of } G \textit{ and the Peter-Weyl theorem}}$

Now we are going to investigate the properties of the *coordinate functions of representations* as functions on G. More precisely, let ρ be an irreducible representation of dimension d. For $1 \leq j, k \leq d$ we define the functions $f^\rho_{jk} : G \to \mathbb{C}$ by $f^\rho_{jk}(g) :=$ the entry in the j'th row and the k'th column of $\rho(g)$. The preceding preparations enable us to study the f^ρ_{jk} as elements of the Hilbert space X_G.

Lemma 16.6 *The coordinate functions satisfy the following orthogonality properties as elements of X_G:*

(i) *If ρ_1 and ρ_2 are irreducible representations which are not equivalent, then $\langle f^{\rho_1}_{jk}, f^{\rho_2}_{lm} \rangle_G = 0$ for arbitrary indices j, k, l, m.*

(ii) *Let ρ be irreducible. Then each two different functions f^ρ_{jk} are orthogonal, and $\langle f^\rho_{jk}, f^\rho_{jk} \rangle_G = 1/d_\rho$ (with $d_\rho =$ the dimension of ρ).*

Proof. (i) Denote by d_1 and d_2 the dimensions of ρ_1 and ρ_2, respectively. Fix j, k, l, m and define a $d_2 \times d_1$-matrix $A = (a_{st})$ by

$$a_{lj} = 1, \text{ and the other } a_{st} \text{ vanish.}$$

Then, with $x =$ the k'th unit vector of $\mathbb{C}^{d_1}$ resp. $y =$ the m'th unit vector in $\mathbb{C}^{d_2}$, the $\mathbb{C}^{d_2}$-scalar product $\langle A\rho_1(g)x, \rho_2(g)y \rangle$ equals $f^{\rho_1}_{jk}(g)\overline{f^{\rho_2}_{lm}(g)}$. Since $(\rho_2(g))^* = \rho_2(g^{-1})$, the scalar product is just $\langle \rho_2(g^{-1})A\rho_1(g)x, y \rangle$ so that summation over g leads to

$$\langle \tilde{A}x, y \rangle = \langle f^{\rho_1}_{jk}, f^{\rho_2}_{lm} \rangle_G,$$

with $\tilde{A}$ as in lemma 16.5. But $\tilde{A}$ is the zero matrix by this lemma, hence the result.
(ii) We apply once more the preceding argument, now with $\rho = \rho_1 = \rho_2$. If $j \neq l$, then the trace of A is zero and therefore $\tilde{A}$ is the zero matrix in this case. In the case $j = l$ the trace is one, and this time $\tilde{A}$ is $(1/d_\rho)Id$. It follows that – with the above notation –

$$\langle f^\rho_{jk}, f^\rho_{jm} \rangle_G = \frac{1}{d_\rho}\langle x, y \rangle.$$

Since $\langle x, y \rangle = \delta_{km}$ (with Kronecker's delta) the proof is complete. $\qquad\square$

Now we know that there are "not too many" irreducible representations: if we select precisely one representative ρ from every equivalence class, then $\sum d_\rho^2$ is bounded by the dimension of X_G which is just the cardinality of G. We will show that the coordinate functions are in fact an orthogonal basis of X_G so that there are also "sufficiently many" such ρ.

Definition 16.7 Any finite collection of representations which contains precisely one representative from each equivalence class of irreducible representations is called *a dual of G*. In general there are many duals, nevertheless one uses the symbol $\widehat{G}$. (Only in the commutative case there is only one $\widehat{G}$, it can be identified with the character group.)

Any dual $\widehat{G}$ of G suffices to describe the structure of G completely. This is the main content of the *Peter-Weyl theorem*, more precisely it states that the (normalized) coordinate functions which are associated with a dual $\widehat{G}$ are an orthonormal basis of X_G. (Note that the theorem can be thought of as a generalization of lemma 15.3.) Once this result is established everything which is needed to study rapid mixing can easily be obtained.

In the proof of the Peter-Weyl theorem we will use some special properties of representations, in particular some facts concerning the left regular representation. It is convenient to prepare the proof by dealing with these facts separately.

Preparations I: some general properties of representations

For commutative groups we know that the complex conjugate of a character is again a character. There is an analogue for representations of arbitrary groups:

Lemma 16.8 *Let $M = (m_{jk})$ be unitary and ρ a representation of G.*

  (i) *Denote by $\overline{M}$ the matrix $(\overline{m_{jk}})$, that is the component-wise complex conjugate. Then $\overline{M}$ is unitary.*

  (ii) *Let ρ be d-dimensional. If we define $\overline{\rho} : G \to \mathcal{U}_d$ by $g \mapsto \overline{\rho(g)}$, then $\overline{\rho}$ is also a d-dimensional representation.*

  (iii) *With ρ also $\overline{\rho}$ is irreducible.*

Proof. (i) and (ii) are obvious, and (iii) is easy with the help of Schur's lemma: if a matrix A commutes with all $\overline{\rho}(g)$, then – if ρ is irreducible – $\overline{A}$ lies in $\mathbb{C}\,Id$ since it commutes with the $\rho(g)$; therefore A is a constant multiple of the identity as well. $\qquad\square$

Every representation "contains" an irreducible one:

Lemma 16.9 *Any representation is irreducible or is equivalent with a representation $\rho_1 \oplus \rho_2$ with an irreducible ρ_1.*

Proof. We argue by induction on the dimension d_ρ of ρ.

In the case $d_\rho = 1$ nothing has to be shown. Suppose that $d_\rho > 1$, that ρ is reducible and that for representations with smaller dimensions the statement has been verified.

By assumption there are a unitary M and ρ', ρ'' such that $M\rho(g)M^{-1} = \rho'(g) \oplus \rho''(g)$. If ρ' is irreducible, we are done, otherwise we find by our induction hypothesis an irreducible ρ_1 such that $\rho'_{\tilde{M}} = \rho_1 \oplus \rho_2$ with suitable $\tilde{M}$ and ρ_2. It should be clear that

$$\hat{M}\rho_M\hat{M}^{-1} = \rho_1 \oplus (\rho_2 \oplus \rho''),$$

where $\hat{M} = (\tilde{M} \oplus Id)M$. $\qquad\qquad\qquad\qquad\qquad\qquad\qquad\qquad\qquad\square$

Preparations II: some properties of the left-regular representation

Let W be a nonzero subspace of X_G which is invariant with respect to the left-regular representation:

$$f \in W \Rightarrow T_{g_0}f \in W$$

for all g_0; recall that $T_{g_0}f : g \mapsto f(g_0^{-1} \circ g)$.

Since T_{g_0} is unitary not only on X_G but on every reducing subspace such a situtation gives rise to a representation of G: fix any orthonormal basis $f_1,\ldots,f_d$ of W and define $\rho(g_0)$ as the matrix

$$\begin{pmatrix}
\langle T_{g_0}f_1, f_1\rangle_G & \langle T_{g_0}f_2, f_1\rangle_G & \cdots & \langle T_{g_0}f_d, f_1\rangle_G \\
\langle T_{g_0}f_1, f_2\rangle_G & \langle T_{g_0}f_2, f_2\rangle_G & \cdots & \langle T_{g_0}f_d, f_2\rangle_G \\
\vdots & \vdots & & \vdots \\
\langle T_{g_0}f_1, f_d\rangle_G & \langle T_{g_0}f_2, f_d\rangle_G & \cdots & \langle T_{g_0}f_d, f_d\rangle_G
\end{pmatrix}.$$

As in the case of the left-regular representation (which corresponds to the case $W = X_G$) $g \mapsto \rho(g)$ is a representation of G. How are the coordinate functions $g_0 \mapsto \langle T_{g_0}f_k, f_j\rangle_G$ of ρ related with the functions f_j? In the case of irreducibility the answer is at follows, it will be crucial for the proof of the Peter-Weyl theorem:

Lemma 16.10 *Suppose that the preceding ρ is irreducible. Further, let ρ' be an irreducible representation which is equivalent with $\bar{\rho}$. Then it is not true that all coordinate functions of ρ' are orthogonal with all $f_1,\ldots,f_d$ (i.e., not all $f_{jk}^{\rho'}$ lie in $W^\perp$).*

Proof. Suppose that the $f_{jk}^{\rho'}$ lie in $W^\perp$. Then the $f_{jk}^{\bar{\rho}}$ – by assumption they are linear combinations of the $f_{jk}^{\rho'}$ – also lie in this space:

$$\langle f_l, f_{jk}^{\bar{\rho}}\rangle_G = 0 \text{ for } j,\, k,\, l = 1,\ldots,d.$$

By the definition of ρ and the scalar product this means that

$$\begin{aligned}
0 &= \sum_{g_0} f_l(g_0)\overline{f_{jk}^{\bar{\rho}}(g_0)} \\
&= \sum_{g_0} f_l(g_0)f_{jk}^{\rho}(g_0) \\
&= \sum_{g_0} f_l(g_0)\langle T_{g_0}f_k, f_j\rangle_G.
\end{aligned}$$

On the other hand, we know that

$$T_{g_0}f_k = \sum_j \langle T_{g_0}f_k, f_j\rangle_G f_j \qquad\qquad (16.1)$$

since the $f_1,\ldots,f_d$ are an orthonormal basis. We evaluate this equation at $g = g_0$:

$$(T_{g_0}f_k)(g_0) = f_k(g_0^{-1} \circ g_0) = f_k(e) = \sum_j \langle T_{g_0}f_k, f_j\rangle_G f_j(g_0).$$

This holds for all g_0 and thus, if we sum up all these equations with various g_0, we get

$$N f_k(e) = \sum_{g_0,j} f_j(g_0)\langle T_{g_0}f_k, f_j\rangle_G.$$

But we have shown above that this expression is zero, and therefore all f_k vanish at zero. Then – by an evaluation of (16.1) at $g = e$ – it follows that the f_k are zero at g_0^{-1} for arbitrary g_0, that is they all vanish identically. This is impossible for an orthonormal family, and we can conclude our proof with this contradiction. $\qquad\square$

Now we are ready for

Theorem 16.11 (The Peter-Weyl theorem)
Let $\widehat{G}$ be a dual of the finite group G (we continue to denote the dimension of a $\rho \in \widehat{G}$ by d_ρ). Then the family

$$\{\sqrt{d_\rho}\, f_{jk}^\rho \mid \rho \in \widehat{G},\ j,k = 1,\ldots,d_\rho\}$$

is an orthonormal basis of X_G.
It follows that
   (i) $\sum_{\rho\in\widehat{G}} d_\rho^2 = N$ *(= the cardinality of G),*
   (ii) $f = \sum_{\rho,j,k} d_\rho \langle f, f_{jk}^\rho\rangle_G f_{jk}^\rho$ *for every $f \in X_G$.*
   (iii) *For every $g \neq e$ there is an irreducible representation $\rho \in \widehat{G}$ such that $\rho(g)$ is not the identity matrix.*
   (iv) *G is commutative iff all irreducible ρ are one-dimensional.*

Proof. Let V be the linear span of the f_{jk}^ρ. Since the orthogonality properties are already established we only have to show that V is all of X_G or, equivalently, that the orthogonal complement $W := V^\perp$ of V vanishes. We assume the contrary, and we will derive a contradiction.
Claim 1: V is an invariant subspace for ρ_{regular}.
Let $\rho \in \widehat{G}$ be arbitrary. For fixed g_0 we know that $\rho(g_0^{-1} \circ g) = \rho(g_0^{-1})\rho(g)$ holds for arbitrary g. Therefore every translate $T_{g_0}f_{jk}^\rho$ of every coordinate function f_{jk}^ρ is a certain linear combination of the f_{lm}^ρ and thus lies in V. It follows that $T_{g_0}V \subset V$, i.e., V is invariant for ρ_{regular}.
Claim 2: W is invariant as well.
It is a general fact that orthogonal complements of invariant subspaces are also invariant. For a proof let $f_1 \in W$ be given, we have to show that $T_{g_0}f_1 \in W$. To this end, let $f_2 \in V$ be arbitrary, the claim is that $\langle T_{g_0}f_1, f_2\rangle_G = 0$.
Since V is invariant, we know that $f_3 := T_{g_0^{-1}}f_2$ belongs to V so that $\langle f_1, f_3\rangle_G = 0$. Now the unitarity of the T_{g_0} comes into play, it leads to

$$0 = \langle f_1, f_3\rangle_G = \langle T_{g_0}f_1, T_{g_0}f_3\rangle_G = \langle T_{g_0}f_1, f_2\rangle_G.$$

It's time to apply the above preparations. First we consider the representation which is induced by the left-regular representation on W. It will not be irreducible in general, but nevertheless we find orthonormal $f_1,\ldots,f_d$ in W such that the linear span is invariant and the induced representation is irreducible. This is what has been shown in lemma 16.9.

Summing up, we are precisely in the situation of lemma 16.10: we have $f_1, \ldots, f_d$ which give rise to an irreducible representation ρ. On the one hand, $\widehat{G}$ contains a ρ' which is equivalent with $\bar{\rho}$, and all coordinate functions $f_{jk}^{\rho'}$ lie in $V = W^{\perp}$. On the other hand, lemma 16.10 just states that such a situation never occurs.

This contradiction proves that necessarily $W = \{0\}$, or $V = X_G$ as claimed.

The consequences (i), (ii), (iii) are obvious. One implication in (iv) has already been shown in corollary 16.3, the other follows from (iii): if all ρ are one-dimensional, then $\rho(g \circ h) = \rho(h \circ g)$ so that – by (iii) – $g \circ h = h \circ g$ for arbitrary g, h. $\square$

It will be important in the sequel to know – for a given G – a dual $\widehat{G}$. The construction of such a dual can be an extremely demanding problem, it cannot be our aim here to provide a complete list of available techniques. We will confine ourselves to a discussion of some examples. The following observation is very helpful:

Proposition 16.12 *Let C be a finite collection of representations of G such that the family*

$$\{\sqrt{d_\rho}\, f_{jk}^\rho \mid \rho \in C, \; j, k = 1, \ldots, d_\rho\}$$

is an orthonormal basis of X_G; the f_{jk}^ρ are defined as in the Peter-Weyl theorem.
Then C contains precisely one representative from every class of irreducible representations, i.e., C is a dual of G.

Proof. Let ρ be a d-dimensional representation such that the coordinate functions are orthogonal and have norm $1/\sqrt{d}$; also let $M = (m_{jk}) \in \mathcal{U}_d$ be given. How are the coordinate functions of ρ related with those of ρ_M? If we denote the former by f_{jk} and the latter by $\tilde{f}_{jk}$, then

$$\tilde{f}_{jk} = \sum_{l,n} m_{jn} \overline{m_{kl}}\, f_{nl}$$

by the definition of ρ_M. Therefore

$$
\begin{aligned}
\langle \tilde{f}_{jk}, \tilde{f}_{j'k'} \rangle_G &= \sum_{l,n,l',n'} \langle f_{nl}, f_{n'l'} \rangle_G\, m_{jn} \overline{m_{kl}}\, m_{k'l'} \overline{m_{j'n'}} \\
&= \frac{1}{d} \sum_{l,n} \overline{m_{kl}}\, m_{k'l}\, m_{jn} \overline{m_{j'n}} \\
&= \frac{1}{d} \Big(\sum_{l} \overline{m_{kl}}\, m_{k'l}\Big) \Big(\sum_{n} m_{jn} \overline{m_{j'n}}\Big) \\
&= \frac{1}{d} \delta_{kk'}\, \delta_{jj'};
\end{aligned}
$$

here we have used the orthogonality relations for the f_{jk} and the equations $\sum_l m_{jl} \overline{m_{kl}} = \delta_{jk}$ (which are a restatement of $MM^* = Id$).

Now let $\rho \in C$ with dimension d be arbitrary, we claim that ρ is irreducible. If this were not the case there would be ρ_1, ρ_2 (with dimensions $d_1, d_2 > 0, d_1 + d_2 = d$) such that ρ_1 is irreducible, and $\rho_M = \rho_1 \oplus \rho_2$ for a suitable unitary M. Then, by the preceding considerations, the coordinate functions of ρ_1 would have norm $1/\sqrt{d}$. On the other hand, by lemma 16.6, they have norm $1/\sqrt{d_1} \neq 1/\sqrt{d}$.

Thus $\mathcal{C}$ solely *consists of irreducible representations*. We claim that every two ρ_1, ρ_2 *are not equivalent*: by assumption the coordinate functions of ρ_1 are orthogonal to the coordinate functions of ρ_2; in the case $\rho_2 = (\rho_1)_M$, however, the coordinate functions of ρ_2 are linear combinations of the coordinate functions of ρ_1, and thus orthogonality would not be possible.

It remains to show that $\mathcal{C}$ contains a representative for each irreducible ρ'. In fact, if an irreducible ρ' existed which were *not* equivalent with any $\rho \in \mathcal{C}$, then the coordinate functions of ρ' would be orthogonal to the coordinate functions associated with the $\rho \in \mathcal{C}$ (lemma 16.6), which by assumption are an orthogonal basis of X_G. Therefore all $f_{jk}^{\rho'}$ would vanish simultaneously, and this is surely not possible for entries of unitary matrices. $\Box$

$\boxed{\textit{Some examples of duals}}$

By the preceding proposition it suffices to produce irreducible representations until the coordinate functions exhaust X_G: finally a $\hat{G}$ is found.

Example 1: The permutation group S_3
We abbreviate the six elements of S_3 as follows:

$$0 = \begin{pmatrix} 1\,2\,3 \\ 1\,2\,3 \end{pmatrix},\ 1 = \begin{pmatrix} 1\,2\,3 \\ 3\,1\,2 \end{pmatrix},\ 2 = \begin{pmatrix} 1\,2\,3 \\ 2\,3\,1 \end{pmatrix},\ 3 = \begin{pmatrix} 1\,2\,3 \\ 1\,3\,2 \end{pmatrix},\ 4 = \begin{pmatrix} 1\,2\,3 \\ 3\,2\,1 \end{pmatrix},\ 5 = \begin{pmatrix} 1\,2\,3 \\ 2\,1\,3 \end{pmatrix},$$

and multiplication is defined such that the permutation on the left-hand side is applied first (e.g., $1 \cdot 3 = 5$).

Since the group is not commutative, there must be an irreducible representation with a dimension strictly larger than one. By theorem 16.11(i) this is only possible if there are one irreducible representation of dimension two and two one-dimensional ones.

The latter are easily identified as the trivial representation ρ_{triv} and the sign representation ρ_{sign} which we will regard as mappings χ_{triv} and χ_{sign} from S_3 to $\mathbb{C}$ (cf. page 143). To find the remaining two-dimensional candidate it is useful to remember that motions in the plane which fix the origin correspond to unitary matrices. Therefore one could try to model permutations by such motions. This in fact works: if we label the vertices of an equilateral triangle by $1, 2, 3$, then the motions which leave the triangle invariant give rise to permutations of the vertices and thus to certain elements of S_3. This induces a representation ρ.

The unitary matrices which are associated with the group elements $0, 1, 2, 3, 4, 5$ are

$$\begin{pmatrix} 1 & 0 \\ 0 & 1 \end{pmatrix},\ \begin{pmatrix} -1/2 & \sqrt{3/4} \\ -\sqrt{3/4} & -1/2 \end{pmatrix},\ \begin{pmatrix} -1/2 & -\sqrt{3/4} \\ \sqrt{3/4} & -1/2 \end{pmatrix},$$

$$\begin{pmatrix} -1 & 0 \\ 0 & 1 \end{pmatrix},\ \begin{pmatrix} 1/2 & -\sqrt{3/4} \\ -\sqrt{3/4} & -1/2 \end{pmatrix},\ \begin{pmatrix} 1/2 & \sqrt{3/4} \\ \sqrt{3/4} & -1/2 \end{pmatrix}.$$

In the following table we have collected the values of all coordinate functions:

$g:$	0	1	2	3	4	5
χ_{triv}	1	1	1	1	1	1
χ_{sign}	1	1	1	-1	-1	-1
f_{11}^{ρ}	1	$-1/2$	$-1/2$	-1	$1/2$	$1/2$
f_{12}^{ρ}	0	$\sqrt{3/4}$	$-\sqrt{3/4}$	0	$-\sqrt{3/4}$	$\sqrt{3/4}$
f_{21}^{ρ}	0	$-\sqrt{3/4}$	$\sqrt{3/4}$	0	$-\sqrt{3/4}$	$\sqrt{3/4}$
f_{22}^{ρ}	1	$-1/2$	$-1/2$	1	$-1/2$	$-1/2.$

It is routine to check that ρ is a representation, that the six coordinate functions are orthogonal and that the f^ρ_{jk} have norm $1/\sqrt{2}$. Therefore

$$\widehat{\mathcal{S}_3} := \{\chi_{\text{triv}}, \chi_{\text{sign}}, \rho\}$$

is a dual of $\mathcal{S}_3$.

Example 2: The quaternion group Q

This group consists of certain distinguished elements of the skew-field of *quaternions*, namely of the eight elements

$$\pm\underline{1}, \pm\underline{i}, \pm\underline{j}, \pm\underline{k};$$

those who don't know the quaternions might think of an arbitrary set of eight elements which are called $+\underline{1}, -\underline{1}, \ldots$.

The group operation "$\circ$" is written multiplicatively, usually the dot is omitted. It is defined by the following rules:

- $\underline{1}\,g = g\,\underline{1} = g$;

- $\underline{i}^2 = \underline{j}^2 = \underline{k}^2 = -\underline{1}$;

- $\underline{i}\,\underline{j} = \underline{k}, \underline{j}\,\underline{k} = \underline{i}, \underline{k}\,\underline{i} = \underline{j}, \underline{j}\,\underline{i} = -\underline{k}, \underline{k}\,\underline{j} = -\underline{i}, \underline{i}\,\underline{k} = -\underline{j}$;

- the remaining 40 definitions of the group multiplication are evident if one applies the usual rules for calculations with ± 1 and $\pm$ (e.g., $(\pm\underline{1})\,g = g\,(\pm\underline{1}) = \pm g$ for all g, where $+g := g$; $\underline{k}\,(-\underline{i}) = -\underline{k}\,\underline{i} = -\underline{j}, \ldots$)

It is straigthforward to show that this multiplication gives rise to a group, we want to find a dual.

Q is not commutative, at least one irreducible representation of dimension greater than one is to be expected. Since the squares of the dimensions sum up to eight and since there is at least one one-dimensional candidate – the trivial representation – we can conclude:

Q admits one two-dimensional and four one-dimensional irreducible representations.

The one-dimensional representations are easy to be found, for simplicity we write them as mappings $\chi = Q \to \Gamma$ (that is we identify a one-dimensional representation with its coordinate function). Such a χ necessarily maps $\underline{1}$ to 1 and $-\underline{1}$ to -1 or $+1$. The value -1, however, is not possible: in this case $\underline{i}, \underline{j}, \underline{k}$ would be mapped to i (the complex number!) or $-i$, and this is not compatible with the equation $\underline{i}\,\underline{j} = \underline{k}$.

Similarly it turns out that there are not many choices for the χ-values of $\underline{i}, \underline{j}, \underline{k}$, finally one arrives at three possibilities for multiplicative mappings from Q to Γ. They are denoted $\chi_{\underline{i}}, \chi_{\underline{j}}, \chi_{\underline{k}}$, the definitions are as follows:

$$\chi_{\underline{i}} : \pm\underline{1}, \pm\underline{i} \mapsto 1, \ \pm\underline{j}, \pm\underline{k} \mapsto -1,$$

$$\chi_{\underline{j}} : \pm\underline{1}, \pm\underline{j} \mapsto 1, \ \pm\underline{i}, \pm\underline{k} \mapsto -1,$$

$$\chi_{\underline{k}} : \pm\underline{1}, \pm\underline{k} \mapsto 1, \ \pm\underline{i}, \pm\underline{j} \mapsto -1.$$

To find a two-dimensional representation ρ means to model the group Q by unitary 2×2-matrices. This is considerably more difficult than to determine the above one-dimensional representations, we only give the result:
Define ρ by

$$\pm\underline{1} \mapsto \pm E, \ \pm\underline{i} \mapsto \pm I, \ \pm\underline{j} \mapsto \pm J, \ \pm\underline{k} \mapsto \pm K,$$

where E, I, J, K denote the following unitary matrices:

$$E = \begin{pmatrix} 1 & 0 \\ 0 & 1 \end{pmatrix}, \ I = \frac{1}{\sqrt{2}} \begin{pmatrix} i & i \\ i & -i \end{pmatrix}, \ J = \begin{pmatrix} 0 & 1 \\ -1 & 0 \end{pmatrix}, \ K = \frac{1}{\sqrt{2}} \begin{pmatrix} -i & i \\ i & i \end{pmatrix}.$$

This is in fact a unitary representation, the collection

$g:$	$\underline{1}$	$-\underline{1}$	$\underline{i}$	$-\underline{i}$	$\underline{j}$	$-\underline{j}$	$\underline{k}$	$-\underline{k}$
χ_{triv}	1	1	1	1	1	1	1	1
$\chi_{\underline{i}}$	1	1	1	1	-1	-1	-1	-1
$\chi_{\underline{j}}$	1	1	-1	-1	1	1	-1	-1
$\chi_{\underline{k}}$	1	1	-1	-1	-1	-1	1	1
f_{11}^{ρ}	1	-1	$i/\sqrt{2}$	$-i/\sqrt{2}$	0	0	$-i/\sqrt{2}$	$i/\sqrt{2}$
f_{12}^{ρ}	0	0	$i/\sqrt{2}$	$-i/\sqrt{2}$	1	-1	$i/\sqrt{2}$	$-i/\sqrt{2}$
f_{21}^{ρ}	0	0	$i/\sqrt{2}$	$-i/\sqrt{2}$	-1	1	$i/\sqrt{2}$	$-i/\sqrt{2}$
f_{22}^{ρ}	1	-1	$-i/\sqrt{2}$	$i/\sqrt{2}$	0	0	$i/\sqrt{2}$	$-i/\sqrt{2}$

of coordinate functions satisfies the conditions of proposition 16.12, and thus

$$\widehat{Q} := \{\chi_{\mathrm{triv}}, \chi_{\underline{i}}, \chi_{\underline{j}}, \chi_{\underline{k}}, \rho\}$$

is a dual of Q.

The Fourier transform of functions and measures

From now on we fix a group $(G, \circ)$ having N elements together with a dual $\widehat{G}$.
The definitions and results which follow are generalizations of what has been done in the
preceding chapter for the commutative case. Everything – naturally – is technically more
involved. For example, in the commutative case the Fourier transform of a function is
also a function, here the appropriate definition is

Definition 16.13 Let $f : G \to \mathbb{C}$ be any function. The *Fourier transform* $\hat{f}$ of f is a
family of matrices $(\hat{f}(\rho))_{\rho \in \widehat{G}}$, where $\hat{f}(\rho)$ is defined by

$$\hat{f}(\rho) := \frac{1}{N} \sum_{g} f(g)\rho(g) \ = \ \left(\left(\frac{1}{N} \sum_{g} f(g) f_{jk}^{\rho}(g) \right)_{j,k=1,\ldots,d_\rho} \right).$$

Similarly, for any measure $\mathbb{P}_0$ on G, $\widehat{\mathbb{P}_0}$ is the family $(\widehat{\mathbb{P}_0}(\rho))_{\rho \in \widehat{G}}$, with

$$\widehat{\mathbb{P}_0}(\rho) := \sum_{g} \mathbb{P}_0(\{g\})\rho(g).$$

Thus, formally, $\hat{f}$ and $\widehat{\mathbb{P}_0}$ are elements of

$$\prod_{\rho \in \widehat{G}} \mathcal{M}_{d_\rho},$$

where $\mathcal{M}_d$ denotes the space of complex $d \times d$-matrices. Note that this product is an N-dimensional linear space in a natural way, it can be considered as a certain space of "mappings" defined on $\widehat{G}$, where the range possibly varies for different ρ.

As **examples** consider

$$f : \mathcal{S}_3 \to \mathbb{C}, \quad \begin{pmatrix} 1\,2\,3 \\ a\,b\,c \end{pmatrix} \mapsto a,$$

and $\mathbb{P}_0$ defined on $\mathcal{Q}$ by

$$\mathbb{P}_0(\{\underline{1}\}) = \mathbb{P}_0(\{\underline{i}\}) = \mathbb{P}_0(\{\underline{j}\}) = \mathbb{P}_0(\{\underline{k}\}) = 1/4.$$

Then, with respect to the above duals,

$$\hat{f}(\chi_{\text{triv}}) = \frac{1}{6}(1 + 3 + 2 + 1 + 3 + 2) = 2,$$

$$\hat{f}(\chi_{\text{sign}}) = \frac{1}{6}(1 + 3 + 2 - 1 - 3 - 2) = 0,$$

$$\hat{f}(\rho) = \frac{1}{6}\left[1\begin{pmatrix} 1 & 0 \\ 0 & 1 \end{pmatrix} + 3\begin{pmatrix} -1/2 & \sqrt{3/4} \\ -\sqrt{3/4} & -1/2 \end{pmatrix} + \cdots \right] = \begin{pmatrix} 0 & 0 \\ -\sqrt{1/12} & -1/2 \end{pmatrix},$$

and

$$\widehat{\mathbb{P}_0}(\chi_{\text{triv}}) = \frac{1}{4}(1 + 1 + 1 + 1) = 1,$$

$$\widehat{\mathbb{P}_0}(\chi_{\underline{i}}) = \frac{1}{4}(1 + 1 - 1 - 1) = 0,$$

$$\widehat{\mathbb{P}_0}(\chi_{\underline{j}}) = \frac{1}{4}(1 - 1 + 1 - 1) = 0,$$

$$\widehat{\mathbb{P}_0}(\chi_{\underline{k}}) = \frac{1}{4}(1 - 1 - 1 + 1) = 0,$$

$$\widehat{\mathbb{P}_0}(\rho) = \frac{1}{4}\left[\begin{pmatrix} 1 & 0 \\ 0 & 1 \end{pmatrix} + \begin{pmatrix} i/\sqrt{2} & i/\sqrt{2} \\ i/\sqrt{2} & -i/\sqrt{2} \end{pmatrix} + \cdots \right] = \frac{1}{4}\begin{pmatrix} 1 & 1 + i\sqrt{2} \\ -1 + i\sqrt{2} & 1 \end{pmatrix}.$$

It is obvious that $f \mapsto \hat{f}$ is a linear map, and therefore it is natural to try to identify the Fourier transform of a suitable basis. This is surprisingly simple for the coordinate functions:

Lemma 16.14

   (i) *Let $\overline{f_{jk}^{\rho_0}}$ be the complex conjugate of a coordinate function $f_{jk}^{\rho_0}$ associated with any $\rho_0 \in \widehat{G}$. Then $\widehat{\overline{f_{jk}^{\rho_0}}}(\rho)$ is the zero matrix whenever $\rho \neq \rho_0$, and $\widehat{\overline{f_{jk}^{\rho_0}}}(\rho_0)$ is the following matrix $E_{jk}^{\rho_0} := (a_{lm})_{l,m=1,\ldots,d_{\rho_0}} : a_{jk} = 1/d_{\rho_0}$, and the other a_{lm} vanish.*

   (ii) *The Fourier transform of the uniform distribution U on G vanishes at any $\rho \neq \rho_{\mathrm{triv}}$. At ρ_{triv} it has the value 1.*

Proof. This is nothing but a restatement of lemma 16.6. $\qquad\qquad\square$

By the definition of the scalar product we have to work with $\overline{f_{jk}^{\rho_0}}$ instead of $f_{jk}^{\rho_0}$. If it happens that $\bar{\rho}$ is in $\widehat{G}$ for some $\rho \neq \rho_0$, then the Fourier transform of $f_{jk}^{\rho_0}$ will vanish there. However, in general only a representation which is *equivalent* with $\bar{\rho}$ will be in $\widehat{G}$; this is, for example, the case for the above dual of Q.

By this observation it is – with the help of the Peter-Weyl theorem – not too hard to invert the Fourier transform. One only has to find an expression with the following properties:

- it assigns to any family in $\prod_{\rho \in \widehat{G}} \mathcal{M}_{d_\rho}$ an element of X_G in a linear way;

- for a suitably chosen basis $f_1, \ldots$ of X_G it is true that $\widehat{f_j}$ is mapped to f_j.

Since the f_{jk}^{ρ} are a basis of X_G the $\overline{f_{jk}^{\rho}}$ are a basis as well; this is more than obvious. Therefore we only have to try to find a "linear" definition which produces $\overline{f_{jk}^{\rho}}$ when applied to $\widehat{\overline{f_{jk}^{\rho}}}$. After some trial and error this leads to

Proposition 16.15 (*The Fourier inversion formula*)

   (i) *Let $f \in X_G$ be arbitrary. Then, for $g \in G$,*

$$f(g) = \sum_{\rho \in \widehat{G}} d_\rho \mathrm{tr}(\rho(g^{-1})\hat{f}(\rho));$$

     recall that $\mathrm{tr}(M)$ denotes the trace of M for any square matrix.

   (ii) *$f \mapsto \hat{f}$ is a bijection between X_G and $\prod_{\rho \in \widehat{G}} \mathcal{M}_{d_\rho}$.*

   (iii) *If $\mathbb{P}_0$ is any measure on G, then*

$$\mathbb{P}_0(\{g\}) = \frac{1}{N} \sum_{\rho \in \widehat{G}} d_\rho \mathrm{tr}(\rho(g^{-1})\widehat{\mathbb{P}_0}(\rho))$$

     for all g.

Proof. (i) Let $f = \overline{f_{jk}^{\rho_0}}$ be given. Since $\rho(g^{-1}) = (\rho(g))^*$ it follows that $\rho(g^{-1})\hat{f}(\rho)$ is the zero matrix for $\rho \neq \rho_0$ and the matrix

$$\frac{1}{d_{\rho_0}}\begin{pmatrix} 0 & \cdots & 0 & \overline{f_{j1}^{\rho_0}}(g) & 0 & \cdots & 0 \\ 0 & \cdots & 0 & \overline{f_{j2}^{\rho_0}}(g) & 0 & \cdots & 0 \\ \vdots & & \vdots & \vdots & \vdots & & \vdots \\ 0 & \cdots & 0 & \overline{f_{jd_\rho}^{\rho_0}}(g) & 0 & \cdots & 0 \end{pmatrix}$$

for $\rho = \rho_0$ (the f's are in the k'th column). The trace of this matrix is $\overline{f_{jk}^{\rho_0}}(g)/d_{\rho_0}$, and this proves that the assertion holds for $\overline{f_{jk}^{\rho_0}}$. By the linearity of $f \mapsto \hat{f}$ and $(M_\rho)_\rho \mapsto \sum_\rho d_\rho \mathrm{tr}(\rho(g^{-1})M_\rho)$ it holds for *all* $f \in X_G$.

(ii) By the first part $f \mapsto \hat{f}$ is one-to-one. That this mapping is onto follows from a dimension argument: both X_G and $\prod_{\rho \in \widehat{G}} \mathcal{M}_{d_\rho}$ have dimension $N = \sum_\rho d_\rho^2$.

(iii) With $f_{\mathbb{P}_0}(g) := N\mathbb{P}_0(\{g\})$ we have $\widehat{\mathbb{P}_0} = \hat{f}$, and the claim follows from (i) with $f = f_{\mathbb{P}_0}$. $\qquad\square$

Similarly one can derive a general version of the Plancherel formula (proposition 15.7):

Proposition 16.16 (The "non-commutative" Plancherel formula)
Let $f_1, f_2 \in X_G$ be arbitrary. Then

$$\langle f_1, f_2 \rangle_G = \sum_{\rho \in \widehat{G}} d_\rho \mathrm{tr}[\hat{f}_1(\rho)(\hat{f}_2(\rho))^*].$$

In particular we have

$$\frac{1}{N} \sum_g |f(g)|^2 = \sum_\rho d_\rho \mathrm{tr}[\hat{f}(\rho)(\hat{f}(\rho))^*]$$

for every function f, and

$$\sum_g (\mathbb{P}_0(\{g\}))^2 = \frac{1}{N} \sum_\rho d_\rho \mathrm{tr}[\widehat{\mathbb{P}_0}(\rho)(\widehat{\mathbb{P}_0}(\rho))^*]$$

for every measure $\mathbb{P}_0$.

Proof. $f \mapsto \hat{f}$ is a linear map, and both $(f_1, f_2) \mapsto \langle f_1, f_2 \rangle_G$ and

$$((M_\rho)_\rho, (N_\rho)_\rho) \mapsto \sum_\rho d_\rho \mathrm{tr}(M_\rho N_\rho^*)$$

are linear in the first and conjugate linear in the second argument. Therefore it suffices to prove the claim for f_1, f_2 running through a basis of X_G.

As in the proof of the inversion formula we work with the complex conjugates of the coordinate functions, that is we start with

$$f_1 = \overline{f_{jk}^{\rho'}}, \quad f_2 = \overline{f_{lm}^{\rho''}}$$

with $\rho', \rho'' \in \widehat{G}$. Then the left-hand side of the Plancherel formula is zero unless $\rho' = \rho''$, $j = l$, $k = m$ in which case it is $1/d_{\rho'}$. For the evaluation of the right-hand side we recall that $\hat{f}_1(\rho)$ is zero for the $\rho \neq \rho'$ and equals $E_{jk}^{\rho'}$ at $\rho = \rho'$ (see lemma 16.14). Therefore the resulting sum is different from zero precisely when $\rho' = \rho''$, $j = l$, $k = m$, and then it equals $1/d_{\rho'}$. As we have already noted this proves the result for *all* f_1, f_2.

The Plancherel formula for measures follows as in the preceding proof by considering the special case $f = f_{\mathbb{P}_0}$. $\qquad\square$

As an illustration consider the measure $\mathbb{P}_0$ on the quaternion group on page 155. For this example we have

$$\sum_g (\mathbb{P}_0(\{g\}))^2 = 4\left(\frac{1}{4}\right)^2 = \frac{1}{4}.$$

The right-hand side coincides with this number:

$$\frac{1}{8}\left[1^2 + 0^2 + 0^2 + 0^2 + 2\operatorname{tr}\left(\frac{1}{4}\begin{pmatrix} 1 & 1+i\sqrt{2} \\ -1+i\sqrt{2} & 1 \end{pmatrix}\frac{1}{4}\begin{pmatrix} 1 & -1-i\sqrt{2} \\ 1-i\sqrt{2} & 1 \end{pmatrix}\right)\right]$$

$$= \frac{1}{8}\left(1 + 2\cdot\frac{1}{16}\cdot 8\right) = \frac{1}{4}.$$

With the results proved so far we can describe in terms of Fourier transforms how close measures are to the uniform distribution. The following lemma should be compared with lemma 15.8:

Lemma 16.17 *For probability measures* $\mathbb{P}_0$, $\mathbb{P}_1$, $\mathbb{P}_2,\ldots$ *on* G *the following assertions hold:*

(i) $\mathbb{P}_0$ *coincides with the uniform distribution iff* $\widehat{\mathbb{P}_0}$ *is one at the trivial representation and vanishes at the other* $\rho \in \widehat{G}$.

(ii)

$$\sum_g \left(\mathbb{P}_0(\{g\}) - \frac{1}{N}\right)^2 = \frac{1}{N}\sum_{\rho \neq \rho_{\mathrm{triv}}} d_\rho \operatorname{tr}[\widehat{\mathbb{P}_0}(\rho)(\widehat{\mathbb{P}_0}(\rho))^*].$$

(iii) *The variation distance* $\|\mathbb{P}_0 - U\|$ *can be estimated by*

$$\frac{1}{2}\left(\sum_{\rho \neq \rho_{\mathrm{triv}}} d_\rho \operatorname{tr}[\widehat{\mathbb{P}_0}(\rho)(\widehat{\mathbb{P}_0}(\rho))^*]\right)^{1/2}.$$

(iv) *The* $\mathbb{P}_k$ *tend to* U *in variation norm iff the* $\widehat{\mathbb{P}_k}(\rho)(\widehat{\mathbb{P}_k}(\rho))^*$ *tend to zero for all* $\rho \in \widehat{G}$ *with* $\rho \neq \rho_{\mathrm{triv}}$.

Proof. (i) That $\widehat{U}(\rho_{\mathrm{triv}}) = 1$ and that $\widehat{U}$ vanishes at the other ρ has already been noted in lemma 16.14 (ii). The other implication is then clear: $\mathbb{P}_0 \mapsto \widehat{\mathbb{P}_0}$ is one-to-one.
(ii) This follows from (i) and the Plancherel formula[5].
(iii) One only has to relate the L^2-distance (part (ii)) with the variation distance as in the proof of lemma 15.8 (ii).
(iv) This is a consequence of (iii): the $\widehat{\mathbb{P}_k}(\rho)(\widehat{\mathbb{P}_k}(\rho))^*$ are self-adjoint and nonnegative, and therefore, if the traces tend to zero, the matrices will converge to zero as well (cf. exercise 16.22). $\qquad\square$

---

[5] Strictly speaking the statement we need here has *not* been proved in proposition 16.16 since $\mathbb{P}_0 - U$ is not a probability measure. It is only a signed measure, and one has to check our proof to verify that it covers also this slightly more general situation.

Convolutions

Let $\mathbb{P}_1$ and $\mathbb{P}_2$ be probability measures on G. As in the preceding chapter we want to calculate the distribution of the final position of a two-step walk which starts at the neutral element e and for which the two (independent) steps $e \mapsto h$ and $h \mapsto h' \circ h$ are in accordance with the probability laws $\mathbb{P}_1$ and $\mathbb{P}_2$, respectively. The final position will be h_0 with probability

$$\sum_{\{h' \mid h' \circ h = h_0\}} \mathbb{P}_1(\{h\})\mathbb{P}_2(\{h'\}) = \sum_h \mathbb{P}_1(\{h\})\mathbb{P}_2(\{h_0 \circ h^{-1}\}).$$

This leads to

Definition 16.18 The *convolution* $\mathbb{P}_2 * \mathbb{P}_1$ of $\mathbb{P}_1, \mathbb{P}_2$ is defined to be the measure

$$\mathbb{P}_2 * \mathbb{P}_1(\{h_0\}) := \sum_h \mathbb{P}_1(\{h\})\mathbb{P}_2(\{h_0 \circ h^{-1}\}).$$

As in the case of commutative groups we define $\mathbb{P}_0^{(k*)}$ by induction: $\mathbb{P}_0^{(1*)} = \mathbb{P}_0$, and $\mathbb{P}_0^{((k+1)*)} = \mathbb{P}_0^{(k*)} * \mathbb{P}_0$.

> Note that we have to write the measures $\mathbb{P}_1$, $\mathbb{P}_2$ from the right to the left in $\mathbb{P}_2 * \mathbb{P}_1$ if we want to model transitions where $\mathbb{P}_1$ is used first, this is similar to the case of mappings. In contrast to the commutative case the order might be relevant.

We can now repeat what has been said in chapter 15:

> **If the one-step transitions are governed by $\mathbb{P}_0$, then one will observe k-step transitions of the form $g_0 \mapsto h_0 \circ g_0$ with probability $\mathbb{P}_0^{(k*)}(\{h_0\})$. Consequently the problem of how fast the chain converges to its equilibrium is equivalent with the question of how fast the $\mathbb{P}_0^{(k*)}$ tend to the uniform distribution.**

One of the main reasons to study Fourier transforms is the fact that convolution is transformed to multiplication. The following proposition generalizes a similar assertion for commutative groups (see proposition 15.10):

Proposition 16.19 *Let $\mathbb{P}_1$, $\mathbb{P}_2$ be probability measures on $(G, \circ)$. Then the Fourier transform of $\mathbb{P}_2 * \mathbb{P}_1$ is component-wise the (matrix-)product of the Fourier transforms of $\mathbb{P}_2$ and $\mathbb{P}_1$:*

$$\widehat{\mathbb{P}_2 * \mathbb{P}_1}(\rho) = \widehat{\mathbb{P}_2}(\rho)\widehat{\mathbb{P}_1}(\rho)$$

for every $\rho \in \widehat{G}$.

Consequently, for any probability $\mathbb{P}_0$ and any ρ, $\widehat{\mathbb{P}_0^{(k*)}}(\rho)$ is the k'th power of $\widehat{\mathbb{P}_0}(\rho)$.

Proof. The proof is similar to that of proposition 15.10:

$$\begin{aligned}
\widehat{\mathbb{P}_2 * \mathbb{P}_1}(\rho) &= \sum_{g_0} \rho(g_0)\mathbb{P}_2 * \mathbb{P}_1(\{g_0\}) \\
&= \sum_{g_0} \rho(g_0) \sum_g \mathbb{P}_1(\{g\})\mathbb{P}_2(\{g_0 \circ g^{-1}\})
\end{aligned}$$

$$= \sum_{g_0,g} \rho(g_0 \circ g^{-1} \circ g)\mathbb{P}_1(\{g\})\mathbb{P}_2(\{g_0 \circ g^{-1}\})$$

$$= \sum_{g_0,g} \rho(g_0 \circ g^{-1})\rho(g)\mathbb{P}_1(\{g\})\mathbb{P}_2(\{g_0 \circ g^{-1}\})$$

$$= \sum_{g',g} \rho(g')\rho(g)\mathbb{P}_1(\{g\})\mathbb{P}_2(\{g'\})$$

$$= \left(\sum_{g'} \rho(g')\mathbb{P}_2(\{g'\})\right)\left(\sum_{g} \rho(g)\mathbb{P}_1(\{g\})\right)$$

$$= \widehat{\mathbb{P}_2}(\rho)\widehat{\mathbb{P}_1}(\rho)$$

holds for arbitrary ρ. $\square$

So far our approach to generalize the techniques which worked so successfully in the commutative case looks rather promising. There is, however, *a fundamental difference between the commutative and the general situation.*

In both cases we want to study how fast the $\mathbb{P}_0^{(k*)}$ tend to the uniform distribution. For *commutative groups* this has been transformed to the question:

How fast do the $(\widehat{\mathbb{P}_0}(\chi))^k$ tend to zero for the nontrivial characters?

The answer is simple, one only has to check the absolute value of $\widehat{\mathbb{P}_0}(\chi)$.

For *arbitrary groups*, however, we are faced with the following much more difficult problem, *it will be resumed in the next section:*

Consider, for all nontrivial $\rho \in \widehat{G}$, the matrix $A := \widehat{\mathbb{P}_0}(\rho)$. Does $A^k(A^*)^k$ tend to zero, and if so, how fast?

Now we study a general version of proposition 15.11, the characterization of irreducibility by means of the support of the measure under consideration. As a preparation we introduce a *definition:* whenever Δ is a nonvoid subset of G, then G_Δ denotes the set of all $g \in G$ such that there are an even number $g_1, \ldots, g_{2r}$ of elements of G with $g = g_1 \circ \ldots \circ g_{2r}$ such that r of the g_i's lie in Δ and the others lie in $\Delta^{-1}(= \{g^{-1} \mid g \in \Delta\})$; thus, for example, if g_1, g_2, g_3, g_4 are elements of Δ, then $g_1 \circ g_2 \circ g_3^{-1} \circ g_4^{-1}$ lies in G_Δ, but $g_2 \circ g_3^{-1} \circ g_4^{-1}$ is possibly *not* contained in this set. G_Δ has the following properties:

Lemma 16.20

    (i) G_Δ *is a subgroup of G, it lies between the subgroup and the normal subgroup which are generated by*

$$\Delta \circ \Delta^{-1} \ (= \{g \circ h^{-1} \mid g, h \in \Delta\}).$$

        Both inclusions might be proper.

    (ii) *If Δ contains the neutral element e, then G_Δ is the subgroup generated by Δ.*

    (iii) G_Δ *is all of G iff there is a k such that every g can be written as a product of k elements each lying in Δ.*

Proof. (i) It is clear that with g, h also $g \circ h$ lies in G_Δ. This already implies that G_Δ is a subgroup since G is finite. $\Delta \circ \Delta^{-1}$ – and thus the subgroup generated by this set – surely is contained in G_Δ so that it remains to show that $G_\Delta \subset N_\Delta(:=$ the normal subgroup generated by $\Delta \circ \Delta^{-1})$.
We first note that $\Delta^{-1} \circ \Delta \subset N_\Delta$, this follows from the identity

$$h^{-1} \circ g = g^{-1} \circ g \circ h^{-1} \circ g.$$

For arbitrary $g \in G_\Delta$ we proceed by induction on r (with r as in the above definition). The case $r = 1$ has just been settled, now suppose that $r > 1$ and that the claim is proved for all smaller r'. Let a $g = g_1 \circ \cdots \circ g_{2r}$ be given were r of the g_i are in Δ and the others come from Δ^{-1}.

Case 1: g_1, g_{2r} lie in Δ.
Then there is an $l < r$ such that both $g' = g_1 \circ \cdots \circ g_{2l}$ and $g'' = g_{2l+1} \circ \cdots \circ g_{2r}$ contain the same number of factors from Δ and Δ^{-1}.

For a proof of this assertion define $\varepsilon_i := +1$ (resp. -1) if $g_i \in \Delta$ (resp. $\in \Delta^{-1}$) for $i = 1, \ldots, 2r$. Then $\varepsilon_1 = \varepsilon_{2r} = 1$ and $\sum \varepsilon_i = 0$. Thus there must be an $l' < 2r$ with $\sum_{i=1}^{l'} \varepsilon_i = 0$, and l' is necessarily an even number.
By the induction hypothesis g' and g'' lie in N_Δ, and therefore $g \in N_\Delta$ as well.

Case 2: g_1, g_{2r} lie in Δ^{-1}.
This case can be is treated in a similar way.

Case 3: g_1, $g_{2r}^{-1} \in \Delta$ or g_1^{-1}, $g_{2r} \in \Delta$.
Then $g' := g_2 \circ \cdots \circ g_{2r-1}$ lies in G_Δ (and thus in N_Δ), and it suffices to note that

$$g = (g_1 \circ g' \circ g_1^{-1}) \circ (g_1 \circ g_{2r})$$

with $g_1 \circ g' \circ g_1^{-1} \in N_\Delta$ and $g_1 \circ g_{2r} \in (\Delta \circ \Delta^{-1}) \cup (\Delta^{-1} \circ \Delta) \subset N_\Delta$.
It can happen that G_Δ is strictly larger than the group generated by $\Delta \circ \Delta^{-1}$: consider in $\mathcal{S}_3$ the set $\Delta = \{1, 5\}$ (the notation is as on page 152); then $\Delta \circ \Delta^{-1} = \{0, 4\}$, and this is a subgroup which does not contain $3 = 5^{-1} \circ 1 \in G_\Delta$.
In order to prove that the second inclusion also might be strict it suffices to choose Δ as a subgroup which is not normal.

(ii) The group generated by Δ is the collection of the $g_1 \circ \cdots \circ g_l$ with $g_i \in \Delta$, and every such product can be written as $g_1 \circ \cdots \circ g_l \circ e^{-1} \circ \cdots \circ e^{-1}$.

(iii) Suppose that $G_\Delta = G$. We fix any g_0, and we write g_0 as $g_1 \circ \cdots \circ g_{2r}$, where r of the g_i are in Δ and r are elements of Δ^{-1}; note that r will depend on g_0. Replace those $g_i = h_i^{-1}$ which lie in Δ^{-1} by h_i^{N-1}, where – as usual – N stands for the order of G. The h_i lie in Δ, and this shows that g_0 can be written as a product of $k_{g_0} := r + (N-1)r = rN$ elements of Δ. The element e can also be written in this way, e.g., as g^{rN} with an arbitrary $g \in \Delta$. It follows that $G = \Delta \circ \cdots \circ \Delta$, where Δ occurs $k = \sum_{g_0} k_{g_0}$ times.

If, conversely, $G = \Delta \circ \cdots \circ \Delta$ with k factors, then we can write any given g as $g_1 \circ \cdots \circ g_k$ and also e as $e = h_1 \circ \cdots \circ h_k$. Thus

$$g = g_1 \circ \cdots \circ g_k \circ h_k^{-1} \circ \cdots \circ h_1^{-1}$$

lies in G_Δ. $\qquad\square$

Proposition 16.21 *Let $\mathbb{P}_0$ be a probability measure on the finite group $(G, \circ)$ with support $\Delta = \{g \mid \mathbb{P}_0(\{g\}) > 0\}$. The following conditions are equivalent:*

(i) $G_\Delta = G$.

(ii) *There is a k such that every $g \in G$ can be written as a product $g = g_1 \circ \cdots \circ g_k$ with $g_1, \ldots, g_k \in \Delta$.*

(iii) *For every irreducible representation $\rho \neq \rho_{\mathrm{triv}}$ the matrix $A := \widehat{\mathbb{P}_0}(\rho)$ satisfies $\|A^k(A^*)^k\| \to 0$ for any matrix norm $\| \cdot \|$.*

(iv) *The measures $\mathbb{P}_0^{(k*)}$ converge to the uniform distribution on G with respect to the variation norm.*

Proof. (i) is equivalent with (ii) by the preceding lemma, and (ii) is a restatement of the fact that the associated chain is irreducible and aperiodic; this proves the equivalence with (iv). That (iii) and (iv) are equivalent follows immediately from lemma 16.17 and proposition 16.19.   □

Rapid mixing

A combination of lemma 16.17 with proposition 16.19 also gives rise to a quantitative version of the preceding equivalence (iii)⇔(iv):

Lemma 16.22 (The upper-bound lemma)

$$\|\mathbb{P}_0^{(k*)} - U\|^2 \leq \frac{1}{4} \sum_{\rho \neq \rho_{\mathrm{triv}}} d_\rho \mathrm{tr}[(\widehat{\mathbb{P}_0}(\rho))^k((\widehat{\mathbb{P}_0}(\rho))^*)^k].$$

In order to apply this lemma one needs to know whether, for a given square matrix A which is a convex combination of unitary matrices, one has $A^k(A^*)^k \to 0$. We provide *three techniques* to deal with this question.

$\boxed{\textit{Matrix norms}}$

Let $\| \cdot \|$ be a matrix norm on the space of $d \times d$-matrices; this means that $\|AB\| \leq \|A\|\,\|B\|$ holds for arbitrary A, B. As an example one can regard A as a linear map on $\mathbb{C}^d$, the *operator norm* $\|A\|_{op}$ of this map has the desired properties.

> $\|A\|_{op}$ is defined as the maximum of the numbers $\|Ax\|$ ($=$ the euclidean norm of Ax) with $\|x\| = 1$. It can be shown that $\|A\|_{op}$ is the square root of the maximum of the numbers $|\lambda|$, where λ runs through the eigenvalues of AA^*.

Then $\|A^k(A^*)^k\| \leq \|A\|^k\|A^*\|^k$, and one might hope to apply lemma 16.22 successfully with this estimate.

Unfortunately it can happen that this gives very weak results. Consider for example the following convex combination of unitary matrices:

$$A = \frac{1}{2}\left(\begin{pmatrix} 0 & 1 \\ 1 & 0 \end{pmatrix} + \begin{pmatrix} 0 & 1 \\ -1 & 0 \end{pmatrix}\right) = \begin{pmatrix} 0 & 1 \\ 0 & 0 \end{pmatrix}.$$

The matrix A has operator norm one, but nevertheless A^k tends to zero remarkably fast (already A^2 vanishes).

$\boxed{\textit{Selfadjoint matrices}}$

Let us recall some definitions and facts from linear algebra:

- A square matrix A is called *normal* if $AA^* = A^*A$; note that self-adjoint matrices are trivially normal.

- AA^* is always a self-adjoint matrix.

- If A is normal, then $\|A\|_{op}$ is the supremum of the numbers $|\lambda|$, where λ runs through the eigenvalues of A.

Therefore, if it happens that $A = \widehat{\mathbb{P}_0}(\rho)$ is normal, then $\|A^k(A^*)^k\|_{op} = \alpha^k$, where α is the square of the maximum of the moduli of the eigenvalues of A. Thus the convergence we want to investigate solely depends on the size of α.

There seems to be no simple characterization of the $\mathbb{P}_0$ for which all $\widehat{\mathbb{P}_0}(\rho)$ are normal. A sufficient criterion for self-adjointness, however, can easily be found:

Lemma 16.23 *Let $\mathbb{P}_0$ be a symmetric measure:* $\mathbb{P}_0(\{g\}) = \mathbb{P}_0(\{g^{-1}\})$ *for all g. Then all* $\widehat{\mathbb{P}_0}(\rho)$, $\rho \in \widehat{G}$, *are self-adjoint.*

Proof. This follows from $(\rho(g))^* = \rho(g^{-1})$, that is from the unitarity of the $\rho(g)$:

$$
\begin{aligned}
(\widehat{\mathbb{P}_0}(\rho))^* &= \Big(\sum_g \mathbb{P}_0(\{g\})\rho(g)\Big)^* \\
&= \sum_g \mathbb{P}_0(\{g\})\rho(g^{-1}) \\
&= \sum_g \mathbb{P}_0(\{g^{-1}\})\rho(g^{-1}) \\
&= \sum_g \mathbb{P}_0(\{g\})\rho(g) \\
&= \widehat{\mathbb{P}_0}(\rho).
\end{aligned}
$$

$\square$

$\boxed{\textit{Class functions}}$

Sometimes it happens that $\mathbb{P}_0$ is such that

$$\mathbb{P}_0(\{g \circ h\}) = \mathbb{P}_0(\{h \circ g\})$$

for all $g, h \in G$. This means that $\mathbb{P}_0(\{g\}) = \mathbb{P}_0(\{h \circ g \circ h^{-1}\})$, i.e., $\mathbb{P}_0$ is *constant on conjugacy classes*. We will speak of a *class measure* if this is the case.

Class measures share many features with measures on commutative groups. Their Fourier transforms are particularly simple to determine:

Lemma 16.24 *Let $\mathbb{P}_0$ be a class measure. Then $\widehat{\mathbb{P}_0}(\rho)$ is a multiple of the identity for every $\rho \in \widehat{G}$. More precisely:*

$$\widehat{\mathbb{P}_0}(\rho) = \Big[\frac{1}{d_\rho} \sum_g \mathbb{P}_0(\{g\})\chi_\rho(g)\Big] Id.$$

Here χ_ρ stands for the character associated with ρ, it is defined by $\chi_\rho(g) :=$ the trace of $\rho(g)$.

Proof. Let $h \in G$ be arbitrary. Then

$$
\begin{aligned}
\rho(h)\widehat{\mathbb{P}_0}(\rho) &= \rho(h) \sum_g \mathbb{P}_0(\{g\})\rho(g) \\
&= \sum_g \mathbb{P}_0(\{g\})\rho(h \circ g) \\
&= \sum_{g'} \mathbb{P}_0(\{h^{-1} \circ g'\})\rho(g') \\
&= \sum_{g'} \mathbb{P}_0(\{g' \circ h^{-1}\})\rho(g') \\
&= \sum_g \mathbb{P}_0(\{g\})\rho(g \circ h) \\
&= \Big(\sum_g \mathbb{P}_0(\{g\})\rho(g)\Big)\rho(h) \\
&= \widehat{\mathbb{P}_0}(\rho)\rho(h).
\end{aligned}
$$

Hence the first part of the assertion follows from Schur's lemma 16.2.

Now write $\widehat{\mathbb{P}_0}(\rho)$ as $a\,Id$. Then, on the one hand, the trace of this matrix is $a\,d_\rho$; on the other hand it equals $\sum_g \mathbb{P}_0(\{g\})\chi_\rho(g)$, and this completes the proof. $\qquad\square$

Remarks:

1. Note that in the case of commutative groups "character" means a group homomorphism from G to the complex numbers of modulus one. For general groups the characters are the traces of the irreducible representations. This extends the previous definition since in the commutative case all irreducible representations are one-dimensional.

One could argue that the definition might depend on the particularly chosen dual (so that one should speak of "a character with respect to $\widehat{G}$"). This is not the case: if ρ is an irreducible d-dimensional representation and $M \in \mathcal{U}_d$, then $\rho(g)$ and $M\rho(g)M^{-1}$ have the same trace, and therefore ρ and ρ_M give rise to the same character.

2. Characters play an important role in harmonic analysis, they are sufficient to describe the "commutative" aspect of G in that they are the most general class functions:

- Every χ_ρ satisfies $\chi_\rho(g \circ h) = \chi_\rho(h \circ g)$; this follows at once from $\mathrm{tr}(AB) = \mathrm{tr}(BA)$ for $d \times d$-matrices A, B. As a consequence all linear combinations f of characters are *class functions*, i.e., they satisfy $f(g \circ h) = f(h \circ g)$ for all g, h.

- Conversely, let f be any class function. Then it follows as in the preceding proof that $\hat{f}(\rho)$ is the identity matrix multiplied by $\sum_h f(h)\chi_\rho(h)/Nd_\rho$. The inverse Fourier transform (proposition 16.15) then provides the formula

$$
f(g) = \frac{1}{N} \sum_{\rho,h} f(h)\chi_\rho(h)\chi_\rho(g^{-1}),
$$

and if one notes that $\chi_\rho(g^{-1}) = \chi_{\bar{\rho}}(g)$ with $\bar{\rho}$ as in lemma 16.8, then it follows that f lies in the linear span of the χ_ρ[6].

---

[6] In fact one has to argue a little bit more subtly since $\bar{\rho}$ needs not be an element of $\widehat{G}$. However, there is a ρ' in this dual which is equivalent with $\bar{\rho}$, and both $\bar{\rho}$ and ρ' give rise to the same character.

Sometimes conjugacy classes are rather big. In the case of the *symmetric group* S_r, for example, the collection of all transpositions (= the permutations which only exchange two elements i, j with $i \neq j$) is a conjugacy class. To make use of this fact let us consider a deck of r cards, we want to analyse the *random-transposition shuffle*. This shuffle is slightly different from the random-to-random shuffle we have met in example 5 of chapter 2. In the language of the present chapter the random-to-random shuffle corresponds to a $\mathbb{P}_0$ which assigns mass $1/r$ to the identity, $2/r^2$ to the transpositions which exchange adjacent elements, and $1/r^2$ to the remaining transpositions. This $\mathbb{P}_0$ is *not* a class measure. In the case of the random-transposition shuffle we use a different rule:

> Select j, k in $\{1, \ldots, r\}$ independently according to the uniform distribution and exchange the j'th and the k'th card.

The associated measure (which we continue to call $\mathbb{P}_0$) has mass $1/r$ on the identity and mass $2/r^2$ on each of the $r(r-1)/2$ transpositions so that it is a class measure.

In order to discuss this example further it would be necessary to provide a dual for S_r. In fact, here much less is essential. Since we are dealing with a particularly simple class measure we only need to know, for every irreducible ρ, the value of χ_ρ at the trivial permutation and at any transposition τ (the value will be independent of τ). The first number is d_ρ, the trace of the $d_\rho \times d_\rho$-identity matrix, let us denote the second by χ_ρ^τ.

Since there are $r(r-1)/2$ transpositions and each one has weight $2/r^2$ it follows from lemma 16.24 that

$$\widehat{\mathbb{P}_0}(\rho) = \Big(\frac{1}{r} + \frac{r-1}{r} \cdot \frac{\chi_\rho^\tau}{d_\rho}\Big) Id.$$

Therefore the mixing rate of the associated chain is determined by the number

$$M_r := \max_{\rho, \rho \neq \rho_{\mathrm{triv}}} \left| \frac{1}{r} + \frac{r-1}{r} \cdot \frac{\chi_\rho^\tau}{d_\rho} \right|.$$

Since the χ_ρ^τ are traces of unitary matrices it follows that their absolute values are bounded by d_ρ, but this observation only provides the poor bound $M_r \leq 1$. Better results necessitate to put into action the machinery of advanced group theory. In the literature on this subject one finds tables of the numbers χ_ρ^τ. For example, if $r = 10$, page 354 of [47] contains the information that there are precisely 42 irreducible representations with dimensions ranging from 1 (for the trivial and the sign representation) to 768. On this page also the χ_ρ^τ can be found so that it is possible to derive M_{10} explicitly[7].

It is considerably more difficult to provide results concerning this shuffle *for arbitrary* r. Diaconis proves in theorem 5 of [24] the following assertion. The proof depends on deep properties of characters of the symmetric group which enable one to estimate the M_r and thus to apply the upper-bound lemma 16.22.

Theorem 16.25 *There is a constant $a > 0$ such that for any $c > 0$ one has*

$$\|\mathbb{P}_0^{(k*)} - U\| \leq a\, e^{-2c},$$

provided that $k \geq (r \log r)/2 + cr$.

---

[7] The worst case happens for a representation where $d_\rho = 9$ and $\chi_\rho^\tau = 7$; this leads to

$$M_{10} = \frac{1}{10} + \frac{9}{10} \cdot \frac{7}{9} = \frac{4}{5}.$$

This roughly means that a k of order $r \log r$ suffices to guarantee that after k random-transposition shuffles all permutations of the cards are (approximately) equally likely.

Exercises

($(G, \circ)$ will be a finite group in the following exercises.)

16.1: There exists a G which does not admit any non-trivial one-dimensional representation. (Hint: consider a suitable subgroup of the symmetric group $\mathcal{S}_r$, where r is not too small.)

16.2: Let $\mathbb{P}_0$ be a probability measure on G such that

$$\mathbb{P}_0(A) = \mathbb{P}_0(\{g \circ g_0 \mid g \in A\})$$

for all $A \subset G$ and all g_0; such measures are called *translation invariant*. Prove that the uniform distribution is the only translation invariant probability measure on G.

16.3: Let d be an integer, $d > 1$. Prove that every d-dimensional representation ρ such that all $\rho(g)$ are diagonal can be written as a product of one-dimensional representations.

16.4: Let δ_g and δ_h be Dirac measures associated with two elements g and h of G. Calculate the convolution $\delta_g * \delta_h$.

16.5: G is commutative iff $\mathbb{P}_1 * \mathbb{P}_2 = \mathbb{P}_2 * \mathbb{P}_1$ holds for arbitrary probability measures $\mathbb{P}_1, \mathbb{P}_2$ on G.

16.6: Let G_1 and G_2 be finite groups and ρ_1 and ρ_2 representations of G_1 und G_2, respectively. Prove that

$$(g_1, g_2) \mapsto \begin{pmatrix} \rho_1(g_1) & 0 \\ 0 & \rho_2(g_2) \end{pmatrix}$$

defines a representation of the product group $G_1 \times G_2$.

16.7: Let g_0 be a fixed element of G and δ_{g_0} the Dirac measure associated with g_0. Can you give, for an arbitrary probability measure $\mathbb{P}_0$ on G, an explicit description of $\delta_{g_0} * \mathbb{P}_0$?

16.8: Prove that the convolution is an associative operation.

16.9: Let $\chi : G \to \Gamma$ be multiplicative on G and ρ a d-dimensional representation. Define $\chi\rho$ by

$$g \mapsto \chi(g)\rho(g).$$

a) Prove that $\chi\rho$ is a d-dimensional representation of G.

b) Is $\chi\rho$ irreducible if ρ is?

c) What problems arise if one wants to define, for given d-dimensional representations ρ_1, ρ_2, a new one by

$$g \mapsto \rho_1(g)\rho_2(g)?$$

16.10: The Peter-Weyl theorem implies that $\overline{\rho}$ is equivalent with some $\rho' \in \widehat{G}$ for every ρ in a dual $\widehat{G}$ of G. Verify this fact for the above two-dimensional representation ρ of $\mathcal{Q}$.

16.11: Can there be a d-dimensional irreducible representation on a group with d^2 elements?

16.12: Let H be a normal subgroup of G and g_0 an element which does not belong to H. Prove that there exists an irreducible representation ρ such that $\rho(g)$ is the identity matrix for all $g \in H$, but $\rho(g_0) \neq Id$.

16.13: Let a measure $\mathbb{P}_0$ on the quaternion group Q be defined by

$$\mathbb{P}_0(\{\underline{1}\}) = \mathbb{P}_0(\{\underline{i}\}) = 1/3, \ \mathbb{P}_0(\{\underline{j}\}) = \mathbb{P}_0(\{\underline{k}\}) = 1/6.$$

Calculate $\widehat{\mathbb{P}_0}$ and reconstruct $\mathbb{P}_0$ from $\widehat{\mathbb{P}_0}$ with the help of the Fourier inversion formula.

16.14: For a probability measure $\mathbb{P}_0$ on G define a new measure $\mathbb{P}_0^*$ by

$$\mathbb{P}_0^*(\{g\}) := \mathbb{P}_0(\{g^{-1}\}).$$

a) $\mathbb{P}_0^*$ is a probability measure.

b) If $\mathbb{P}_0 * \mathbb{P}_0^* = U$, the uniform distribution, then $\mathbb{P}_0 = U$.

It is important to consider $\mathbb{P}_0 * \mathbb{P}_0^*$ and *not* $\mathbb{P}_0 * \mathbb{P}_0$ here since there are groups for which measures $\mathbb{P}_0$ different from U exist which nevertheless satisfy $\mathbb{P}_0 * \mathbb{P}_0 = U$. (See [15], a complete discussion of this problem can be found in [27] and [72].) This is in marked contrast to the commutative case, cf. exercise 15.9.

16.15: Let U, V be distinct unitary $d \times d$-matrices. Prove that $\|(U + V)/2\|_{\mathrm{op}}$ is strictly less than one if $d = 1$, but that for $d > 1$ there are examples such that $\|(U + V)/2\|_{\mathrm{op}} = 1$ holds.

16.16: Let f be a complex valued function on a group G with N elements. Define $\tilde{f}$ by

$$\tilde{f}(g_0) := \frac{1}{N} \sum_g f(g \circ g_0 \circ g^{-1}).$$

Prove that $\tilde{f}$ is a class function and that $f \mapsto \tilde{f}$ is a linear projection on the space X_G.

16.17: Prove that the notion "equivalence" for representations of a group has the properties of an equivalence relation: it is reflexive, symmetric and transitive.

16.18: Let ρ be a d-dimensional representation of G. Prove that ρ is irreducible iff the subspace spanned by the $\rho(g)$ is d^2-dimensional.

16.19: How is the Fourier transform of a function f related with that of the functions $g \mapsto f(g^{-1})$ and $g \mapsto f(g_0 \circ g)$?

16.20: We have introduced characters resp. representations such that the $\chi(g)$ resp. the $\rho(g)$ are *complex* numbers resp. unitary matrices. Where was this important, what goes wrong if one restricts oneself to real numbers and matrices?

16.21: Characterize the probability measures $\mathbb{P}_0$ such that the associated chain on G is reversible.

16.22: Let A_k be a self-adjoint $N \times N$-matrix with nonnegative eigenvalues for $k = 1, \ldots$. Prove that the A_k tend to zero (component-wise) iff $\lim_k \mathrm{tr}(A_k) = 0$.

17   Notes and remarks

The results presented in *chapter 9* have been known since the "classical" period of Markov chain theory, the approach presented here emphasizes the use of convexity arguments. The proof of proposition 9.2(iv) is from [69], readers who want to learn more on the algebraic theory of stochastic matrices are referred to this book.

Also all results in *chapter 10* are folklore. The structure of our proof of theorem 10.3 follows chapter 2.2 in [70]. The simple idea to treat certain renewal problems as presented in the text seems to have no counterpart in the literature.

The material of *chapter 11* is mainly from [70], there one also finds extensive comments on the development of conductance techniques. Our proof of theorem 11.3 is similar to the approach in chapter 2.2 of [70]. (Despite considerable effort there seems to be at present no really elegant and simple proof of this result. For competing or similar variants see [18] or [74].) The observation which we have called proposition 11.8 seems to be new.

Chapter 12 covers standard material, in the present context (finite state space, discrete time) the proofs remain rather simple.

As already noted in the text, the couplings in *chapter 13* have a long history (see [54]). Aldous ([1]) seems to be the first who has applied couplings systematically to get bounds of rapid mixing, most of our results are from this paper.

The standard reference for *chapter 14* is [4], further results can be found in [3] and [24]. Our contribution is only to emphasize the property which we have called "$\mathbb{T}$ respects the Markov property".

Who is responsible for the results in *chapter 15* and *chapter 16*? The development of harmonic analysis on finite groups was the work of many mathematicians, it was completed at the beginning of the twentieth century. Here I have tried to find an approach which is self-contained and elementary. My way to the Peter-Weyl theorem is based on lemma 16.10, more common is an application of the Stone-Weierstraß theorem at this point.

I did not find any remark in the literature who used harmonic analysis for the first time to investigate mixing properties of chains on groups. At present the standard reference is [24], it contains an abundance of applications of group theory to probability and statistics. Most of our theorems – in particular the last one – can be found there (as an exception I mention the properties of G_Δ and proposition 16.21 which seem to be due to the author). The results of this chapter can only be applied if one has mastered the problem to exhibit sufficiently many irreducible representations of a given group. For this [19], [34] and [47] might be helpful.

It should be noted that the same technique has been applied similarly successfully to certain *infinite groups:* the approach is the same, the concrete calculations, however, are much more involved (see, e.g., [65]).

This Notes-and-Remarks chapter closes with **some supplements**. The first one concerns *couplings.* The idea with coupled Markov chains is to observe two copies of a random walk until they meet. A variant has been proposed in [63]. There one simulates a chain backwards, more precisely, one tries to find a time step $-k_0$ in such a way that now, at time 0, all walks which have been started at time $-k_0$ have met at the same state, say i. The surprising feature is that a particular state i is found in this way *precisely* with probability π_i. This "coupling from the past" has been applied successfully to treat various problems, see [32] or [64]. Other methods which also provide *exact simulation* – and not only outputs with a distribution arbitrarily close to the equilibrium – can be found in [75].

Next we want to mention a phenomenon which has attracted the attention of several mathematicians. Imagine an irreducible and aperiodic chain with equilibrium π^T and suppose that it starts deterministically. Then often the following happens: for a certain number of steps the probabilities to find the walk at some state i are "far away" from π_i, and then – not much later – they approximate π_i very well. This *"cut-off phenomenon"* has been studied in a number of papers (see, e.g., [1], [3], [24], [55], or [65]), it can be observed at many concrete chains. A theoretical understanding, however, which covers arbitrary chains has not yet been proposed.

And finally it has to be remarked that only a selection of known mixing methods has been treated here. For example, we have completely omitted the so-called L^2-*methods* where Hilbert space methods come systematically into play[1]; see the article of L. Saloff-Coste in [37]. For a *survey of other methods* cf. [55], and in this connection it is also necessary to mention [5] (which hopefully sooner or later will manage the transformation from mere electronic existence into a real book).

---

[1] In section 21, however, in the proof of proposition 21.3, we will make use of such techniques.

Part III

Rapidly mixing chains: applications

In part II we have developed a number of techniques which enable us to determine how fast a given chain converges to its equilibrium. In particular we are now in a position to generate random elements from a finite set according to a prescribed distribution provided it can be thought of as the equilibrium of a chain to which our methods apply.

Part III will contain some examples to demonstrate how one can profit from this idea. In *chapter 18* we describe the connection between *approximate counting* and *random generation*: for a certain class of of sets, the solution sets of self-reducible problems, it is essentially equivalent to be able to count up to a prescribed accuracy or to have access to a (nearly) uniform random generator. Next, in *chapter 19* we introduce *Markov random fields* which can be thought of as a natural generalization of Markov chains. It will be shown how it is possible to generate samples from such a field by using the Markov chain techniques developed earlier in this book. For a special class of random fields, the class of *Gibbs fields*, the probability measure is defined by certain functions, the potential functions. Gibbs fields are studied in *chapter 20*, we prove that they are Markov random fields. A celebrated example of a Gibbs field, the *Ising model*, is investigated more closely. At the end of this chapter we show how one can obtain samples from a Gibbs field, to this end we provide concrete bounds for the mixing rate of the *Gibbs sampler*.

There is a variant of the Gibbs sampler: also with the *Metropolis sampler* it is possible to sample from a finite space for which the probability distribution is given by an energy function. This is studied at the beginning of *chapter 21*. The second half of this chapter is devoted to *simulated annealing*, a stochastic optimization technique which has found numerous applications in various areas of applied mathematics. Finally, the last chapter of this book (*chapter 22*) contains some *Notes and remarks*.

18   Random generation and counting

Sometimes it happens that one is dealing with a set S for which it is easy to check that it is finite but for which there seems to be no simple way to determine the number of elements within reasonable time. There are even situations where this problem is NP-hard so that exact counting is in a sense impossible. However, by using Markov chains one can treat the weaker problem of *approximate counting*, a connection which has been systematically studied by Sinclair and others (see [70] and the literature cited there).

We start our discussion with some sample problems in *section 1*. It will then be important to note that some of these are of a particular type: they can be reduced to "few" simpler ones which in turn give rise to others which are even more tractable and so on. They will be called *self-reducible*, it is this class of problems to which Markov chain techniques apply. Next, in *section 2*, we indicate how the possibility of exact counting gives rise to the possibility of exact uniform simulation. More interesting is the converse: if we have uniform random generators at our disposal we can count the number of solutions of self-reducible problems approximately. This will be presented in *section 3*.

Finally, in *section 4*, we complement the investigations of section 2 in that we describe an approximately uniform random generator over the solutions of a self-reducible problem without the assumption of knowing the numbers associated with the reduced problems.

Self-reducible problems

Most of the ideas we are dealing with in this chapter can be illustrated with the "problem" of determining the number $N(r)$ of all permutations over r elements. Usually one argues as follows: the collection of *all* permutations is the disjoint union of r subsets the ρ'th of which contains the permutations which send 1 to ρ ($\rho = 1, \ldots, r$); each of the subsets has $N(r{-}1)$ elements, this gives rise to the recursion $N(r) = rN(r{-}1)$, and with $N(1) = 1$ one gets $N(r) = r!$. The crucial point in the argument is to split the problem into a "small" number of simpler ones. Before we try to be more formal we consider further

Examples:

1. Let $n > 0$ be an integer. What is the number of *partitions* of n, that is how many families $n_1 \geq \cdots \geq n_k \geq 1$ exist such that $n = n_1 + \cdots + n_k$? (The number 5, e.g., admits the 7 partitions $5, 4 + 1, 3 + 2, 3 + 1 + 1, 2 + 2 + 1, 2 + 1 + 1 + 1, 1 + 1 + 1 + 1 + 1$.)

2. Let $F(u_1, \ldots, u_r)$ be a Boolean formula in the Boolean variables $u_1, \ldots, u_r$, that is a well-defined logical expression which contains the u's and which is built up using the logical operations $\wedge$ (= "and"), $\vee$ (= "or") and $\neg$ (= "not").

Here are examples in the three variables u, v, w:

$$(u \vee v) \wedge (\neg w), \quad ((u \vee v) \wedge (u \vee w)) \vee (\neg v), \ldots$$

There are 2^r possibilities to give the variables the values "true" or "false". How many of these give rise to the value "true" for the expression F? For example, for the case $(u \lor v) \land (\neg w)$, it is easy to check that there are precisely 3 possibilities, but how can the number be determined for larger r? Also it could be important to have an answer to the more modest question whether or not there exist *any* truth values for the u's which make F a true expression[1].

3. Let Δ be a collection of r points in the d-dimensional euclidean space $\mathbb{R}^d$; we assume that $r \geq d + 2$. Then Carathéodory's theorem asserts that for every x in the convex hull of Δ there are elements $x_0, \ldots, x_d$ of Δ such that x lies already in the convex hull of these $d + 1$ elements. There are $\begin{pmatrix} r \\ d+1 \end{pmatrix}$ possibilities to choose these elements, for how many choices will x be in the convex hull?

4. Let n be an integer, one wants to know the number of nontrivial divisors m of n. A particular instance is the problem whether there exist any such m, that is whether or not n is prime.

5. Let (V, E) be a graph (see page 96). A *perfect matching of size n* is a subset M of the edges E such that M has n elements and each two different e_1, e_2 in M have no vertex in common. Denote by $\mathcal{M}_n$ the collection of these M, how many elements does this set have?

A particular case has attracted the attention of many mathematicians. Assume that (V, E) is of the special form that V is the disjoint union of two subsets V_1, V_2 each containing r elements and that there are only edges which join vertices of V_1 to those of V_2 (a *bipartite graph*). Obviously there are precisely as many such graphs as there are $r \times r$-matrices $A = (a_{ij})$ with $a_{ij} \in \{0, 1\}$, one simply has to translate a "1" at position i, j into an edge joining i to j and vice versa. Then a matching of size r for the graph corresponds to the choice of r positions i, j in the matrix with $a_{ij} = 1$ in such a way that every row and every column contains precisely one of these specified positions. The number of such choices can be written more compactly as

$$\operatorname{per}(A) := \sum_{\sigma} \prod_{i=1}^{r} a_{i\sigma(i)},$$

where the summation runs over all permutations σ of $\{1, \ldots, r\}$. One cannot fail to observe the similarity of this expression with that of the determinant of A, surprisingly the calculation of $\operatorname{per}(A)$, which is called *the permanent of* A, is much harder than that of the determinant[2].

We want to emphasize here that some of these enumeration problems share with our introducing example of the permutations the possibility that counting can be reduced to the counting problem for simpler situations. For example:

- Suppose that you know, for all $m < n$, the number $\pi(m, k)$ of partitions of m such that all summands are bounded by k. (For example, $\pi(4, 2) = 3$, since we have to take into account the three partitions $2 + 2$, $2 + 1 + 1$, $1 + 1 + 1 + 1$). A moment's reflection shows that $\pi(n, n)$ – *this* is the number we are interested in – is just the sum

---

[1] For more comments on this *satisfiability problem* and its connections to the $P = NP$ circle of ideas see chapter 22.

[2] A standard reference concerning the permanent is [58].

$$1 + \pi(1, n - 1) + \pi(2, n - 2) + \cdots + \pi(n - 1, 1). \qquad (18.1)$$

- In the last example fix an edge e of the bipartite graph under investigation joining a fixed vertex x_0 to another vertex y_0 and consider the graph which is obtained from the original one by erasing e, the vertices x_0, y_0 and all edges starting at x_0 or y_0. Suppose that you know the number P_e of perfect matchings of size $r - 1$ of this reduced graph for every such e. Then the number of matchings of size r is just the sum over the P_e.

  To phrase it in the language of the permanent: the permanent of an $r \times r$-matrix A with 0-1-entries is the sum of the permanents of the $\leq r$ matrices which are derived from A by erasing a fixed line and a column where this line contains a "1". This corresponds to a similar technique for determinants.

For the other problems it is not obvious how to choose an appropriate reduction. For problem 4, the number of divisors, it is at present even unknown whether a similar simplification is possible.

Now we are going to argue a little bit *more formally*. However, in order to avoid the danger of hiding the relevant ideas behind technicalities our approach will not be perfectly rigorous[3]. We are given a problem which can be posed by prescribing a certain mathematical object x. One knows in advance that there exists a finite set $R(x)$ of solutions to this problem, and we are interested in the cardinality of $R(x)$. In the very first example, e.g., x is the number r and $R(x)$ is the set of all permutations of r elements.

We suppose that we can associate with every such situation a nonnegative integer $l(x)$ which can be thought of as *a measure of difficulty* to treat the problem; in the permutation example surely $l(x) = r$ is a natural choice. And the precise meaning of "the problem can be reduced to 'few' simpler problems" then is that it is possible to find subproblems $x_1, \ldots, x_s$ of a similar kind such that

- $l(x_i) < l(x)$ for $i = 1, \ldots, s$;

- the cardinality of $R(x)$ is the sum of the cardinalities of the $R(x_i)$ (or, more generally, $R(x)$ can simply be counted if all $R(x_i)$ are known);

- the number s of subproblems is not too large: there is a polynomial Q such that $s \leq Q(l(x))$.

Also it is assumed that problems with $l(x) = 0$ have a unique solution which can be found in constant time.

Readers who are not familiar with theoretical investigations of complexity might wonder why polynomials occur here. The reason is that it is now a generally accepted idea to regard only those problems as tractable for which the amount of work to solve them can be bounded by a polynomial in the number of bytes which are necessary to pose the problem. E.g., as everybody knows, the calculation of the product of two numbers *is* tractable, but there are many problems for which this is unlikely (a well-known example is the travelling salesman problem).

---

[3] Cf. [70], chapter 1, for a more extended presentation.

In view of these remarks we hasten to complement our assumptions: it is tacitly understood that the calculation of $l(x)$ and the determination of the $x_1,\ldots,x_s$ have a polynomially bounded running time, also $l(x)$ must be bounded by a polynomial in the number of bytes which are necessary to describe x. If these conditions are met we will speak of a *self-reducing problem.*

It should be clear how the above examples fit into this framework, we omit to identify the $l(x)$ and the $x_1,\ldots,x_s$ in these special cases. Rather we want to recommend the following *visualization of self-reducible problems.* We associate with each such problem a *tree* with root $R(x)$, it is depicted at level $l = l(x)$. The $R(x_i)$ are connected by edges with $R(x)$, they occupy certain levels which are smaller than l (for simplicity we have placed all of them at level $l-1$ in our picture). The individual $R(x_i)$ give rise to new sub-subproblems at even lower levels, and so on until we reach the level zero. There we find the "leaves" of our tree, certain $R(y)$ which can rapidly be determined. This information then gives rise – by working backwards – to the number of elements in $R(x)$.

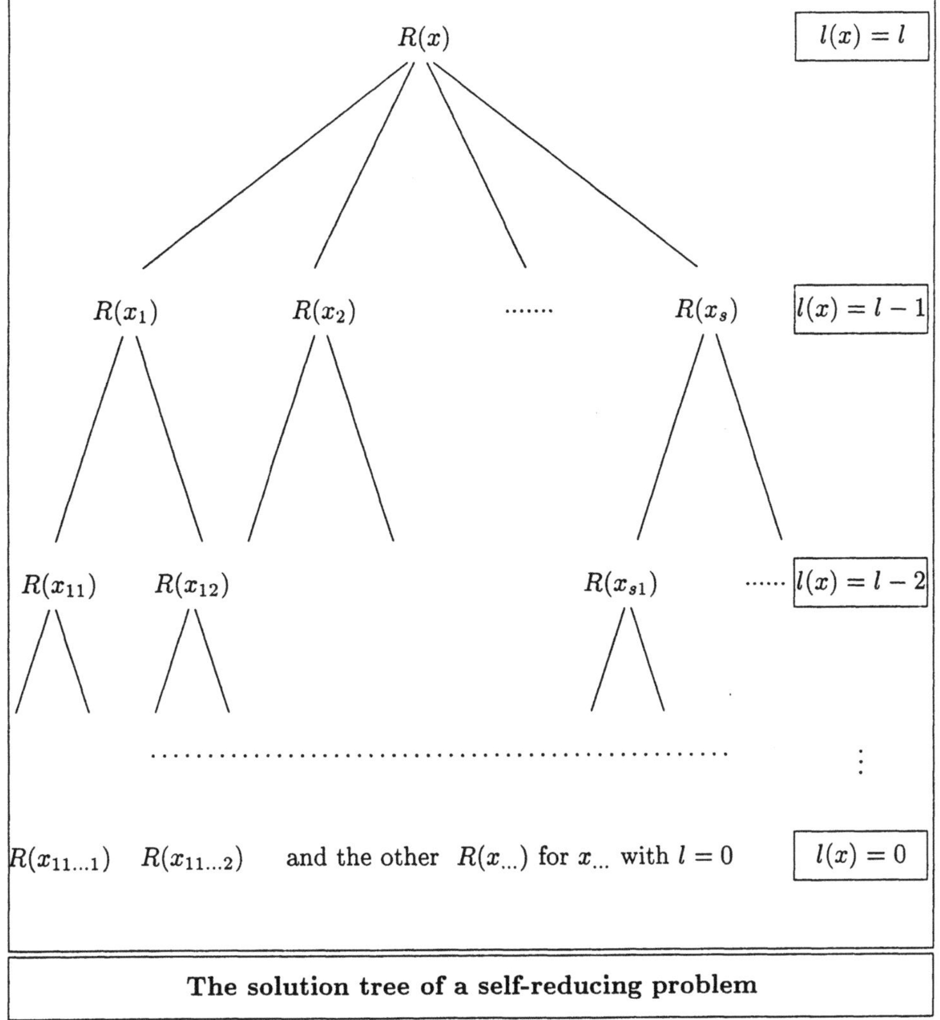

The solution tree of a self-reducing problem

Counting plus self-reducibility implies simulation

We resume our permutation example. Suppose that we want to provide a random permutation $\tau = (\tau_1, \ldots, \tau_{100})$ of $\{1, \ldots, 100\}$. There are

$$
\begin{aligned}
100! \quad = \quad &93326215443944152681699238856266700490715\\
&96826438162146859296389521759999322991560\\
&89414639761565182862536979208272237582511\\
&8521091686400000000000000000000000000
\end{aligned}
$$

of them, a finite though rather big number. Therefore there is no hope to approach the problem by enumerating all candidates and then to generate a random number between 1 and 100!. It is much more natural to start with a random choice in $\{1, \ldots, 100\}$ to obtain the first element τ_1, then to select uniformly τ_2 among the remaining 99 numbers and so on. In this way we can get a uniformly distributed τ from all permutations of n elements after n random choices among at most n elements. It should be obvious that this method reduces the complexity of the problem drastically (otherwise it would not even be tractable).

The reason why the resulting permutation is in fact uniformly distributed is simple, one only has to apply repeatedly the following argument[4]:

> Suppose a finite set M is written as a disjoint union of sets $M_1, \ldots, M_s$ each having t elements. If one chooses $\sigma \in \{1, \ldots, s\}$ uniformly at random and then x in M_σ independently and also uniformly, then x will be equidistributed in M.

Only a little modification of this idea is necessary to treat arbitrary self-reducing problems in a similar way:

Proposition 18.1 *Let $R(x)$ be the solution set associated with a self-reducing problem. Suppose that this class of problems is such that it is easy[5] to determine the cardinality of $R(y)$ for all y.*
Then one can generate a uniformly distributed z in $R(x)$ as follows:

- *Denote by $n, n_1, \ldots, n_s$ the cardinalities of $R(x), R(x_1), \ldots, R(x_s)$, respectively; here $x_1, \ldots, x_s$ are the subproblems associated with x (note that $n = n_1 + \cdots + n_s$ by the definition of self-reducibility). Choose $\sigma \in \{1, \ldots, s\}$ according to the probabilities $n_1/n, \ldots, n_s/n$ and continue to work with x_σ.*

- *Pass similarly from x_σ to a sub-subproblem, from there to a sub-sub-subproblem and so on until you arrive at a problem y of level $l = 0$.*

- *Find the unique z' in the leave $R(y)$ and use this – by working backwards – to get a z in $R(x)$. This z is the output of our generator.*

Proof. The justification is easy: if $U_1, \ldots, U_s$ denote the uniform distributions on sets $M_1, \ldots, M_s$ which are a disjoint partition of a set M, then

---

[4] In technical terms it is the trivial statement that the product of two uniform distributions is also uniform.

[5] "Easy", of course, means "in polynomial time".

$$\frac{\mathrm{card}(M_1)}{\mathrm{card}(M)} U_1 + \cdots + \frac{\mathrm{card}(M_1)}{\mathrm{card}(M)} U_s$$

is uniform on M. This is applied here to $M(x)$ which – up to isomorphism – is the disjoint union of the $R(x_i)$. $\qquad\qquad\qquad\qquad\qquad\qquad\qquad\qquad\qquad\qquad\qquad\qquad\quad$ $\square$

As a *variant of this idea* suppose that we don't know the $R(x)$ exactly but that it is possible to get approximations as close as we wish within reasonable time. More rigorously this can be expressed by saying that we have an algorithm at our disposal with the following two properties:

- For given $\varepsilon > 0$ and x the algorithm provides a number r such that

$$\frac{|r - \mathrm{card}(R(x))|}{\mathrm{card}(R(x))} \leq \varepsilon$$

  with a probability of at least $1 - \varepsilon$.

- The running time to get r is bounded by a polynomial in $1/\varepsilon$ and the number of bytes to describe x.

Then we can modify the simulation procedure in proposition 18.1 to get a random generator for the $z \in R(x)$ which has distribution ε-close to the uniform distribution and a polynomially bounded running time (with a polynomial in $1/\varepsilon$ and the "length" of x).

The proof is not difficult, one only has to glue together the various polynomials. Nevertheless, a rigorous argument is technically cumbersome, the reader is referred to chapter 1.2 in [70].

Simulation offers the possibility of counting

Surprisingly, it is possible to reverse the idea. To motivate the approach let's consider again the permutations $\rho = (\rho_1, \ldots, \rho_r)$ of $\{1, \ldots, r\}$. We pretend not to have the slightest idea of how big the number $N(r)$ of the ρ might be. However, we suppose that we are clever enough to *simulate* them with respect to the uniform distribution. With the help of such a generator we will observe after some time that roughly a $1/r$-fraction of the samples satify $\rho_1 = i$ for $i = 1, \ldots, r$. We repeat the experiments with permutations over $\{1, \ldots, r - 1\}$, and also this time the first entry is uniformly distributed. In this way we proceed until we arrive at a level which can be treated directly, say $r = 3$. Then we argue as follows: $N(3) = 6$, and among the permutations of *four* elements it is (roughly) equally likely to get "1 followed by a permutation of $\{2, 3, 4\}$" or "2 followed by a permutation of $\{1, 3, 4\}$" or $\ldots$. Thus $N(4)$ should be close to $4N(3) = 24$.

> Note that this is a variant of the following elementary fact: if an urn contains an unknown number m of balls precisely k of which are white, then you can estimate m provided that k and the probability of drawing a white ball are known.

Working backwards we really arrive at the estimate $N(r) \approx r!$, a guess which of course should be complemented by an error analysis.

This can be made precise for arbitrary self-reducing problems: the number of elements in the set $R(x)$ can be counted approximately in polynomially bounded time if one has access to a generator which provides these elements nearly uniformly distributed in polynomial time. A rigorous formulation and the details of the canonical proof are omitted here, we refer the reader to chapter 1.4 in [70].

A celebrated example where this counting method plays an important role is the *permanent* which we have met at the beginning of this chapter: we are given a matrix A containing solely 1's and 0's, and the problem is to determine per (A), the number of permutations σ such that all $a_{i\sigma(i)}$, $i = 1, \ldots, r$, are one. It has been shown by Valiant ([73]) that the calculation of per (A) is "difficult": this calculation is $\#$P-complete, a notion from the zoo of complexity definitions which essentially states that there is no hope for a polynomially bounded algorithm. On the other hand, the permanent can be calculated approximately in polynomial time, and thus we have an example of a situation where Monte-Carlo techniques are provably superior to exact methods.

The idea is to combine the following facts:

1. The permanent of a matrix is just the the number M_n of perfect matchings of size n of a suitable bipartite graph (V, E) with $2n$ vertices.

2. The determination of the permanent is a self-reducing problem.

3. It is possible to generate a random perfect matching of size n in polynomial time[6].

This has first been sketched by Broder in [21], the result is described in full detail in chapter 3 of [70].

Simulation without counting

Let us finally remark that one can *always* simulate elements of $R(x)$ in the case of self-reducing problems. It has to be admitted, however, that the method performs rather poorly.

The idea is to apply the Markov chain techniques for chains which are defined by graphs from chapter 11 to the graph which is associated with the problem (see page 177).

We want to produce elements of $R(x)$, or, equivalently, vertices at the level $l = 0$ (these are the leaves of our tree); they all should have (approximately) the same probability.

On the other hand, *we know* how to produce a random vertex of the *whole* graph, this has been described in the second half of chapter 11. From the results we have proved there we even can derive concrete bounds for the mixing rate, it will depend on the edge magnification μ of the concrete graph under consideration.

And to bridge the gap between our "we want" and the "we know" one simply produces a random vertex of the whole graph with the restriction that it will be used as the output of our random generator only if it is at level $l = 0$. Then the outputs will clearly be (nearly) equidistributed in this subset.

---

[6] More precisely: the samples are nearly uniformly distributed, and the distance ε between the uniform and the real distribution contributes with a polynomial in $\log \varepsilon$ to the running time.

Let us analyse what happens if we are going to produce random permutations τ of $\{1,\ldots,r\}$ in this way. The various levels $l = r,\ r-1,\ \ldots,\ 0$ can be thought of as the number of components of τ which remain to be specified. It is reasonable to identify a vertex with the sequence of components we already know, and thus the root of our tree is the empty set $\emptyset$, let us start there.

We proceed as described in chapter 11 (cf. definition 11.4; for simplicity we will work with $\beta = 1$). The maximal number d of edges in our graph is $r+1$, and thus we stay at $\emptyset$ or we will move to 1, to 2, $\ldots$, or to r each with probability $1/(r+1)$. Suppose that we arrive at 2. Then, with equal probability, we will return to $\emptyset$ or continue to one of 21, 23, $\ldots$, or $2r$. In this way we perform a walk on the possible selections of 0, 1, $\ldots$, or r numbers (without repetition) out of $\{1,\ldots,r\}$, a walk which will from time to time be in $\Pi :=$ the selection of r numbers $=$ the set of permutations. We stop the walk after some time, and only if we are at a state in Π this is used as an output.

Two natural questions arise. The first is the question of *efficiency*, how often will it happen that we stop at a state in Π? The graph has $g_r := 1 + r + r(r-1) + \cdots + r!$ elements $r!$ of which are favourable. The quotient is close to $1/e \approx 0.38$, and this ratio is surely not too bad: the random walk will produce roughly 38 outputs out of 100 runs.

And what about the *running time*? Unfortunately, the graph under consideration is particularly unsuitable for rapid mixing because of the bottleneck at the root; to pass from one leave (= a permutation) to another one with a different first component one has to climb up to the very top of the graph. Let us assume for simplicity that r is even, by T we denote the vertices which belong to the left half of the graph (not including the root). The capacity C_T is very close to $1/2$, the ergodic flow F_T, however, is rather small. There are $r/2$ edges from T to its complement each with transition probability $1/(r+1)$, and therefore $F_T = r/[2(r+1)g_r]$. It follows that Φ_T and thus also the conductance Φ of our chain is as tiny as $1/g_r$, and since the mixing rate is bounded by $1 - 2\Phi$ it is impossible to guarantee a good mixing rate within reasonable time.

Moral of the story: Simulation without counting is – at least in the case where one works with the solution tree of a self-reducing problem – mainly interesting for theoretical reasons, this method can be used only for very restricted examples.

Exercises

18.1: Let $(M, \leq)$ be a finite ordered space. Prove that the problem of finding all totally ordered subsets of M is self-reducing.

18.2: Let A be an $N \times N$-matrix with integer coefficients. Verify that the problem of calculating the determinant of A is self-reducing.

18.3: Prove formula (18.1).

18.4: Calculate the permanent of

$$A = \begin{pmatrix} 0 & 1 & 1 & 0 \\ 1 & 1 & 0 & 0 \\ 0 & 0 & 1 & 1 \\ 1 & 1 & 1 & 1 \end{pmatrix}$$

and sketch the associate bipartite graph.

18.5: Let A_1 and A_2 be square matrices. Give a formula of the permanent of

$$\left(\begin{array}{cc} A_1 & 0 \\ 0 & A_2 \end{array}\right)$$

in terms of the permanents of A_1 and A_2. What does this result mean for the associated bipartite graphs if A_1 and A_2 have 0-1-entries?

18.6: Let N be an integer with r digits.

a) Prove that the number of elementary calculations to determine N^2 is bounded by a polynomial of degree 2 in r.

b) What degree belongs to the calculation of N^3?

c) Suppose that N is a square, $N = M^2$. Is the number of calculations which are necessary to determine M bounded by a polynomial in r?

18.7: Let A be a finite set which is written as the disjoint union of subsets $A_1, \ldots, A_r$. Suppose that one has the information that the number of elements in A_ρ is n_ρ for $\rho = 1, \ldots, r$. However, this is known to be true in each of the r cases only with probability $1 - \varepsilon$ (where "true" or "false" are independent for the various ρ). Also n_ρ is possibly not precisely the cardinality of A_ρ, there might be a relative error δ:

$$\frac{|\text{card}(A_\rho) - n_\rho|}{\text{card}(A_\rho)} \leq \delta.$$

What can be said about the number of elements in A?

19 Markov random fields

Imagine a set S of people, the inhabitants of your home town, say. For every $s \in S$ there is a subset $\mathcal{N}_s$ of S: the people whom s knows, his or her neighbours, friends or colleagues. It happens that some people are infected by a dangerous disease D, the probability that a particular person s has D will naturally depend on the number of $t \in \mathcal{N}_s$ with D. What can be said about the distribution of infected people? Will D eventually disappear or will everybody be infected sooner or later?

It is easy to find similar situations, "s has D" can be replaced by "s has heard of the rumour R" or by "s is in favour of the political party P". Of great interest are also examples from physics. It is known, for example, that the orientation of a magnetic dipole d depends stochastically on the orientation of its neighbours: if they all have the same orientation O then there is a strong tendency that O is also assumed by d.

In the present chapter we introduce a model to deal with such probabilities which are (solely) influenced by the "neighbours". For many years it has been used in various areas, ranging from physics over sociology, medicine and biology to applications in image reconstruction.

The appropriate setting will be *Markov random fields*. Such a field is a family $(X_s)_s$ of random variables indexed by a set S, where the term "Markov" refers to the fact that "little" information (namely the values of X_t for the "neighbours" t of s) is as good as the knowledge of *all* X_t, $t \neq s$ if one wants to predict X_s. We start with rigorous definitions: random fields, neighbourhood systems, Markov random fields, local characteristics. The connection between a local and a global view is discussed in some detail in the second section, there Markov chains and their equilibrium distributions will play an important role.

Markov random fields: definitions and examples

Let us first fix *notation*. We need a finite set S, the *sites*, and a finite set Λ of *states*. To avoid trivialities we will assume throughout that Λ has at least two elements. Usually Λ is small, but even then the collection Λ^S of all mappings from S to Λ can be incredibly large, *this* is the space we are interested in[1].

To illustrate these abstract notions we consider some

Examples: 1. Let S be as at the beginning of this chapter. With $\Lambda := \{0,1\}$, a mapping $x : S \to \Lambda$ can be thought of as a description of the distribution of the disease D at a fixed moment: one only has to translate $x(s) = 1$ (resp. $= 0$) into "s has D" (resp. not).

2. Let $T_1, \ldots, T_r$ be the football teams (political parties, preferred restaurants, $\ldots$) of a town. We put $\Lambda = \{0, 1, \ldots, r\}$, and we describe the football preferences at a particular moment by a function $x : S \to \Lambda$; $x(s) = \rho$ of course means that T_ρ is the favourite team of s, with the interpretation "s has no favourite team" in case $x(s) = 0$.

---

[1] The $x : S \to \Lambda$ are sometimes called *configurations*.

3. Now we turn to pictures. With $S = \{1, \ldots, 256\} \times \{1, \ldots, 256\}$ and $\Lambda = \{0, 1\}$ a black-and-white picture in a 256×256-resolution corresponds to a map $x : S \to \Lambda$. It is easy to introduce colours by passing to a bigger Λ.

Note that even in the black-and-white case and even with this moderate number of pixels we have to deal with a space Λ^S containing $2^{256 \cdot 256} \approx 10^{19,660}$ elements.

This was the set theoretical part, *we now turn to probability considerations*. By a **random field** we mean a space Λ^S together with a probability measure $\mathbb{P}$; if it is necessary to emphasize the roles of S and Λ it is more precise to speak of *a random field on S with state space Λ*. As before S and Λ are finite, and it is usually assumed that $\mathbb{P}$ is strictly positive at every point; we will follow this convention.

By $X_s : \Lambda^S \to \Lambda$, $s \in S$, we will denote the evaluation maps $x \mapsto x(s)$. If $(\Lambda^S, \mathbb{P})$ is a random field, then the X_s are random variables, we are mainly interested in *stochastic dependencies* between them. With the $s \in S$ there are associated the conditional probabilities $\mathbb{P}(X_s \mid X_t, \ t \neq s)$. For our purposes it will be convenient to work with the numbers

$$\mathbb{P}(X_s = \lambda \mid X_t = \lambda_t \text{ for } t \in S, \ t \neq s), \tag{19.1}$$

where λ and the λ_t are arbitrary elements of Λ. (Or, equivalently, with the

$$\mathbb{P}(X_s = x(s) \mid X_t = x(t) \text{ for } t \in S, \ t \neq s),$$

for $x \in \Lambda^S$.)

> Suppose that you know at a given moment the state of health of all persons who are different from a fixed person s. What is the probability that s has the disease D?

These quantities can always be defined. In many cases, however, it is not necessary to have access to *all* X_t with $t \neq s$ in order to deal with the probabilities in (19.1) since already the t in a "small" subset of S contain the relevant information.

> If our person s lives alone with his family then it might be sufficient to know whether or not his wife or his children have D.

This leads to the following fundamental definition, the starting point of the investigations to come:

Definition 19.1 Let S be as above. By a *neighbourhood system* we mean a family $\mathcal{N} = (\mathcal{N}_s)_{s \in S}$ such that

    (i) for $s \in S$, $\mathcal{N}_s$ is a (possibly empty) subset of S which does *not* contain s.

    (ii) $t \in \mathcal{N}_s$ yields $s \in \mathcal{N}_t$ for all s, t.

Now let $(X_s)_s$ be a random field with state space Λ. It is called a *Markov random field with respect to the neighbourhood system $\mathcal{N}$* provided that

$$\mathbb{P}(X_s = \lambda \mid X_t = \lambda_t \text{ for } t \in S, \ t \neq s) = \mathbb{P}(X_s = \lambda \mid X_t = \lambda_t \text{ for } t \in \mathcal{N}_s)$$

for arbitrary $s \in S$, $\lambda, \lambda_t \in \Lambda$.

Some *remarks* are in order. First we emphasize that all conditional probabilities are defined since we are dealing with a strictly positive $\mathbb{P}$; note that, if $\mathcal{N}_s = \emptyset$, the conditional probability on the left-hand side is just $\mathbb{P}(X_s = \lambda)$. And we also want to stress that it is important always to have in mind the dependency of the Markov property on $\mathcal{N}$: if $\mathcal{N}_{\max}$ denotes the *maximal neighbourhood system* – where the neighbours of s are *all* t with $t \neq s$ – then *every* random field is Markov with respect to $\mathcal{N}_{\max}$. The other extreme is the case where all $\mathcal{N}_s$ are empty. Now "Markov random field" translates into "the $(X_s)_s$ are independent random variables" (cf. exercise 19.6).

How are the Markov processes which we have studied throughout *related to random fields?* To explain the connection we fix a (strictly positive) initial distribution $(p_i)_{i=1,\ldots,N}$ and a strictly positive $N \times N$-matrix P.

> Maybe you have noted that we are considering a chain on $\{1,\ldots,N\}$ and not on a general state space S as often before. The reason is that the letter S now has a different meaning than before. For most of the book "S" was used for the *states*, and we have investigated which states are occupied at time steps $k = 0, 1, \ldots$. In this chapter, however, "S" abbreviates *site*, the states are now the elements of Λ. Hopefully you are not too much confused by these different notational preferences of the Markov chain versus the random field community.

Let k_0 be a fixed integer: in order to arrive at a finite probability space we will have to restrict ourselves to a fixed number of steps. The *sites* of our random field are the numbers $k = 0, 1, \ldots, k_0$, with every site k we associate the random variable $X_k =$ "the position of the walk at step number k". So far this is not new, we have a random field with $S = \{0,\ldots,k_0\}$ and $\Lambda = \{1,\ldots,N\}$, and we also know how to calculate all probabilities in connection with this field: a particular element of Λ^S, that is a path $i_0, i_1, \ldots, i_{k_0}$, will be observed with probability

$$p_{i_0} p_{i_0 i_1} p_{i_1 i_2} \cdots p_{i_{k_0-1} i_{k_0}}; \tag{19.2}$$

this has been observed in (1.2) in chapter 1.

Now some care is needed. From the very beginning of this book until the previous chapter the Markov property of a process $X_0, X_1, \ldots$ was synonymous with the fact that, in order to predict X_k, the information $X_0 = i_0, \ldots, X_{k-1} = i_{k-1}$ is precisely as good as $X_{k-1} = i_{k-1}$.

With $(X_k)_{k=0,\ldots,k_0}$, considered as a random field, the situation is different, there we are faced with the following problem:

> For a k between 0 and k_0, what are the subsets $\Delta \subset \{0,\ldots,k-1,k+1,\ldots,k_0\}$ such that
>
> $$\mathbb{P}(X_k = i_k \mid X_l = i_l \text{ for } l \neq k) = \mathbb{P}(X_k = i_k \mid X_l = i_l \text{ for } l \in \Delta)$$
>
> for all $i_0, \ldots, i_{k_0}$; what is the smallest such Δ?

In view of the preceding remarks it is tempting to try $\Delta = \{k-1\}$. This, however, is not successful, since sometimes the knowledge of *all* X_l with $l \neq k$ is strictly better than that of X_{k-1} alone. (If, for example, an ordinary random walk on $\{0,\ldots,9\}$ is at position 5 at time $k-1 = 7$, it might be at 4 or 6 with equal probability at $k = 8$; the additional information, however, that $X_{k+1} = 3$ implies that $X_k = 4$ with probability one.) A better choice is to use both "neighbours" of k instead:

Lemma 19.2 *With the preceding notation we define $\mathcal{N}_k := \{k-1,\, k+1\}$ for $k = 1, \ldots,$ $k_0 - 1$, $\mathcal{N}_0 := \{1\}$, $\mathcal{N}_{k_0} := \{k_0 - 1\}$. With respect to this neighbourhood system the Markov chain $(X_k)_{k=0,\ldots,k_0}$ is a Markov random field.*

Proof. Fix k as well as states $i_0, \ldots, i_k$, we assume that k lies strictly between 0 and k_0 (the cases $k = 0$ and $k = k_0$ can be treated similarly). Consider the events

$$
\begin{aligned}
A &:= \{X_k = i_k\}, \\
B &:= \{X_0 = i_0,\, X_1 = i_1, \ldots,\, X_{k-1} = i_{k-1},\, X_{k+1} = i_{k+1}, \ldots,\, X_{k_0} = i_{k_0}\}, \\
C &:= \{X_{k-1} = i_{k-1},\, X_{k+1} = i_{k+1}\}.
\end{aligned}
$$

By (19.2), the respective probabilities are

$$
\begin{aligned}
\mathbb{P}(A \cap B) &= p_{i_0} p_{i_0 i_1} p_{i_1 i_2} \cdots p_{i_{k_0-1} i_{k_0}}, \\
\mathbb{P}(B) &= \sum_{i_k'} p_{i_0} p_{i_0 i_1} p_{i_1 i_2} \cdots p_{i_{k-1} i_k'} p_{i_k' i_{k+1}} \cdots p_{i_{k_0-1} i_{k_0}}, \\
\mathbb{P}(A \cap C) &= \sum_{i_0', i_1', \ldots, i_{k-2}'} p_{i_0'} p_{i_0' i_1'} p_{i_1' i_2'} \cdots p_{i_{k-2}' i_{k-1}} p_{i_{k-1} i_k} p_{i_k i_{k+1}}, \\
\mathbb{P}(C) &= \sum_{i_0', i_1', \ldots, i_{k-2}', i_k'} p_{i_0'} p_{i_0' i_1'} p_{i_1' i_2'} \cdots p_{i_{k-2}' i_{k-1}} p_{i_{k-1} i_k'} p_{i_k' i_{k+1}},
\end{aligned}
$$

and it follows that

$$
\begin{aligned}
\mathbb{P}(A \mid B) &= \frac{\mathbb{P}(A \cap B)}{\mathbb{P}(B)} \\
&= \frac{p_{i_{k-1} i_k} p_{i_k i_{k+1}}}{\sum_{i_k'} p_{i_{k-1} i_k'} p_{i_k' i_{k+1}}} \\
&= \frac{\mathbb{P}(A \cap C)}{\mathbb{P}(C)} \\
&= \mathbb{P}(A \mid C).
\end{aligned}
$$

This is just the Markov condition for the neighbourhoods under consideration. $\qquad\square$

The global versus the local approach

Now fix S, Λ, and a neighbourhood system $\mathcal{N} = (\mathcal{N}_s)_s$. There are *two possible approaches* to deal with the random fields which possibly have the Markov property with respect to $\mathcal{N}$. The *first one* is to prescribe a strictly positive measure $\mathbb{P}$ and then to check whether or not the Markov property is satisfied; this will be called the *global approach* here.

If we have defined a Markov random field with respect to $\mathcal{N}$, then surely the numbers $\mathbb{P}(X_s = \lambda \mid X_t = \lambda_t \text{ for } t \in \mathcal{N}_s)$ will play an important role, they are called the *family of local characteristics associated with* $\mathbb{P}$.

Now we turn to the *second approach*, the *local* one. Our starting point is introduced in the following

Definition 19.3 With S, Λ and $\mathcal{N}$ as before a *family of local characteristics* is a family $(\Pi(s;\lambda;y)_{s,\lambda,y})$ of strictly positive numbers; here s runs through S, λ is an arbitrary state and y denotes a map from $\mathcal{N}_s$ to Λ. We assume that

$$\sum_{\lambda \in \Lambda} \Pi(s;\lambda;y) = 1$$

for all s and all y.

The significance of these numbers is the following: $\Pi(s;\lambda;y)$ is a candidate of a conditional probability, namely of the probability that the site s is in state λ under the assumption that the neighbours are in states given by y. Note that it will depend on the size of the $\mathcal{N}_s$ whether there are many or few local characteristics.

It is clear that, given a Markov random field with respect to $\mathcal{N}$ by a measure $\mathbb{P}$ on Λ^S, the family of local characteristics associated with $\mathbb{P}$ is in fact a family of local characteristics. It is not obvious, however, whether such families always arise in this way.

But this is an extremely important problem since, in view of the examples we have introduced at the beginning of this chapter, it is surely very natural *to start with local probability assumptions.*

> For example, we do *not* know how large the probability of a particular distribution of the disease D is (and in most cases this is not even of interest). However, it is not too hard to invent models which specify how likely it is that a person s has D under the assumption that a certain percentage of the neighbours is infected. This observation applies similarly to the other examples: the preference of a football team, the orientation of a magnetic dipole and so on.

To phrase it more formally, we want to know:

Let a family $(\Pi(s;\lambda;y)_{s,\lambda,y})$ of local characteristics be given. Does there exist a probability $\mathbb{P}$ on Λ^S which gives rise to a Markov random field such that the $\Pi(s;\lambda;y)$ satisfy

$$\Pi(s;\lambda;y) = \mathbb{P}(X_s = \lambda \mid X_t = y(t) \text{ for } t \in \mathcal{N}_s)$$

for all s, λ, y?

If $\mathbb{P}$ does exist, is it uniquely determined?

In order to investigate this problem we will work with *Markov chains on* Λ^S. To understand the underlying idea we resume once more the very first example. How will the disease D be distributed?

We can think of a mechanism which works as follows. A person s is selected at random, and then his or her neighbours are inspected whether or not they have D; let the result of this inspection be a function $y : \mathcal{N}_s \to \Lambda = \{0,1\}$. Now the family of local characteristics is consulted, of interest are the numbers $\Pi(s;\lambda;y)$, $\lambda \in \Lambda$, for *this* site s and *this* function y. They sum up to one, and therefore we can regard them as probabilities according to which we select $\lambda \in \Lambda$. *This* is the new state of s. The procedure can be applied "very often", every second, say. Each single step is some kind of update of the distribution of D, and it is natural to assume that "in the limit" our model reflects somehow the real situation.

After this heuristic consideration we are going to be more rigorous. Let S, Λ, $\mathcal{N}$ and $(\Pi(s; \lambda; y)_{s,\lambda,y}$ be given. We fix a strictly positive probability $q = (q_s)_s$ on S, and we define a Markov chain by the following rules:

- The state space of our chain is $\tilde{S} := \Lambda^S$; one must not confuse the elements of $\tilde{S}$ with those of Λ which also are called states.

- The chain starts deterministically at a fixed $x_0 \in \tilde{S}$. This configuration can be defined arbitrarily, for example as a constant function.

- Let $x \in \tilde{S}$ be the actual state of the walk. To find the next position, first select an $s \in S$ in accordance with the distribution q. Then put $y :=$ "the restriction of x to $\mathcal{N}_s$" and choose λ by using the distribution $\Pi(s; \lambda; y)$, $\lambda \in \Lambda$; the q_s- and the λ-choice are assumed to be independent. Then the new position will be the state $z \in \tilde{S}$, where z is defined by

$$z(t) := \begin{cases} x(t) & : \quad \text{if } t \neq s \\ \lambda & : \quad \text{if } t = s. \end{cases}$$

Theorem 19.4 *The previously defined chain has the following properties:*
  (i) *It is irreducible and aperiodic so that there is a unique equilibrium distribution $\pi = (\pi_x)_{x \in \Lambda^S}$ on Λ^S.*
  (ii) *For $s \in S$ and $x \in \Lambda^S$ let $\pi(x; S\backslash\{s\})$ be the sum over all π_z, where z runs through the elements of Λ^S which coincide with x on $S \setminus \{s\}$. Then*

$$\pi_x = \sum_s q_s \Pi(s; x(s); y) \pi(x; S\backslash\{s\}),$$

  with $y =$ the restriction of x to $\mathcal{N}_s$.
  (iii) *Suppose that the local characteristics are such that the equilibrium π is independent of (q_s). Then Λ^S, provided with the measure associated with π, is a Markov random field with respect to $\mathcal{N}$ for which the local characteristics are just the $\Pi(s; \lambda; y)$.*
  Also, the chain is reversible.

Proof. (i) This follows from the assumption that all q_s and all $\Pi(s; \lambda; y)$ are strictly positive: with some luck one can pass from any x to any z in $\mathrm{card}(S)$ many steps, and there is also a positive probability to pause.
(ii) The assertion is a reformulation of the equilibrium condition.
(iii) Denote by $\mathbb{P}_\pi$ the measure associated with π, i.e., $\mathbb{P}_\pi(A) := \sum_{x \in A} \pi_x$. (Of course one could identify π with $\mathbb{P}_\pi$, but we have introduced π as a vector. Also, the measure notation is more convenient for our purposes.) The claim is that

$$\mathbb{P}_\pi(X_s = x(s) \mid X_t = x(t) \text{ for } t \neq s) \quad = \quad \mathbb{P}_\pi(X_s = x(s) \mid X_t = x(t) \text{ for } t \in \mathcal{N}_s)$$
$$= \quad \Pi(s; x(s); x|_{\mathcal{N}_s})$$

for all $x \in \Lambda^S$ and all s.
  We start our investigations with the observation that

$$\pi_x = \Pi(s; x(s); x|_{\mathcal{N}_s}) \, \pi(x; S\backslash\{s\}) \tag{19.3}$$

for all x and all s; this equation follows immediately from the fact that by assumption the equation in (ii) holds for *all* choices of the (q_s).

Now let us fix x and s. If we divide (19.3) by $\pi(x; S\backslash\{s\})$, we get

$$\mathbb{P}_\pi(X_s = x(s) \mid X(t) = x(t) \text{ for } t \neq s) = \Pi(s; x(s); x|_{\mathcal{N}_s}).$$

To prove the second half of the claim we introduce the following notation: for $z \in \Lambda^S$ the function z' will be defined by

$$z'(t) := \left\{ \begin{array}{lll} z(t) & : & \text{if } t \neq s \\ x(s) & : & \text{if } t = s. \end{array} \right.$$

Let z be arbitrary such that z coincides with x on $\mathcal{N}_s$. From (19.3) we conclude that

$$\begin{array}{rcl} \pi_z & = & \Pi(s; z(s); z|_{\mathcal{N}_s})\pi(z; S\backslash\{s\}) \\ & = & \Pi(s; z(s); x|_{\mathcal{N}_s})\pi(z; S\backslash\{s\}) \\ & = & \Pi(s; z(s); x|_{\mathcal{N}_s})\pi(z'; S\backslash\{s\}), \end{array}$$

and summation over these probabilities for all z leads us to

$$\begin{array}{rcl} \mathbb{P}_\pi(X(t) = x(t) \text{ for } t \in \mathcal{N}_s) & = & \displaystyle\sum_{z|_{\mathcal{N}_s} = x|_{\mathcal{N}_s}} \pi_z \\[2em] & = & \displaystyle\sum_{\lambda \in \Lambda} \sum_{z|_{\mathcal{N}_s} = x|_{\mathcal{N}_s}, z(s) = \lambda} \pi(z) \\[2em] & = & \displaystyle\sum_{\lambda \in \Lambda} \sum_{z|_{\mathcal{N}_s} = x|_{\mathcal{N}_s}, z(s) = \lambda} \Pi(s; \lambda; x|_{\mathcal{N}_s})\pi(z'; S\backslash\{s\}) \\[2em] & = & \displaystyle\sum_{\lambda \in \Lambda} \Pi(s; \lambda; x|_{\mathcal{N}_s}) \sum_{z|_{\mathcal{N}_s} = x|_{\mathcal{N}_s}, z(s) = \lambda} \pi(z'; S\backslash\{s\}) \\[2em] & = & \displaystyle\sum_{\lambda \in \Lambda} \Pi(s; \lambda; x|_{\mathcal{N}_s}) \sum_{z|_{\mathcal{N}_s \cup \{s\}} = x|_{\mathcal{N}_s \cup \{s\}}} \pi(z; S\backslash\{s\}) \\[2em] & = & \displaystyle\sum_{z|_{\mathcal{N}_s \cup \{s\}} = x|_{\mathcal{N}_s \cup \{s\}}} \pi(z; S\backslash\{s\}) \sum_{\lambda \in \Lambda} \Pi(s; \lambda; x|_{\mathcal{N}_s}) \\[2em] & = & \displaystyle\sum_{z|_{\mathcal{N}_s \cup \{s\}} = x|_{\mathcal{N}_s \cup \{s\}}} \pi(z; S\backslash\{s\}); \end{array}$$

in the last step the normalization from definition 19.3 has come into play.

Also we have, by (19.3),

$$\begin{array}{rcl} \mathbb{P}_\pi(X(t) = x(t) \text{ for } t \in \mathcal{N}_s \cup \{s\}) & = & \displaystyle\sum_{z|_{\mathcal{N}_s \cup \{s\}} = x|_{\mathcal{N}_s \cup \{s\}}} \pi_z \\[2em] & = & \displaystyle\sum_{z|_{\mathcal{N}_s \cup \{s\}} = x|_{\mathcal{N}_s \cup \{s\}}} \Pi(s; x(s); x|_{\mathcal{N}_s})\pi(z; S\backslash\{s\}), \end{array}$$

and together with the preceding calculations this implies

$$\mathbb{P}_\pi(X_s = x(s) \mid X(t) = x(t) \text{ for } t \in \mathcal{N}_s) = \Pi(s; x(s); x|_{\mathcal{N}_s})$$

as claimed.

It remains to verify that the chain is reversible, here once more equation (19.3) is helpful. Let x, z be arbitrary elements of Λ^S such that a transition from x to z or vice versa is possible. We may assume that $x \neq z$, and hence there is a unique s such that x and z coincide on $S \setminus \{s\}$. By the definition of the chain the probability p_{xz} (resp. p_{zx}) for a jump from x to z (resp. from z to x) is $q_s\Pi(s; z(s); x|_{\mathcal{N}_s})$ (resp. $q_s\Pi(s; x(s); x|_{\mathcal{N}_s})$). Thus, by (19.3) and since $\pi(x; S\setminus\{s\}) = \pi(z; S\setminus\{s\})$, it follows that

$$
\begin{aligned}
\pi_x p_{xz} &= \Pi(s; x(s); x|_{\mathcal{N}_s})\pi(x; S\setminus\{s\})q_s\Pi(s; z(s); x|_{\mathcal{N}_s}) \\
&= \Pi(s; z(s); z|_{\mathcal{N}_s})\pi(z; S\setminus\{s\})q_s\Pi(s; x(s); x|_{\mathcal{N}_s}) \\
&= \pi_z p_{zx}.
\end{aligned}
$$

This completes the proof. $\qquad\qquad\qquad\qquad\qquad\qquad\qquad\qquad\qquad\qquad\qquad\square$

The space Λ^S, provided with the the equilibrium π, will in general *not* be a Markov random field with respect to $\mathcal{N}$, condition (ii) of the theorem is much weaker than the Markov property. Let's analyse an

Example: Consider $S = \{a, b, c\}$ and $\Lambda = \{0, 1\}$. The elements of Λ^S will be denoted by 000, 001, 010, 011, ..., 111, the element 101, e.g., is the element which maps a to 1, b to 0 and c to 1. In the description of the transition matrix which we will give shortly they will occur in exactly this order (000 $\in \Lambda^S$ corresponds to state 1 and so on).

A neighbourhood system $\mathcal{N}$ is defined by $\mathcal{N}_a := \{b\}$, $\mathcal{N}_b := \{a, c\}$, and $\mathcal{N}_c := \{b\}$. With $\varepsilon = $ "a small positive number" we define the local characteristics as follows:

- For a, we have to prescribe the $\Pi(a; \lambda; y)$ for $\lambda \in \Lambda$ and $y : \mathcal{N}_a \to \Lambda$. We can identify y with 0 or 1, here is the definition:

$$\Pi(a; 0; 0) := \Pi(a; 1; 0) := \Pi(a; 0; 1) := \Pi(a; 1; 1) := 1/2;$$

  this means that – regardless of the state of b –, the state of a will be set to 0 or 1 with equal probability.

- The neighbourhood of b has the two elements a and c. If the function y is 0 on both a and c, then we put

$$\Pi(b; 1, y) := 1 - \varepsilon, \ \Pi(b; 0, y) := \varepsilon.$$

For the other three possible y the definition is

$$\Pi(b; 1, y) := \varepsilon, \ \Pi(b; 0, y) := 1 - \varepsilon.$$

- As in the case of a also for c the function y (= the state of b) is 0 or 1. We set

$$\Pi(c; 0; 0) := 1 - \varepsilon, \ \Pi(c; 1; 0) := \varepsilon, \ \Pi(c; 0; 1) := \varepsilon, \ \Pi(c; 1; 1) := 1 - \varepsilon.$$

These probabilities are designed such that:

- For $x := 000$ and $z := 100$ one has $p_{xz} = p_{zx}$.
- By the definition of the Π's one tries to let the chain run in such a way that it is more often in z than in position x; this is done by the choice of the local characteristics at b and c.

- If one really succeeds with this idea then $\pi_x < \pi_z$ will hold. Therefore the chain is *not* reversible and thus Λ^S cannot be a Markov random field by the next theorem.

The number ε is only included to meet the condition of strictly positive $\Pi(s; \lambda; y)$.

It remains to fix the (q_s), we choose the uniform distribution. Then it is easy to calculate the transition matrix, for the special case $\varepsilon = 1/100$ it has the form

$$P = \frac{1}{300} \begin{pmatrix} 150 & 1 & 99 & 0 & 50 & 0 & 0 & 0 \\ 99 & 52 & 0 & 99 & 0 & 50 & 0 & 0 \\ 1 & 0 & 150 & 99 & 0 & 0 & 50 & 0 \\ 0 & 99 & 1 & 150 & 0 & 0 & 0 & 50 \\ 50 & 0 & 0 & 0 & 248 & 1 & 1 & 0 \\ 0 & 50 & 0 & 0 & 99 & 150 & 0 & 1 \\ 0 & 0 & 50 & 0 & 99 & 0 & 52 & 99 \\ 0 & 0 & 0 & 50 & 0 & 99 & 1 & 150 \end{pmatrix}.$$

Why, for example, does the transition $010 \to 010$ have the probability $1/2$?

With probability $1/3$ the state of a is (possibly) changed. The (conditional) probability that a keeps its state is $1/2$, and thus this possibility contributes with $1/6$. If, however, the state of b is concerned (this also will happen with probability $1/3$), then one has to look at the states of a and c: both are 0, and thus with (conditional) probability $99/100$ the state 1 survives at b; therefore the second contribution for the transition $010 \to 010$ is $99/300$. Finally, if the (q_s)-sampler chooses c, the chance that state 0 is chosen again at c is only $1/100$, that is we have to add $1/300$ to the already determined $1/6 + 99/300$. In this way $1/2$ has been obtained, and similarly all other transition probabilities can be derived.

With the help of a computer the equilibrium π is easily calculated as the positive normalized solution of $\pi^\top P = \pi^\top$, here is the result:

$$\pi^\top = (0.162,\ 0.074,\ 0.117,\ 0.148,\ 0.338,\ 0.070,\ 0.025,\ 0.066).$$

In particular, we see that really $\pi_{000} = 0.162 < 0.338 = \pi_{100}$ so that the chain is *not* reversible. Consequently, by the following theorem, (Λ^S, π) is not a Markov random field.

This, of course, can also be calculated directly. We have, e.g.,

$$\mathbb{P}_\pi(X_a = 0 \mid X_b = X_c = 0) = \frac{\pi_{000}}{\pi_{000} + \pi_{100}} = \frac{0.162}{0.162 + 0.338} = 0.324,$$

this is the conditional probability under *complete information* on the complement of a. If, however, the states are only known on $\mathcal{N}_a$ we have to calculate

$$P_\pi(X_a = 0 \mid X_b = 0) = \frac{\pi_{000} + \pi_{001}}{\pi_{000} + \pi_{100} + \pi_{001} + \pi_{101}} = 0.366.$$

What happens if we consider a chain on Λ^S which is defined not by arbitrary local characteristics but rather by those associated with a Markov random field? The following assertion is not too surprising:

Theorem 19.5 *With S, Λ and $\mathcal{N}$ as above let $\mathbb{P}$ be a measure on Λ^S which gives rise to a Markov random field.*
Denote by $\Pi_{\mathbb{P}}(s; \lambda; y)$ the associated local characteristics, that is

$$\Pi_{\mathbb{P}}(s; \lambda; y) := \mathbb{P}(X_s = \lambda \mid X_t = y(t) \text{ for } t \in \mathcal{N}_s).$$

If the Markov chain which we have defined preceding theorem 19.4 is run with arbitrary $q_s > 0$ and with the $\Pi(s; \lambda; y) := \Pi_{\mathbb{P}}(s; \lambda; y)$, then the equilibrium of this chain coincides with $\mathbb{P}$. Also, the chain is reversible.

Proof. Rather than to compare $\mathbb{P}$ and the unique equilibrium directly we prefer to start with the detailed balance condition. If x and z are different elements of Λ^S such that transitions are possible – so that they are different at a unique s –, then the probability to come from x to z is $q_s\Pi_{\mathbb{P}}(s; z(s); x|_{\mathcal{N}_s})$. Also, in view of the Markov condition, $\mathbb{P}(\{x\})$ can be replaced by

$$\Pi_{\mathbb{P}}(s; x(s); x|_{\mathcal{N}_s}) \cdot \mathbb{P}(X_t = x(t) \text{ for } t \neq s).$$

Consequently the product "probability of x" times "probability for a jump $x \to z$" is the same as "probability of z" times "probability for a jump $z \to x$": both numbers equal

$$q_s\Pi_{\mathbb{P}}(s; x(s); x|_{\mathcal{N}_s})\Pi_{\mathbb{P}}(s; z(s); x|_{\mathcal{N}_s})\mathbb{P}(X_t = x(t) \text{ for } t \neq s).$$

We have already remarked in chapter 10 (see (10.2)) that then $\pi = (\mathbb{P}\{x\})_{x \in \Lambda^S}$ must be the equilibrium. The remaining assertions now can be read from theorem 19.4. $\square$

Corollary 19.6 *If $(\Pi(s; \lambda; y)_{s,\lambda,y})$ is a family of local characteristics, then there is at most one measure $\mathbb{P}$ on Λ^S such that*

    (i) *Λ^S, together with $\mathbb{P}$, is a Markov random field with respect to $\mathcal{N}$;*
    (ii) *the local characteristics associated with $\mathbb{P}$ are just the $\Pi(s; \lambda; y)$.*

Proof. This follows at once from the preceding theorem and the uniqueness of the equilibrium distribution.

$\square$

Corollary 19.7 *Under the assumptions of theorem 19.5 one can produce samples from Λ^S with probabilities given by $\mathbb{P}$ up to arbitrary precision; it is only necessary to run the above chain on Λ^S for "sufficiently many" steps and to use the position obtained in this way as an output.*

Proof. This follows from theorem 19.4(i), theorem 7.4 and theorem 19.5. (In order to apply this corollary a more detailed analysis of the mixing rate will be necessary; see, e.g., the discussion of the Gibbs sampler at the end of the next chapter.) $\square$

As an *illustration* we resume the Markov random field induced by a Markov chain which we have studied in lemma 19.2. In the proof of this lemma we have already calculated the local characteristics: if k is in S and the states of $k-1$ and $k+1$ are i_{k-1} and i_{k+1}, respectively, then k will be in state i_k with probability

$$\frac{p_{i_{k-1}i_k}p_{i_ki_{k+1}}}{\sum_{i'_k} p_{i_{k-1}i'_k}p_{i'_ki_{k+1}}}. \tag{19.4}$$

This provides a *second possibility to simulate ordinary Markov chains*. If a sample of the chain is needed for the time steps $k = 0, 1, \ldots, k_0$, start with an arbitrary sequence $i_0, i_1, \ldots, i_{k_0}$. Then update this element of Λ^S "very often" by choosing a k at random and changing the state i at k in accordance with the probabilities in (19.4). It is plain that this is much less effective than the usual procedure where one chooses i_0 according to (p_i) then i_1 by using the i_0'th row of the transition matrix and so on. This sampling method is extremely faster and provides the output even with the exact probabilities.

Exercises

19.1: Let $(\Lambda^S, \mathbb{P})$ be a Markov random field and $s \in S$.

a) Prove that there is a minimal subset Δ of $S \setminus \{s\}$ such that

$$\mathbb{P}(X_s = x(s) \mid X_t = x(t) \text{ for } t \in \Delta) = \mathbb{P}(X_s = x(s) \mid X_t = x(t) \text{ for } t \in S, \ t \neq s).$$

b) Give an example to show that in general Δ is not unique.

c) It can happen that the minimal set Δ coincides with $S \setminus \{s\}$ for all s.

19.2: Let (S, d) be a finite metric space.

a) Fix $R > 0$ and put $\mathcal{N}_s^{d,R} = \{t \mid 0 < d(s,t) \leq R\}$ for $s \in S$. Prove that $\mathcal{N}^{d,R} = (\mathcal{N}_s^{d,R})_{s \in S}$ defines a neighbourhood system.

b) Let $\mathcal{N}$ be an arbitrary neighbourhood system on a finite set S. Prove that there is a metric d on S such that, for a suitable R, $\mathcal{N}$ is of the form $\mathcal{N}^{d,R}$. In particular the minimal and the maximal neighbourhood systems $\mathcal{N}_{\min}$ and $\mathcal{N}_{\max}$ can be represented in this way.

c) Let $R_s > 0$, $s \in S$, be arbitrary numbers, we put

$$\mathcal{N}_s := \{t \mid 0 < d(s,t) \leq R_s\}.$$

Is $(\mathcal{N}_s)_s$ a neighbourhood system?

19.3: Consider an ordinary cyclic random walk on $\{0, \ldots, 9\}$ which starts deterministically at 5, we observe this walk at "times" $k = 0, \ldots, 100$. Now we regard this random walk as a Markov random field. What are the local characteristics at $k = 15$?

19.4: Similarly to the case of ordinary Markov chains one can treat processes with a short memory. To be specific, we consider the stochastic process of example 6 in chapter 2, we observe this process for the steps $k = 0, \ldots, k_0$. As in the present chapter this gives rise to a random field: there are two states (a and b), and the set of sites is $\{0, \ldots, k_0\}$. Find natural candidates for neighbourhoods in order to have a Markov random field and determine the local characteristics.

19.5: Neighbourhood systems have been introduced in definition 19.1. By the second condition they have a certain kind of symmetry: $t \in \mathcal{N}_s$ is equivalent with $s \in \mathcal{N}_t$. In which of the arguments of the present chapter was this property of importance?

19.6: Prove that a random field $(\Lambda^S, \mathbb{P})$ is Markov with respect to the minimal neighbourhood systems (= empty neighbourhoods) iff $(X_s)_{s \in S}$ is a family of independent random variables.

19.7: Provide – with arbitrary finite sets S and Λ – the set Λ^S with the uniform distribution. Determine all neighbourhood systems such that this random field is Markov.

19.8: Let $\mathcal{N}$ be a neighbourhood system for S. Is it possible to find a probability $\mathbb{P}$ on Λ^S such that the random field $(\Lambda^S, \mathbb{P})$ is Markov with respect to $\mathcal{N}$?

19.9: Let a Markov random field $(\Lambda^S, \mathbb{P})$ be given (Markov with respect to $\mathcal{N}$). Now let $\widehat{\mathcal{N}}$ be a second neighbourhood system such that $\mathcal{N}_s \subset \widehat{\mathcal{N}_s}$ for every s. Is $(\Lambda^S, \mathbb{P})$ also Markov with respect to $\widehat{\mathcal{N}}$?

19.10: Let a family of local characteristics be given such that $\Pi(s; \lambda; y) = r_\lambda$ for all s, λ and y and suitable $r_\lambda > 0$ with $\sum_\lambda r_\lambda = 1$. Prove that the equilibrium of the chain of theorem 19.4 gives rise to a Markov random field on Λ^S.

19.11: In the example preceding theorem 19.5, calculate the numbers

$$\mathbb{P}_\pi(X_a = 0 \mid X_b = X_c) \text{ and } \mathbb{P}_\pi(X_c = 1 \mid X_a = 1).$$

19.12: In lemma 19.2 we have considered a Λ-valued Markov process as a random field, and it turned out that this field is Markov with respect to the neighbourhood system $\mathcal{N}$ defined in this lemma. Prove the following more general assertion: if an *arbitrary* Λ-valued stochastic process $X_0, X_1, \ldots$ is considered as a random field, then the Markov property of this field with respect to $\mathcal{N}$ is equivalent with the Markov property of the process (X_k).

20 Potentials, Gibbs fields, and the Ising model

Let the set of sites S, the state space Λ and a neighbourhood system $\mathcal{N}$ be given as in the preceding chapter[1]. Sometimes a probability measure on Λ^S is given in closed form by a *potential*, this will lead us to the *Gibbs fields*. We will show that such fields are Markov random fields, the celebrated *Ising model* will serve as a simple example. In the final section we describe the *Gibbs sampler*, a method by which one can produce samples from Gibbs fields with (approximately) correct probabilities.

The energy function and potentials

We begin with an elementary observation: if $\mathbb{P}$ is any strictly positive probability measure on an arbitrary finite set $\tilde{S}$, then there is a real-valued function $\mathcal{H}$ such that

$$\mathbb{P}(\{x\}) = e^{-\mathcal{H}(x)}$$

holds for every x. Conversely, if $\mathcal{H} : \tilde{S} \to \mathbb{R}$ is arbitrary, then

$$\mathbb{P}_{\mathcal{H}}(\{x\}) := \frac{e^{-\mathcal{H}(x)}}{Z} \qquad (20.1)$$

will be a strictly positive probability on $\tilde{S}$ if we define Z by

$$Z := \sum_{y \in \tilde{S}} e^{-\mathcal{H}(y)}.$$

Since $\mathcal{H}$ and Z have their origin in statistical physics they are usually called the *energy function* and the *partition function*, respectively (see [51] or [66] for the physical background). Note that two energy functions $\mathcal{H}$ and $\mathcal{H}'$ induce the same $\mathbb{P}$ iff $\mathcal{H} - \mathcal{H}'$ is constant.

In the applications we have in mind the probabilities $\mathbb{P}_{\mathcal{H}}$ are usually introduced by (20.1) with a more or less easy-to-calculate energy function $\mathcal{H}$. At first glance this seems to be as good as to work with an explicitly defined $\mathbb{P}_{\mathcal{H}}$, but this is far from being true. The reason is that in most cases the set $\tilde{S}$ is that huge that it is hopeless to calculate Z. This has *a remarkable consequence:*

If a probability $\mathbb{P}$ is defined by an energy function, then the $\mathbb{P}(\{x\})$ are in many cases practically unknown. Easy to determine, however, is usually the ratio $\mathbb{P}(\{x\})/\mathbb{P}(\{z\})$ – which is just $e^{-\mathcal{H}(x)+\mathcal{H}(z)}$ – for arbitrary x, z. Similarly simple is the calculation of conditional probabilities $\mathbb{P}(A \mid B)$ for "small" sets A and B.

---

[1] The definitions which are not explained in the present chapter can also be found there.

One should have this in mind when discussing whether definitions or methods are merely of theoretical interest or useful for real applications.

Now we turn to the case $\tilde{S} = \Lambda^S$. The energy functions we are going to study will be defined such that for the calculation of $\mathcal{H}(x)$ the neighbourhood system $\mathcal{N}$ plays a crucial role[2]. We have already seen by the example in the preceding chapter that some care is necessary: for the Markov condition to hold it is *not* sufficient to work with definitions which solely use properties of neighbourhoods. Something more is needed, here it is some kind of *symmetry* which is implicitly introduced in the following

Definition 20.1 Let S, Λ and $\mathcal{N}$ be as above.

   (i) A nonempty subset C of S is called a *clique* if it is a singleton or if – for different $s, t \in C$ – one has $s \in \mathcal{N}_t$ and $t \in \mathcal{N}_s$.
       The collection of all cliques will be denoted by $\mathcal{C}$; note that the definition depends on $\mathcal{N}$ so that it would be more precise to write $\mathcal{C}_\mathcal{N}$.

  (ii) By a *Gibbs potential* we mean a family $V = (V_C)_{C \in \mathcal{C}}$, where each V_C is a map $V_C : \Lambda^C \to \mathbb{R}$.

 (iii) Let $V = (V_C)_{C \in \mathcal{C}}$ be a Gibbs potential, the induced energy function $\mathcal{H}_V$ is defined by

$$\mathcal{H}_V(x) := \sum_C V_C(x|_C).$$

 (iv) Let the measure $\mathbb{P}_V$ be induced by $\mathcal{H}_V$ as in (20.1), it is called the *Gibbs measure*. Λ^S, together with $\mathbb{P}_V$, is the *Gibbs field* associated with V.

To illustrate these notions let us consider some *examples*.

1. Let $\mathcal{N}^{\max}$ be the system of maximal neighbourhoods: $\mathcal{N}_s := S \setminus \{s\}$ for every s. Then every nonempty subset of S is a clique. The other extreme case, $\mathcal{N}^{\min}$, occurs when all $\mathcal{N}_s$ are empty: now only the singletons are cliques.

2. In the Markov chain example which has been introduced in lemma 19.2 the cliques are precisely the one-point sets and the $\{k, k+1\}$, where $k = 0, \ldots, k_0 - 1$.

   This is the special case $r = 1$ of the more general situation where S is the set $\{0, \ldots, k_0\}^m$ and $\mathcal{N}_s$ consists of those t for which the l^1-distance[3] to s is precisely one: here the cliques which are not singletons are sets of the form

$$\{(i_1, \ldots, i_{\rho-1}, k, i_{\rho+1}, \ldots, i_m), (i_1, \ldots, i_{\rho-1}, k+1, i_{\rho+1}, \ldots, i_m)\},$$

with $\rho = 1, \ldots, m$, $i_1, \ldots, i_{\rho-1}, i_{\rho+1}, \ldots, i_m \in \{0, \ldots, k_0\}$, and $k = 0, \ldots, k_0 - 1$.

3. More generally, one can start with an S together with *any* metric: for $s \in S$, the neighbourhood of s is defined to be the set of all t for which the distance to s lies in $]0, R]$, where R is a fixed positive number. It is illustrative to identify the cliques for various metrics, the reader is invited to check the case of the euclidean and the maximum metric.

4. Let $\mathcal{N}^{\max}$ be the maximal neighbourhood system from example 1. Then S is a clique, and therefore *every* function $\mathcal{H} : \Lambda^S \to \mathbb{R}$ is of the form $\mathcal{H}_V$: simply define $V_S := \mathcal{H}$ and let all other V_C vanish.

---

[2] Of course this is to be expected if one wants to arrive at a Markov random field with respect to $\mathcal{N}$.
[3] Cf. page 86.

This implies that every measure on Λ^S can be considered as a Gibbs measure for a system of suitably large neighbourhoods.

5. Now suppose that all neighbourhoods are empty. A potential for this situation can be identified with a family of mappings $V_s : \Lambda \to \mathbb{R}$, $s \in S$, and the energy function then has the particularly simple form

$$\mathcal{H}_V(x) = \sum_s V_s(x(s)).$$

It follows rather easily that the evaluation maps from Λ^S to Λ are independent random variables with respect to the measure $\mathbb{P}_V$ which is induced by $\mathcal{H}_V$; also Λ^S is a Markov random field for $\mathcal{N}^{\min}$.

Note that, conversely, every Markov random field relative to $\mathcal{N}^{\min}$ is a Gibbs field, an appropriate definition of the potential functions is $V_s(\lambda) := -\log \mathbb{P}(X_s = \lambda)$ for $\lambda \in \Lambda$ and $s \in S$.

6. In example 2 above we have identified the cliques of the random field associated with an ordinary Markov chain. Is it a Gibbs field?

One has to solve the following problem: an $x = (i_0, \ldots, i_{k_0}) \in \Lambda^S$ has the probability $\mathbb{P}(\{x\}) = p_{i_0} p_{i_0 i_1} \cdots p_{i_{k_0-1} i_{k_0}}$, and one must find functions

$$V_k : \Lambda \to \mathbb{R} \quad \text{and} \quad V_{k,k+1} : \Lambda \times \Lambda \to \mathbb{R}$$

such that $\mathbb{P}(\{x\})$ is – possibly up to a constant – the number

$$\exp\Big(-\sum_{k=0}^{k_0} V_k(i_k) - \sum_{k=0}^{k_0-1} V_{k,k+1}(i_k, i_{k+1})\Big).$$

A moment's reflection shows that this is achieved by the following definition:

- V_k vanishes for $k = 1, \ldots, k_0$, and $V_0(i) := -\log p_i$;

- $V_{k,k+1}(i,j) := -\log p_{ij}$ for $k = 0, \ldots, k_0 - 1$.

Gibbs fields are Markov random fields

Let S, Λ and $\mathcal{N}$ be as before, and again $\mathcal{C}$ will denote the collection of cliques. We fix a Gibbs potential $V = (V_C)_C$, and for the sake of *notational convenience* we agree to write

- $\mathcal{H}$ resp. $\mathbb{P}$ instead of $\mathcal{H}_V$ resp. $\mathbb{P}_V$, and

- $V_C(x)$ instead of the more correct $V_C(x|_C)$.

We then claim:

Theorem 20.2 *The Gibbs field induced by V is a Markov random field, and the associated local characteristics have the form*

$$\mathbb{P}(X_s = x(s) \mid X_t = x(t) \text{ for } t \in \mathcal{N}_s) = \frac{\exp(-\sum_{C, s \in C} V_C(x))}{\sum_{\lambda \in \Lambda} \exp(-\sum_{C, s \in C} V_C(x^{s;\lambda}))}$$

for $s \in S$ and $x \in \Lambda^S$; here $x^{s;\lambda}$ stands for that function from S to Λ which has the value λ at s and coincides with x at the other points of S.

Proof. The only difficulty of the proof is to avoid notational confusion.

Let s and x be given, we have to show that both $\mathbb{P}(X_s = x(s) \mid X_t = x(t)$ for $t \in \mathcal{N}_s)$ and $\mathbb{P}(X_s = x(s) \mid X_t = x(t)$ for $t \in S,\ t \neq s)$ coincide with

$$\exp\left(- \sum_{C,\, s \in C} V_C(x)\right) \bigg/ \sum_{\lambda \in \Lambda} \exp\left(- \sum_{C,\, s \in C} V_C(x^{s;\lambda})\right).$$

s and x will be fixed from now on.

It will be convenient to split the energy into two sums, where the first one measures the contribution of the V_C with $s \in C$; this term is usually called the *local energy at the site s*.

More precisely, we define

$$\mathcal{H}_s(z) := \sum_{C \in \mathcal{C},\, s \in C} V_C(z), \quad \mathcal{H}_s^*(z) := \sum_{C \in \mathcal{C},\, s \notin C} V_C(z)$$

for $z \in \Lambda^S$.

Trivially $\mathcal{H}(z) = \mathcal{H}_s(z) + \mathcal{H}_s^*(z)$ holds, we will need some further facts. The first is the identity

$$\mathcal{H}_s^*(z) = \mathcal{H}_s^*(x) \tag{20.2}$$

for the z which are identical with x at all t, $t \neq s$ (obvious); further,

$$\mathcal{H}_s(z) = \mathcal{H}_s(x) \tag{20.3}$$

holds for all z which coincide with x on $\mathcal{N}_s \cup \{s\}$ (this is true since – by the definition of "clique" – every $C \in \mathcal{C}$ such that $s \in C$ is a subset of $\mathcal{N}_s \cup \{s\}$); and finally, (20.3) has a variant which will also be important:

$$\mathcal{H}_s(z^{s;\lambda}) = \mathcal{H}_s(x^{s;\lambda}), \tag{20.4}$$

for all z such that $x(t) = z(t)$ for $t \in \mathcal{N}_s$.

Now the calculations are straightforward (Z will denote the partition function):

$$\begin{aligned}
Z\mathbb{P}(\{x\}) &= e^{-\mathcal{H}_s(x) - \mathcal{H}_s^*(x)} \\
&= e^{-\mathcal{H}_s(x)}\, e^{-\mathcal{H}_s^*(x)},
\end{aligned}$$

and also, with the help of (20.2),

$$\begin{aligned}
Z\mathbb{P}(\{z \mid z(t) = x(t) \text{ for } t \neq s\}) &= Z \sum_{\lambda \in \Lambda} \mathbb{P}(x^{s;\lambda}) \\
&= \sum_{\lambda} e^{-\mathcal{H}_s(x^{s;\lambda}) - \mathcal{H}_s^*(x^{s;\lambda})} \\
&= \sum_{\lambda} e^{-\mathcal{H}_s(x^{s;\lambda}) - \mathcal{H}_s^*(x)} \\
&= e^{-\mathcal{H}_s^*(x)} \sum_{\lambda} e^{-\mathcal{H}_s(x^{s;\lambda})}.
\end{aligned}$$

This proves that

$$\mathbb{P}(X_s = x(s) \mid X_t = x(t) \text{ for } t \in S,\, t \neq s) = \frac{e^{-\mathcal{H}_s(x)}}{\sum_\lambda e^{-\mathcal{H}_s(x^{s,\lambda})}},$$

the first half of the assertion.

Similarly we calculate, with (20.3),

$$
\begin{aligned}
Z\mathbb{P}(\{z \mid z(t) = x(t) \text{ for } t \in \mathcal{N}_s \cup \{s\}\}) &= Z \sum_{x|_{\mathcal{N}_s \cup \{s\}} = z|_{\mathcal{N}_s \cup \{s\}}} \mathbb{P}(\{z\}) \\
&= \sum_{x|_{\mathcal{N}_s \cup \{s\}} = z|_{\mathcal{N}_s \cup \{s\}}} e^{-\mathcal{H}_s(z) - \mathcal{H}_s^*(z)} \\
&= \sum_{x|_{\mathcal{N}_s \cup \{s\}} = z|_{\mathcal{N}_s \cup \{s\}}} e^{-\mathcal{H}_s(x) - \mathcal{H}_s^*(z)} \\
&= e^{-\mathcal{H}_s(x)} \sum_{x|_{\mathcal{N}_s \cup \{s\}} = z|_{\mathcal{N}_s \cup \{s\}}} e^{-\mathcal{H}_s^*(z)},
\end{aligned}
$$

and with (20.2) and (20.4) we obtain

$$
\begin{aligned}
Z\mathbb{P}(\{z \mid z(t) = x(t) \text{ for } t \in \mathcal{N}_s\}) &= Z \sum_{x|_{\mathcal{N}_s} = z|_{\mathcal{N}_s}} \mathbb{P}(\{z\}) \\
&= \sum_{x|_{\mathcal{N}_s \cup \{s\}} = z|_{\mathcal{N}_s \cup \{s\}}} \sum_\lambda \mathbb{P}(\{z^{s;\lambda}\}) \\
&= \sum_{x|_{\mathcal{N}_s \cup \{s\}} = z|_{\mathcal{N}_s \cup \{s\}}} \sum_\lambda e^{-\mathcal{H}_s(z^{s;\lambda}) - \mathcal{H}_s^*(z^{s;\lambda})} \\
&= \sum_{x|_{\mathcal{N}_s \cup \{s\}} = z|_{\mathcal{N}_s \cup \{s\}}} \sum_\lambda e^{-\mathcal{H}_s(x^{s;\lambda}) - \mathcal{H}_s^*(z)} \\
&= \sum_\lambda e^{-\mathcal{H}_s(x^{s;\lambda})} \sum_{x|_{\mathcal{N}_s \cup \{s\}} = z|_{\mathcal{N}_s \cup \{s\}}} e^{-\mathcal{H}_s^*(z)}.
\end{aligned}
$$

By taking the quotient of these two expressions we arrive at

$$\mathbb{P}(X_s = x(s) \mid X_t = x(t) \text{ for } t \in \mathcal{N}_s) = \frac{e^{-\mathcal{H}_s(x)}}{\sum_\lambda e^{-\mathcal{H}_s(x^{s,\lambda})}},$$

and this completes the proof. $\square$

Let us try to *understand why it was important to work with cliques*. To phrase it otherwise:

Let $\mathcal{D}$ be *any* system of subsets of S and suppose that V_D is a function from Λ^D to $\mathbb{R}$ for every $D \in \mathcal{D}$. Similarly as before we pass from the family (V_D) to $\mathcal{H}_\mathcal{D}$, defined by

$$\mathcal{H}_\mathcal{D}(x) := \sum_D V_D(x|_D),$$

and further to the induced probability measure $\mathbb{P}_\mathcal{D}$. Under what conditions on $\mathcal{D}$ can we mimic the proof of the preceding theorem, when will the random field have the Markov property?

Of course, in the line of the above arguments, one would first introduce the $(\mathcal{H}_D)_s$
and the $(\mathcal{H}_D)_s^*$ as before. Then one would note that equation (20.2) is valid without any
restriction on $\mathcal{D}$. For (20.3) and (20.4) to hold, however, it is essential to know that D is
a subset of $\mathcal{N}_s \cup \{s\}$ whenever $s \in D$ and $D \in \mathcal{D}$. This implies that all $D \in \mathcal{D}$ are cliques,
and we conclude that our approach necessitates to work with this type of subsets.

What about *the converse of the preceding theorem*? Is every Markov random field a
Gibbs field for a suitable potential? In view of the above examples there is some evidence
for this to hold since for all random fields with the Markov property studied above we
have been able to provide a $\mathcal{V}$ with $\mathbb{P} = \mathbb{P}_\mathcal{V}$. In fact, *this is generally true*: a random
field is a Markov random field iff it is a Gibbs field. This is the celebrated theorem of
Hammersley and Clifford from 1968, for a proof we refer the reader to [76], theorem 3.3.2,
or to [20], theorem 7.2.2.

The Ising model[4]

This model was introduced by Ising ([45]) in 1925 to model the phenomenon of phase
transition in ferromagnetic materials. It is a particularly simple example of a Gibbs field,
it has applications in physics (theory of matter), medicine (distribution of epidemies),
biology and sociology.

We will only discuss the *two-dimensional situation*, the modifications which are necessary to generalize the definitions to arbitrarily many dimensions are canonical. It would
be desirable to work with $\mathbb{Z} \times \mathbb{Z}$ as the set of sites, S. This set, however, is infinite so
that one restricts oneself to $S := \{0, \ldots, r-1\} \times \{0, \ldots, r-1\}$ with a sufficiently large
r. But now there is another drawback: different sites might have a different number of
neighbours. To remedy this one uses *the wrap-around trick:* in the set $\{0, \ldots, r-1\}$ one
declares the elements 0 and $r-1$ to be neighbours, similarly to constructions in previous chapters when we have studied Markov chains on the cyclic group $\mathbb{Z}/r\mathbb{Z}$. Then S is
something like a discrete torus, all $s = (i,j) \in S$ have the same neighbourhood structure
if we define the neighbourhood $\mathcal{N}_s$ of s by

$$\mathcal{N}_s := \{(i-1,j),\ (i+1,j),\ (i,j-1),\ (i,j+1)\}$$

(with $i \pm 1$ and $j \pm 1$ modulo r).

Every site s can have one of the two states in $\Lambda = \{-1, +1\}$. In Ising's model "state"
was meant to be the orientation of a magnetic dipole, but you can think of any other
situation where a site has to choose its state among two possibilities. It remains to define
a Gibbs potential in order to arrive at a Markov random field. The special feature of the
Ising model is that, for a clique S, the potential V_C is a particularly simple function of
the $x(s), s \in C$.

To motivate the definition of the V_C we first note that there are *two types of cliques*,
namely the singletons and the four two-point sets

$$\{(i,j),(i-1,j)\},\ \{(i,j),(i+1,j)\},\ \{(i,j),(i,j-1)\},\ \{(i,j),(i,j+1)\}.$$

---

[4] Those who want to pronounce the name "Ising" correctly should know that it is a German name;
therefore the vowel "I" is spoken like the "ea" in "eagle" and *not* like the "i" in "icecream".

Next we recall that the probability of a configuration $x \in \Lambda^S$ will be proportional to $\exp(-\mathcal{H}_V(x))$, and therefore the x with a *high probability* will be those with *a small value of the energy function*. These probabilities can be controlled by assigning appropriate potentials to the cliques, in the Ising model the two different types are treated as follows.

The singletons: Consider a clique $C = \{s\}$. Then V_C could be any function from Λ to the reals. Since $V_C(x)$ is just $V_C(x(s))$, the definition of V_C controls the $x(s)$-entry. By the values of $V_C(\pm 1)$ we can prescribe how likely the state $x(s) = \pm 1$ is: if $V_C(+1)$ is *smaller* than $V_C(-1)$, then $x(s) = +1$ is *more likely* than $x(s) = -1$.

> Such a "tendency to have the value -1 or $+1$" can be a desirable feature of the model. In Ising's original approach this part of the potential stood for an external magnetic field, in other situations it can be an inherent tendency to have a certain opinion, or the disposition to catch a certain disease.

In the Ising model the potentials for the singletons are defined by

$$V_{\{s\}}(x) := -hx(s),$$

where h is a real number (the same for all s). By the *size* of h it is possible to quantify, e.g., the strength of a magnetic field or the disposition to have political opinions. Usually h will be positive so that state $+1$ is favoured, but negative values might also be reasonable.

The two-point cliques: This is the more interesting part, it describes the *interactions*. The underlying idea can be phrased as

Do in Rome what the Romans do.

More seriously: if all neighbours of s are in state $+1$ resp. all are in -1, then there is a more or less strong tendency for s – it will be quantified by a parameter K – also to be in state $+1$ resp. -1; and if some neighbours are in state $+1$ and others are in state -1, then both $x(s) = +1$ and $x(s) = -1$ might be equally likely. Now note that, for a clique $C = \{s, t\}$, the number $x(s)x(t)$ is $+1$ resp. -1 if $x(s) = x(t)$ resp. $x(s) \neq x(t)$. Hence the definition

$$V_{\{s,t\}}(x) := -Kx(s)x(t)$$

favours $x(s) = x(t)$ if K is positive; note that then $x(s) = x(t)$ leads to a smaller energy than $x(s) \neq x(t)$. For a negative K, however, the model enforces $x(s)$ to be different from $x(t)$.

Now we are ready for the definition of the Ising model:

Definition 20.3 Let r be an integer and h and K real numbers. The *two-dimensional toric Ising model* consists of

   (i)   the set of sites $S = \{0, \ldots, r-1\} \times \{0, \ldots, r-1\}$,
   (ii)   the state space $\Lambda = \{-1, +1\}$,
   (iii)   the neighbourhoods $\mathcal{N}_s := \{(i-1, j), (i+1, j), (i, j-1), (i, j+1)\}$ for $s = (i, j)$ (where $i \pm 1$ and $j \pm 1$ are calulated modulo r),
   (iv)   the Gibbs potential $V = (V_C)_C$ which is defined by

$$V_{\{s\}}(x) = -hx(s) \quad \text{and} \quad V_{\{s,t\}}(x) = -Kx(s)x(t)$$

for the one- and two-dimensional cliques, respectively.

It is not hard to visualize this model, one only has to apply corollary 19.7. In order to get a "typical sample", one has to proceed as follows:

- Start with any configuration x_0, often this is chosen by flipping coins at every site s to determine $x_0(s)$.

- Then let a Markov chain run on Λ^S. If the present position is the configuration x, then the next one is obtained as follows: choose an s at random; calculate, for $\varepsilon = \pm 1$, the number

$$a(\varepsilon) := h\varepsilon + K\varepsilon \sum_{t,\ \{s,t\}\ \text{is a clique}} x(t);$$

  select $\eta = +1$ resp. -1 according to the probabilities

$$\frac{e^{a(+1)}}{e^{a(+1)} + e^{a(-1)}} \quad \text{resp.} \quad \frac{e^{a(-1)}}{e^{a(+1)} + e^{a(-1)}};$$

  the next position of the walk is then the configuration which is identical with x on $S \setminus \{s\}$ and for which the state at s is η.

- Let the chain run for "a long time" in this way and use the configuration which is occupied then as an output. *These* samples are distributed approximately in accordance with $\mathbb{P}_\mathcal{V}$ so that you will observe something typical.

Similarly one can treat the case $S = \{0, \ldots, r-1\} \times \{0, \ldots, l-1\}$ with possibly different r and l; e.g., the neighbours of $(0,0)$ are $(1,0)$, $(0,1)$, $(r-1,0)$ and $(0,l-1)$. Here are some samples with $r = 30$ and $l = 20$, the states ± 1 are visualized by little white and black squares in a 30×20-grid:

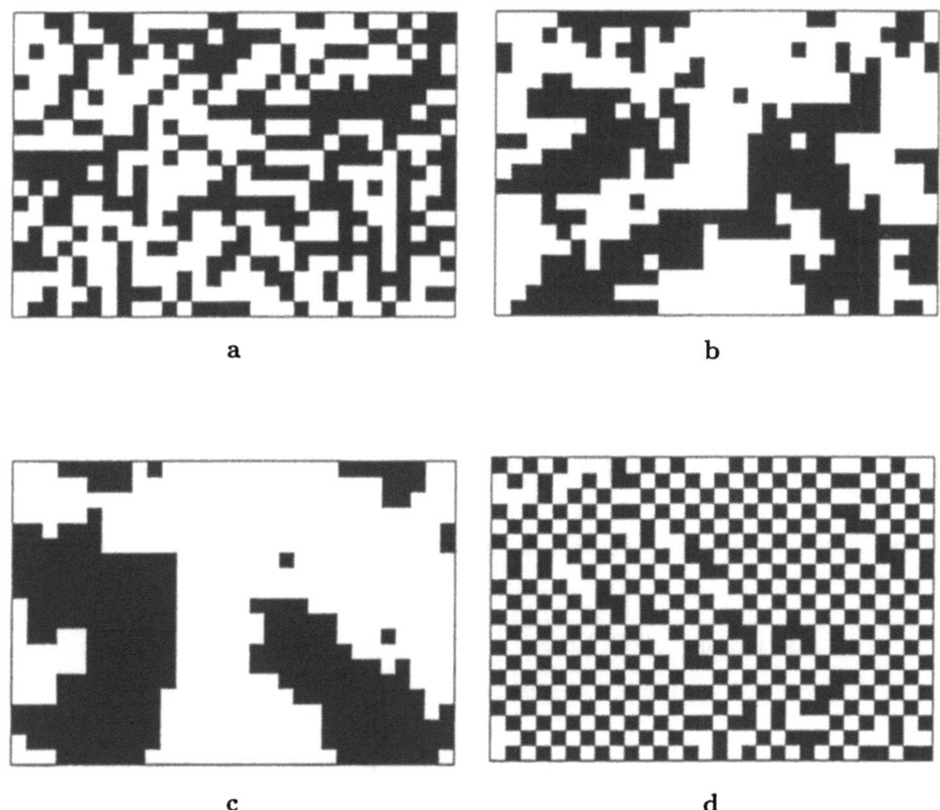

a   b

c   d

Picture "a" is a starting position, the states at the sites are chosen at random with equal probability.

Picture "b" shows the state of the random walk after 10,000 steps; the parameters are $h = 0$ and $K = 0.3$. Since K is rather small, there is weak interaction and the configuration has a lot of structure.

In *picture "c"* one sees the position of the walk after 10,000 further iterations. All these new iterations, however, are with $K = 0.8$ (and $h = 0$ as before). Note that now there is more interaction, the state at a site is now influenced more strongly by the states of the neighbours than before. Not surprisingly, this results in a picture of a different character, there are much fewer black or white "islands".

Finally, to prepare *picture "d"*, we have produced another 10,000 steps of the walk, this time with $K = -0.8$ (and $h = 0$). As it was to be expected, the character of the picture has changed once more: the state of a site tends to be "as opposite as possible" to the states of the neighbours, and thus we see something like a checkerboard pattern.

We close this section on the Ising model with the *calculation of a partition function*, one of the main results in Ising's doctoral thesis. It has been stressed above that it is in most cases impossible to determine $Z = \sum_x e^{-\mathcal{H}(x)}$. For a particular case of the Ising model, the **one-dimensional situation**, one can provide this number explicitly.

Fix an integer N. The *one-dimensional toric Ising model* is defined similarly as above, that is the neighbours of a site $i \in S = \{0, \ldots, N-1\}$ are $i \pm 1$

modulo N. Therefore the cliques are the sets $\{i\}$ and $\{i, i+1\}$, and the energy function has the concrete form

$$\mathcal{H}_V(x) = -h \sum_{i=0}^{N-1} x(i) - K \sum_{i=0}^{N-1} x(i)x(i+1).$$

By definition, the partition function $Z = Z_N$ is the number

$$\begin{aligned}
Z_N &= \sum_{x \in \Lambda^S} e^{-\mathcal{H}_V(x)} \\
&= \sum_x \exp\Big(h \sum_{i=0}^{N-1} x(i) + K \sum_{i=0}^{N-1} x(i)x(i+1)\Big) \\
&= \sum_x \exp\Big(\sum_{i=0}^{N-1} \frac{h}{2}(x(i) + x(i+1)) + Kx(i)x(i+1)\Big).
\end{aligned}$$

The evaluation needs some preparations, we introduce the numbers $Z_N^\pm$, $R^{\pm 1, \pm 1}$ and the *transfer matrix R:*

- $Z_N^\pm := \sum_{x \in \Lambda^S, x(0) = \pm 1} \exp(h \sum_{i=0}^{N-1} x(i) + K \sum_{i=0}^{N-1} x(i)x(i+1))$;

- $R^{\varepsilon, \eta} := \exp(h(\varepsilon + \eta)/2 + K\varepsilon\eta)$, for $\varepsilon, \eta = \pm 1$;

- $R = \begin{pmatrix} R^{+1,+1} & R^{+1,-1} \\ R^{-1,+1} & R^{-1,-1} \end{pmatrix} = \begin{pmatrix} e^{h+K} & e^{-K} \\ e^{-K} & e^{-h+K} \end{pmatrix}$.

Then it is plain that $Z_N = Z_N^+ + Z_N^-$, and also — since

$$Z_N = \sum_{\varepsilon_0, \dots, \varepsilon_{N-1} = \pm 1} R^{\varepsilon_0, \varepsilon_1} R^{\varepsilon_1, \varepsilon_2} \dots R^{\varepsilon_{N-1}, \varepsilon_0}$$

holds — that

$$Z_n^+ = R^{+1,+1} Z_{N-1}^+ + R^{+1,-1} Z_{N-1}^-, \quad Z_n^- = R^{-1,+1} Z_{N-1}^+ + R^{-1,-1} Z_{N-1}^-.$$

This means that

$$\begin{pmatrix} Z_N^+ \\ Z_N^- \end{pmatrix} = R \begin{pmatrix} Z_{N-1}^+ \\ Z_{N-1}^- \end{pmatrix},$$

and by induction on N it follows immediately that Z_N^+ resp. Z_N^- is the top left resp. the bottom right element of R^N (the case $N = 2$ has to be calculated directly). Therefore Z_N is just *the trace of R^N.*

To arrive at our final result it remains to recall that the trace of a self-adjoint matrix is the sum of the eigenvalues and that the eigenvalues of R^N are the N'th powers of the eigenvalues of R:

Proposition 20.4 *Let λ and μ be the eigenvalues of the transfer matrix R. Then Z_N, the partition function of the one-dimensional toric Ising model on $\{0, \dots, N-1\}$, has the value*

$$Z_N = \lambda^N + \mu^N.$$

The Gibbs sampler

We return to the situation of a general Gibbs field which is given by sets S and Λ, a neighbourhood system $\mathcal{N}$ and a Gibbs potential $\mathcal{V}$. We continue to write $\mathcal{H}$ instead of the more correct $\mathcal{H}_{\mathcal{V}}$.

The local characteristics have been determined in theorem 20.2, and in corollary 19.7 we have shown how samples can be produced by running a suitable Markov chain. In this section we want to replace the vague "let the chain run for a sufficiently long time" by concrete bounds.

For these calculations it will be convenient to change the definitions of the associated Markov chain slightly. Originally – cf. page 188 – the choice of the next configuration was started with the selection of a site s in accordance with certain probabilities (q_s). Now we fix once and for all an enumeration of S, that is we identify this set with $\{1, \ldots, N\}$. And the sites where one possibly changes the state by using the probabilities prescribed by the local characteristics are no longer chosen at random. Instead we work systematically: in the first step – the step after the arbitrarily chosen starting position – we work with $s = 1$, then with $s = 2$ and so on until we arrive at $s = N$. Then, in step $N + 1$, we start again at $s = 1$, next we consider $s = 2$, ... Usually the enumeration of S is called a *visiting scheme*, and a *sweep* with respect to this scheme is the result of N consecutive steps (beginning with a – possible – change at $s = 1$).

The Markov chain we are going to analyse has Λ^S as its state space, the starting position is arbitrary but fixed, and the next step is the configuration which is produced after a sweep. The technique to produce samples from a Gibbs field with this chain is called the *Gibbs sampler*, the "output" is the position of the chain after "many" sweeps.

To fix notation, let P_s be the transition matrix which is associated with a (possible) change at s. Then $Q := P_1 P_2 \cdots P_N$ is the matrix which governs the transitions of the new chain.

Lemma 20.5

    (i) *The entries of Q are strictly positive so that the chain is irreducible and aperiodic.*

    (ii) *The equilibrium π of Q is the Gibbs distribution:*

$$\pi_x = \frac{e^{-\mathcal{H}(x)}}{Z}$$

for $x \in \Lambda^S$.

Proof. (i) is obvious: since all local characteristics are strictly positive, one may pass in one step[5] from any x to any z.

(ii) Fix $s \in S$ and $x \in \Lambda^S$. By $T_{s,x}$ we will denote the set of configurations z' such that a transition from z' to x with respect to the transition matrix P_s is possible; $T_{s,x}$ thus contains the z' which coincide with x on $S \setminus \{s\}$. By definition, a transition from a $z \in T_{s,x}$ to x has probability

$$\frac{e^{-\mathcal{H}(x)}}{\sum_{z' \in T_{s,x}} e^{-\mathcal{H}(z')}}.$$

---

[5] Note that a step for Q is a complete sweep.

Thus, with $\pi_x := e^{-\mathcal{H}(x)}/Z$, it follows that $\pi^{\top} P_s = \pi^{\top}$, and this implies that $\pi^{\top} Q = \pi^{\top}$. This completes the proof since there is precisely one equilibrium. $\qquad\square$

We will estimate the rate of convergence of the Q-chain by using theorem 10.5. To apply this result it is necessary to bound the number δ, the *minimal nonzero probability for a transition between two arbitrary configurations*. Let us introduce some further notation:

- the number L will stand for the cardinality of Λ;

- for $s \in S$ and $x \in \Lambda^S$, $m_{s,x}$ resp. $M_{s,x}$ mean the minimum resp. the maximum of the $\{\mathcal{H}(z) \mid z \in T_{s,x}\}$ (for the definition of $T_{s,x}$ see the preceding proof);

- $v_s := \max\{M_{s,x} - m_{s,x} \mid x \in \Lambda^S\}$, for $s \in S$;

- $\Delta := \max_s v_s$.

Now let x and z be arbitrary such that transitions are possible. The probability for a jump from x to z according to Q can be estimated as follows.

Consider first P_1. The probability that x and z will coincide at state 1 after one step of the P_1-chain is

$$\frac{e^{-\mathcal{H}(x')}}{\sum_{z' \in T_{s,x}} e^{-\mathcal{H}(z')}},$$

where $x'(1) = z(1)$ and $x'(s) = x(s)$ for the other s. This number can be estimated by

$$\frac{e^{-\mathcal{H}(x')}}{\sum_{z' \in T_{s,x}} e^{-\mathcal{H}(z')}} \geq \frac{e^{-\mathcal{H}(x')+m_{1,x}}}{\sum_{z' \in T_{1,x}} e^{-\mathcal{H}(z')+m_{1,x}}}$$

$$\geq \frac{e^{-v_1}}{L}.$$

The transitions at $s = 2, \ldots, N$ can be treated similarly, and therefore the probability of a transition $x \to z$ after a complete sweep is at least

$$\prod_{s=1}^{N} \frac{e^{-v_s}}{L} \geq \frac{e^{-N\Delta}}{L^N};$$

this is a lower bound for the number δ from theorem 10.5. It remains to apply this result[6]:

Proposition 20.6 *The convergence rate of the Gibbs sampler can be estimated as follows. If the starting position is arbitrary then a configuration x will be observed as the position of the walk after k sweeps with a probability p which satisfies*

$$|p - \mathbb{P}_{\mathcal{V}}(\{x\})| \leq (1 - e^{-N\Delta})^k.$$

As an *illustration* of this result lets consider the *Ising model* on an $r \times r$-square (considered as above as a discrete torus). For $s = (i,j)$, the local energy is

---

[6] Note that "N" has two different meanings in theorem 10.5 and in the present investigations. There it was the cardinality of the state space, now it is the number of sites, and therefore we have to replace N in 10.5 by L^N.

$$\mathcal{H}_s(x) = hx(s) + K\Big(\sum_{t\in\mathcal{N}_s} x(t)x(s)\Big).$$

We will assume that h and K are positive. Then

$$\mathcal{H}(z) = hz(s) + K\Big(\sum_{t\in\mathcal{N}_s} x(t)x(s)\Big) + \mathcal{H}_s^*(x)$$

for fixed s and the $z \in T_{s,x}$; therefore

$$m_{s,x} = -h - 4K + \mathcal{H}_s^*(x), \;\; M_{s,x} = h + 4K + \mathcal{H}_s^*(x)$$

and consequently

$$v_{s,x} = v_s = \Delta = 2h + 8K.$$

It has to be admitted that only for small values of r, h and K the resulting error bound

$$\leq \left(1 - e^{-(2h+8K)r^2}\right)^k$$

will lead to satisfactory results, some readers will be reminded of the citation on page 78.

Exercises

20.1: Let (S,d) be a finite metric space, the neighbourhood system $\mathcal{N}^{d,R}$ is defined as in exercise 19.2. Prove that the cliques are precisely the subsets of S of diameter at most R.

20.2: Is it possible to reconstruct the neighbourhood system from the collection of cliques? (More precisely: if $\mathcal{N}$ and $\widehat{\mathcal{N}}$ are neighbourhood systems on the same set which give rise to the same cliques, does it follow that $\mathcal{N} = \widehat{\mathcal{N}}$?)

20.3: Let S together with a neighbourhood system be given. Suppose that one knows which of the subsets containing precisely two elements are cliques. Is it then possible to find *all* cliques?

20.4: Let S be a finite set. Characterize the collections $\mathcal{C}$ of subsets which are the cliques for a suitable neighbourhood system.

20.5: Let $\mathcal{V}$ be a potential and $\mathcal{H}_\mathcal{V}$ the associated energy function. Prove or disprove:

a) in the case of the minimal neighbourhood system (i.e., all neighbourhoods are empty) it is possible to reconstruct $\mathcal{V}$ from $\mathcal{H}$;

b) in general, this is not possible.

20.6: Suppose that the neighbourhood system on S is such that every energy function $\mathcal{H}$ can be written as $\mathcal{H}_\mathcal{V}$ for a suitable potential $\mathcal{V}$. Prove that then S is a clique.

20.7: Let $\mathcal{V} = (V_C)_{C\in\mathcal{C}}$ be a potential. Define, for a fixed real number r, another potential $\widehat{\mathcal{V}}$ by $\widehat{V}_C := V_C - r$. Prove that $\mathcal{V}$ and $\widehat{\mathcal{V}}$ give rise to the same Gibbs field.

20.8: We consider the two-dimensional (or, more generally, the r-dimensional) toric Ising model with $h = 0$ and a site s all neighbours of which are in state -1.

a) Suppose that $K = 0.8$. What is the probability that s is also in state -1?

b) Let $p \in [0,1]$ be arbitrary. Is it possible to choose K such that s is in state 1 with probability p?

20.9: Let a, b, c be real numbers such that $|b| + |c| \leq a$. Use the method of the transfer matrix to evaluate the expression

$$\sum_{\varepsilon_0,\ldots,\varepsilon_{N-1}=\pm 1} R^{\varepsilon_0,\varepsilon_1} R^{\varepsilon_1,\varepsilon_2} \cdots R^{\varepsilon_{N-1},\varepsilon_0},$$

where $R^{\varepsilon,\delta} := \sqrt{a + b\varepsilon + c\delta}$.

20.10: Denote by λ and μ the eigenvalues of the transfer matrix as in proposition 20.4. Determine the collection of all (λ, μ) which arise in this way for various h and K. Can it happen, e.g., that $\lambda = \mu$? Or that λ or μ vanishes?

20.11: Consider the Gibbs field induced by a potential $\mathcal{V}$. Denote, for $T \subset S$, by $\mathcal{N}_T$ the collection of all elements in the union of the $\mathcal{N}_s$, $s \in T$, which are not in T. (For obvious reasons, $\mathcal{N}_T$ is called *the neighbourhood of T*.)

a) Calculate $\mathcal{N}_T$ for some T in our standard examples.

b) Prove that the Markov property of Gibbs fields (theorem 20.2) admits the following generalization: whenever $T \subset S$, then $\mathbb{P}(X_s = x(s)$ for $s \in T \mid X_t = x(t)$ for $t \notin T) = \mathbb{P}(X_s = x(s)$ for $s \in T \mid X_t = x(t)$ for $t \in \mathcal{N}_T)$.
(Note that theorem 20.2 corresponds to the case $T = \{s\}$.)

20.12: Let $(\Lambda^S, \mathbb{P}_\mathcal{V})$ be a Gibbs field given by a potential $\mathcal{V}$ and r an integer such that every s lies in at most r cliques. We consider the Markov chain on Λ^S induced by the local characteristics of $\mathbb{P}_\mathcal{V}$. How many calculations (= evaluations of potentials) are necessary to simulate one step of the chain?

21   The Metropolis sampler and simulated annealing

The setting is as at the beginning of the previous chapter: we are given a finite set $\tilde{S}$ and a function[1] $\mathcal{H} : \tilde{S} \to \mathbb{R}$, and this function gives rise to a probability measure $\mathbb{P}_\mathcal{H}$ on $\tilde{S}$ by

$$\mathbb{P}_\mathcal{H}(\{x\}) := \frac{e^{-\mathcal{H}(x)}}{Z}, \quad \text{with } Z := \sum_z e^{-\mathcal{H}(z)}.$$

Since from now on no sites and configurations will be considered, it will cause no confusion if we write S instead of $\tilde{S}$.

The aim of the present chapter is twofold. In the *first section* we present the *Metropolis sampler*, a method to provide samples from S which are distributed like $\mathbb{P}_\mathcal{H}$. This variant of the Gibbs sampler also works without an explicit use of the (generally) unknown number Z. Then, in the *second section*, we present an introduction to *simulated annealing* which is based on the Metropolis sampler. This is a stochastic optimization technique which since many years has found applications in various fields.

The Metropolis sampler

As in chapter 20 we don't know $\mathbb{P}_\mathcal{H}(\{x\})$ explicitly, but nevertheless samples are required with this distribution. In the preceding chapter the relevant notion was that of neighbourhoods: (hopefully) only few neighbours influence the state of a given site. Here the situation is similar in that the set S has an additional structure. It is helpful to visualize S as the set of the vertices of a graph such that

- every vertex is connected with "few" other vertices, and

- there are sufficiently many edges to get from any vertex to any other in "not too many" steps.

(Typical examples are the graphs which we met in chapter 19 or the lattices $\{0,1\}^m$; there edges connect two m-tuples iff they differ at precisely one component.)

The approach will be a little bit more general, we will assume that a Markov chain on S is prescribed by a transition matrix $Q = (q_{xy})_{x,y \in S}$. This matrix Q will be called the *proposal matrix*, the reason for this notion should be clear by the following

Definition 21.1 Let the state space S, the energy function $\mathcal{H}$ and the proposal matrix Q be given. We suppose that Q is symmetric (i.e., $q_{xy} = q_{yx}$) and that the chain defined by Q on S is irreducible and aperiodic.

By the *Metropolis chain* we mean the Markov chain on S which is defined by the following transitions (the starting position is fixed but arbitrary):

- Suppose that the walk is now at position x. Choose a y according to the proposal matrix Q, that is, y is selected with probability q_{xy}.

---

[1] For historical reasons $\mathcal{H}$ is called the "energy function", even in this general approach.

- If $\mathcal{H}(y) \leq \mathcal{H}(x)$, then the next position of the walk will be y.

- If, however, $\mathcal{H}(y) > \mathcal{H}(x)$, an additional Bernoulli experiment is needed: it should provide 1 resp. 0 with probability $\exp(\mathcal{H}(x) - \mathcal{H}(y))$ resp. $1 - \exp(\mathcal{H}(x) - \mathcal{H}(y))$. If the experiment produces a 1, then go to y, otherwise stay at x.

To phrase it otherwise: by the Q-matrix a state y is "proposed" as the possible next position. Only if this does not result in an *increase* of the energy, the proposal is accepted immediately. If $\mathcal{H}$ increases, there is nevertheless a chance to go to y, the probability of this transition depends on the difference between $\mathcal{H}(x)$ and $\mathcal{H}(y)$. If $\mathcal{H}(y)$ is much greater than $\mathcal{H}(y)$, then a jump to y is not to be expected.

By this definition it is likely that the chain has a tendency to occupy positions where $\mathcal{H}$ is low. Even more is true, in the long run the probability to occupy position x tends to $\mathbb{P}_{\mathcal{H}}(x)$:

Proposition 21.2 *The previously defined chain is irreducible, aperiodic and reversible, the equilibrium distribution π coincides with $\mathbb{P}_{\mathcal{H}}$.*

Proof. A transition $x \to y$ has a positive probability with respect to the Metropolis chain iff this is true with respect to the Q-chain. Therefore irreducibility and aperiodicity follow from the assumptions on Q.

Now let $\pi_x := e^{-\mathcal{H}(x)}/Z$ for $x \in S$. If the p_{xy} denote the Metropolis transition probabilities, then we have to show that

$$\pi_x p_{xy} = \pi_y p_{yx} \tag{21.1}$$

for arbitrary x, y; then $(\pi_x)_x$ will be *the* equilibrium, and the proof will be complete. For the proof of (21.1) we may assume that $x \neq y$. If $\mathcal{H}(y) > \mathcal{H}(x)$, then

$$\begin{aligned}
Z\pi_x p_{xy} &= e^{-\mathcal{H}(x)} q_{xy} e^{\mathcal{H}(x) - \mathcal{H}(y)} \\
&= q_{xy} e^{-\mathcal{H}(y)} \\
&= q_{yx} e^{-\mathcal{H}(y)} \\
&= Z\pi_y p_{yx}.
\end{aligned}$$

The proof for the case $\mathcal{H}(x) \leq \mathcal{H}(y)$ is similar. $\qquad\square$

Since the probabilities associated with the k-step transitions of an irreducible and aperiodic chain tend to the equilibrium, one can use the preceding result to produce samples from S which are distributed like $\mathbb{P}_{\mathcal{H}}$; it is only necessary to run the Metropolis chain for "sufficiently many" steps. This is called the **Metropolis sampler**.

But how many steps are "sufficiently many"? A moment's reflection shows that the mixing rate of the Metropolis chain will depend on the mixing rate of the Q-chain – this is obvious – and also on the variation of the function $\mathcal{H}$: if great differences $\mathcal{H}(x) - \mathcal{H}(y)$ are possible, then the chain can be trapped at local minima of $\mathcal{H}$.

We will present a result due to P. Mathé ([56]) by which this observation is quantified:

Proposition 21.3 *Let S and $\mathcal{H}$ be as above, we continue to denote by π the equilibrium of the Metropolis chain. As a measure of the variation of $\mathcal{H}$ we define*

$$a := \min_x \pi_x / \max_x \pi_x.$$

To quantify the mixing rate we use the second largest eigenvalue[2]. By the results of chapter 10 the distance of this eigenvalue to 1 contains all relevant informations, it is called the spectral gap. Let λ_2^P resp. λ_2^Q stand for the second largest eigenvalue of the Metropolis chain and the proposal chain, respectively, and denote by $\mu_P := 1 - \lambda_2^P$ and $\mu_Q := 1 - \lambda_2^P$ the corresponding spectral gaps. Then

$$\frac{a^2}{2}\mu_Q \leq \mu_P \leq \frac{2}{a^2}\mu_Q.$$

Proof. First we have to develop the Hilbert space techniques with which we have already been concerned in chapter 10 a little bit further. We consider an arbitrary irreducible, aperiodic and reversible chain on a set S with N elements, given by a stochastic matrix P and with equilibrium π. The Hilbert space H_π has been introduced in proposition 10.2 as the space $\mathbb{R}^N$ together with the scalar product $\langle \cdot, \cdot \rangle_\pi$ induced by π:

$$\langle f, g \rangle_\pi := \sum_{x \in S} f(x)g(x)\pi_x.$$

As in this proposition we identify P with a map on H_π. Since our chain is reversible by assumption the operator P is self-adjoint, and therefore we may apply proposition 10.4: λ_2^P, the second-largest eigenvalue of P, is the maximum of the numbers $\langle f, Pf \rangle_\pi / \langle f, f \rangle_\pi$, where the f run through all nonzero vectors in H_π which are orthogonal to $(1, \ldots, 1)$. Consequently, the spectral gap $1 - \lambda_2^P$ equals

$$\min\left\{ \frac{\langle f, (Id - P)f \rangle_\pi}{\langle f, f \rangle_\pi} \,\Big|\, f \neq 0,\, f \perp (1, \ldots, 1) \right\}.$$

We have already met the expression in the numerator, it has played an important role in the proof of theorem 11.3. It is usually called the *Dirichlet form* associated with P and written

$$\mathcal{E}_P(f, f).$$

Let g be an arbitrary element of H_π which does not lie in the linear span of $e := (1, \ldots, 1)$. Then $g = \langle e, g \rangle_\pi e + f$ with an f which is orthogonal to e. Since $Id - P$ annihilates e, the Dirichlet form has the same value at g as at f, and therefore we may rewrite the formula for the spectral gap as

$$1 - \lambda_2^P = \min\left\{ \frac{\mathcal{E}_P(g, g)}{\mathrm{Var}_P(g)} \,\Big|\, \mathrm{Var}_\pi(g) > 0 \right\}, \tag{21.2}$$

where $\mathrm{Var}_P(g) := \|g - \langle g, e \rangle_\pi e\|_\pi^2$. This number is called the *variation of g*, note that it is just the ordinary variance of g if g is considered as a random variable from S – provided with π as the probability measure – to $\mathbb{R}$.

Now we turn to the proof. The equilibrium distributions of the proposal chain resp. the Metropolis chain will be denoted by (η_x) resp. (π_x), the corresponding Dirichlet forms, second-largest eigenvalues and variations by $\mathcal{E}_P$ resp. $\mathcal{E}_Q$, λ_2^P resp. λ_2^Q and Var_P resp. Var_Q; note that $\eta_x = 1/N$ for every x since Q is symmetric by assumption.

The number a is defined such that

---

[2] Cf. theorem 10.3. Note that λ_2 is useful to bound the mixing rate only if it coincides with λ^*, the maximum of the eigenvalues which are different from 1. Recall that this can always be achieved by passing from the transition matrix P to $(Id + P)/2$.

$$a\pi_x \leq a \max_y \pi_y = \min_y \pi_y \leq \pi_x$$

holds for every x. Summation leads to

$$a \leq a N \max_y \pi_y = N \min_y \pi_y \leq 1,$$

and thus

$$a\,\eta_x \leq \pi_x \leq \eta_x/a$$

for all x.

With these preparations at hand we investigate the above formula (21.2) for $1 - \lambda_2$. To deal with Var_P and Var_Q we recall from elementary probability theory that the variance of a random variable X is the minimum of the expectations of $(X - c)^2$, with $c \in \mathbb{R}$ (the minimum is assumed at the expectation of X). For our setting this yields, for an arbitrary $g \in H_\pi$,

$$
\begin{aligned}
\mathrm{Var}_P(g) &= \min_c \sum_x (g(x) - c)^2 \pi_x \\
&\geq a \min_c \sum_x (g(x) - c)^2 \eta_x \\
&= a\,\mathrm{Var}_Q(g).
\end{aligned}
$$

To treat the Dirichlet form we first note that $\mathcal{E}_P(g,g) = \sum_{x,y}(g(x) - g(y))^2 p_{xy}\pi_x/2$: this is easy to verify. From this expression one deduces that

$$\mathcal{E}_P(g,g) \leq \sum_{\pi_y \geq \pi_x} (g(x) - g(y))^2 p_{xy}\pi_x.$$

(Here "$\leq$" cannot be replaced by "$=$" in general; this is due to the x, y with $\pi_x = \pi_y$.)

Now the special structure of our chains comes into play: if $\pi_y \geq \pi_x$, then $p_{xy} = q_{xy}$ by definition. We are thus led to

$$
\begin{aligned}
\mathcal{E}_P(g,g) &\leq \sum_{\pi_y \geq \pi_x} (g(x) - g(y))^2 p_{xy}\pi_x \\
&= \sum_{\pi_y \geq \pi_x} (g(x) - g(y))^2 q_{xy}\pi_x \\
&\leq \frac{1}{a} \sum_{\pi_y \geq \pi_x} (g(x) - g(y))^2 q_{xy}\eta_x \\
&\leq \frac{1}{a} \sum_{x,y} (g(x) - g(y))^2 q_{xy}\eta_x \\
&= \frac{2}{a}\,\mathcal{E}_Q(g,g).
\end{aligned}
$$

It remains to put both inequalities together to arrive at

$$\frac{\mathcal{E}_P(g,g)}{\mathrm{Var}_P(g)} \leq \frac{2}{a^2}\frac{\mathcal{E}_Q(g,g)}{\mathrm{Var}_Q(g)},$$

and – by (21.2) – this shows that

$$\mu_P \leq \frac{2}{a^2}\mu_Q.$$

That $a^2\mu_Q/2 \leq \mu_P$ also holds is proved similarly. $\square$

Simulated annealing

As before, let $\mathcal{H}$ be a real valued function on a finite set S. The measure which assigns the probability $\mathbb{P}_{\mathcal{H}}(\{x\}) = \exp(-\mathcal{H}(x))/Z$ to $x \in S$ is large where $\mathcal{H}$ is small. Thus there should be a good chance to get an x with $\mathcal{H}(x)$ close to $\min \mathcal{H}$ if one produces a sample from S with distribution $\mathbb{P}_{\mathcal{H}}$. Now let β be a positive real parameter, for historical reasons[3] it is called the *inverse temperature*.

The idea is to pass from $\mathcal{H}$ to $\beta\mathcal{H}$, the associated partition function and the measure will be denoted by Z_β and $\mathbb{P}_\beta$. What is to be expected? Suppose first that β is close to zero ("high temperature"). Then $\beta\mathcal{H}$ is essentially constant so that $\mathbb{P}_\beta$ is close to the uniform distribution: all $x \in S$ are (nearly) equally likely to be sampled. Now let β be large ("low temperature"). Then, for an x where $\mathcal{H}(x)$ is larger than $\min \mathcal{H}$, the number $\exp(-\beta\mathcal{H}(x))$ will be tiny compared with the $\exp(-\beta\mathcal{H}(y))$ for the y with $\mathcal{H}(y) = \min \mathcal{H}$, and thus it is hardly to be expected that a sample with distribution $\mathbb{P}_\beta$ will produce x.

The preceding observations motivate a stochastic optimization technique which is called *simulated annealing:*

- A function $\mathcal{H}$ on a finite set S is given. One wants to find an x_0 such that $\mathcal{H}(x_0) = \min \mathcal{H}$.

- Fix numbers $0 < \beta_1 < \beta_2 < \cdots$, the *cooling schedule*.

- Let – with an arbitrary starting position – a suitable Markov chain run on S which has $\mathbb{P}_{\beta_1}$ as its equilibrium. Let it run so many steps such that the chain occupies a position, x_1, say, which is approximately distributed in accordance with $\mathbb{P}_{\beta_1}$.

- Now repeat this procedure with x_1 as the new starting position and a chain with $\mathbb{P}_{\beta_2}$ as its equilibrium. Stop after "sufficiently many" steps in state x_2.

- Continue this way with $\mathbb{P}_{\beta_3}$, $\mathbb{P}_{\beta_4}$ and so on. Then, for large m, the x_m should be such that $\mathcal{H}(x_m)$ is close to $\min \mathcal{H}$.

Simulated annealing can be compared with someone who seeks the deepest point in a valley on a foggy day. He starts somewhere, and then he walks around without taking much care about where he is (high temperature $1/\beta_1$). Next – with parameter β_2 — he favours a little, but not too much, to go downwards. For large β_m, at later stages of his search, going upwards is practically not taken into account: he prefers to stay at a (local or global) minimum.

It should be clear that the chance that he will arrive at the global minimum (= the deepest point of the valley) will be influenced by many facts. The geometry of the valley will play an important role, and also it is surely not desirable to choose large values of β_m early: the walk would be trapped in a local minimum.

---

[3] In statistical physics the Gibbs distribution takes the form $\exp(-\mathcal{H}(x)/kT)/Z$, with the Boltzmann constant k and the absolute temperature T; thus β corresponds to $1/kT$.

The underlying idea of simulated annealing is in fact very appealing. Once it is possible to sample from S with prescribed distributions $\mathbb{P}_\beta$ it remains to choose the cooling schedule. This, of course, has to be done carefully. If the (β_m) tend to infinity too fast then it is likely that the procedure only produces a local minimum. On the other hand, if they increase only slowly, then it costs too much time to arrive at an x_m with $\mathcal{H}(x_m)$ close to $\min \mathcal{H}$.

Unfortunately, rigorous results which provide reasonable bounds are rare. This is balanced by the fact that simulated annealing can be applied in various situations where constructive methods are not available. Implementation is easy, even large optimization problems can be treated.

We are now going to study **an example of a cooling schedule**, the Markov chains which will come into play will be Metropolis chains. Our main result will be *theorem 21.5* below, we need a number of preparations.

First we agree to modify our procedure a little bit. Instead of using the Metropolis sampler associated with $\beta_1 \mathcal{H}$ for "sufficiently many" steps, then that for $\beta_2 \mathcal{H}$ for some further steps and so on it will be more convenient to walk *only one step* with the inverse temperature β_1, one with β_2, and similarly for the other β_n. This can easily be achieved by passing from the original β-sequence $\beta_1, \beta_2, \ldots$ to $\beta_1, \ldots, \beta_1, \beta_2, \ldots, \beta_2, \ldots$.

Suppose that the procedure starts at some x. Then the probability to arrive after step number k at a state y can be found in the x-y-position[4] of the product matrix

$$\tilde{P}_k := P_{\beta_1} P_{\beta_2} \cdots P_{\beta_k};$$

this is clear by an argument similar to that from the beginning of chapter 3 which has led to (3.2).

Note that therefore, for the first time in this book, we have to deal with *inhomogeneous Markov chains*. We will prepare the proof of theorem 21.5 by studying some general facts concerning such chains. Let an arbitrary state space $S = \{1, \ldots, N\}$ and a sequence $P_1, P_2, \ldots$ of stochastic matrices be given. $(p_1, \ldots, p_N)$ will denote a starting distribution, and $\tilde{P}_k$ will stand for the matrix $P_1 \cdots P_k$. We are interested in the product $(p_1, \ldots, p_n)\tilde{P}_k$, this vector contains at its j'th component the probability to find the walk at j after the k'th step. In view of the application to simulated annealing we have in mind we have to show that under suitable conditions on the P_k these probabilities converge.

The *norms* which are of importance here have already been used in earlier chapters. We recall that – for a vector $x = (x_1, \ldots, x_N)$ or $x = (x_1, \ldots, x_N)^\top$ – the l^1-norm of x is denoted by $\|x\|_1 = \sum |x_i|$ and that the total variation distance $\|\mu - \nu\|$ of two probability vectors μ and ν is just $\|\mu - \nu\|_1/2$ (see lemma 13.3). With these two norms the Lipschitz property of the operators associated with stochastic matrices can easily be expressed: if P is any such matrix and if C_P stands for the maximum of the total variation distances between the rows of P, then

$$\|(x_1, \ldots x_N)P - (y_1, \ldots, y_N)P\|_1 \leq C_P \|(x_1, \ldots x_N) - (y_1, \ldots, y_N)\|_1 \qquad (21.3)$$

for all probability vectors x, y; this has been shown in chapter 10 (lemma 10.6).

We need another notion in connection with the Lipschitz property: L_P will denote the *best possible Lipschitz constant* of the map $(x_1, \ldots, x_N) \mapsto (x_1, \ldots, x_N)P$, i.e., the minimum of the numbers L such that

---

[4] Those who find this notation confusing should identify S with $\{1, \ldots, N\}$. Then the entry in the "x, y-position" in the transition matrix is – for $x, y \in S$ – the element at the y'th place in the x'th row.

$$\|(x_1,\ldots x_N)P - (y_1,\ldots,y_N)P\|_1 \le L\|(x_1,\ldots x_N) - (y_1,\ldots,y_N)\|_1$$

holds for all probabilities x and y. It is clear that $L_P \le C_P$, and also that $L_{P_1\cdots P_r} \le L_{P_1}\cdots L_{P_r}$ for stochastic matrices $P_1,\ldots,P_r$.

Proposition 21.4 *Suppose that*

- *each of the stochastic matrices P_k, $k = 1,\ldots$ is irreducible and aperiodic, the unique equilibria will be denoted by $(\pi^{(k)})^\top = (\pi_1^{(k)},\pi_2^{(k)},\ldots,\pi_N^{(k)})$;*

- $\sum_k \|\pi^{(k)} - \pi^{(k+1)}\|_1 < \infty$;

- $L_{P_{k'}P_{k'+1}\cdots P_{k'+k}}$ *tends to zero with $k \to \infty$ for every k'.*

This implies:
   (i) *the sequence $(\pi^{(k)})_{k=1,\ldots}$ is convergent to a vector $\pi^\top = (\pi_1,\ldots,\pi_N)$, and*
   (ii) $(p_1,\ldots,p_N)\tilde{P}_k$ *tends to $\pi^\top$ for $k \to \infty$.*

Proof. (i) Fix any i_0. Then

$$
\begin{aligned}
|\pi_{i_0}^{(k+r)} - \pi_{i_0}^{(k)}| &\le \|\pi^{(k+r)} - \pi^{(k)}\|_1 \\
&= \|(\pi^{(k+r)} - \pi^{(k+r-1)}) + \cdots + (\pi^{(k+1)} - \pi^{(k)})\|_1 \\
&\le \sum_{l=k}^{k+r-1} \|\pi^{(l+1)} - \pi^{(l)}\|_1,
\end{aligned}
$$

and therefore, by the second assumption, the sequence $(\pi_{i_0}^{(k)})_k$ is Cauchy.

(ii) As a preparation we prove that $\sup_k \|\pi^\top - \pi^\top P_{k'}P_{k'+1}\cdots P_{k'+k}\|_1$ tends to zero with $k' \to \infty$. The idea is to write the difference under consideration as a telescoping series, for typographical convenience we will use the notation π and π_k instead of the correct but more clumsy $\pi^\top$ and $(\pi^{(k)})^\top$ in the following argument:

$$
\begin{aligned}
\pi P_{k'}P_{k'+1}\cdots P_{k'+k} - \pi &= (\pi - \pi_{k'})P_{k'}\cdots P_{k'+k} + \\
&\quad + \sum_{l=1}^{k}(\pi_{k'-1+l} - \pi_{k'+l})P_{k'+l}\cdots P_{k'+k} + \\
&\quad + (\pi_{k'+k} - \pi) \; ;
\end{aligned}
$$

this is due to the fact that $\pi_k P_k = P_k$. Since all L_P are bounded by 1 it follows that

$$
\begin{aligned}
\|\pi P_{k'}P_{k'+1}\cdots P_{k'+k} - \pi\|_1 &\le L_{P_{k'}\cdots P_{k'+k}}\|\pi - \pi_{k'}\|_1 + \\
&\quad + \sum_{l=1}^{k} L_{P_{k'+l}\cdots P_{k'+k}}\|\pi_{k'-1+l} - \pi_{k'+l}\|_1 + \\
&\quad + \|\pi_{k'+k} - \pi\|_1 \\
&\le 2\sup_{l\ge k'}\|\pi - \pi_l\|_1 + \sum_{l\ge k'}\|\pi_l - \pi_{l+1}\|_1.
\end{aligned}
$$

The first summand is small for large k' by (i), the second as a consequence of

$$\sum_k \|\pi_k - \pi_{k+1}\|_1 < \infty.$$

Now let $(p_1, \ldots, p_N)$ be any starting distribution. Then we have, for arbitrary k',

$$
\begin{aligned}
\|(p_1, \ldots, p_N)\tilde{P}_k - \pi^\top\|_1
&= \|(p_1, \ldots, p_N)P_1 \cdots P_k - \pi^\top\|_1 \\
&= \|((p_1, \ldots, p_N)P_1 \cdots P_{k'-1} - \pi^\top)P_{k'} \cdots P_k + \\
&\qquad + \pi^\top P_{k'} \cdots P_k - \pi^\top\|_1 \\
&\leq L_{P_{k'} \cdots P_k}\|(p_1, \ldots, p_N)P_1 \cdots P_{k'-1} - \pi^\top\|_1 + \\
&\qquad + \|\pi^\top P_{k'} \cdots P_k - \pi^\top\|_1 \\
&\leq 2L_{P_{k'} \cdots P_k} + \|\pi^\top P_{k'} \cdots P_k - \pi^\top\|_1.
\end{aligned}
$$

If now ε is any positive number we can choose a k' with $\|\pi^\top P_{k'} \cdots P_k - \pi^\top\|_1 \leq \varepsilon$ for all $k \geq k'$; for large k, we also have $L_{P_{k'} \cdots P_k} \leq \varepsilon$ by assumption, hence

$$
\|(p_1, \ldots, p_N)P_1 \cdots P_k - \pi^\top\|_1 \leq 3\varepsilon
$$

for these k, and this completes the proof. $\qquad\qquad\qquad\qquad\qquad\qquad\qquad\quad\square$

It remains to choose the inverse temperatures β_k such that the conditions of the preceding proposition are met. We will use the following *notation*:

- $\Delta := \max\{\mathcal{H}(x) - \mathcal{H}(y) \mid x, y \in S,\ q_{xy} > 0\}$; this is the *maximal local increase of* $\mathcal{H}$.

- For $x, y \in S$ we denote by $\sigma(x, y)$ the minimal length of a path from x to y, that is the minimum of the integers r such that there exist $x_0, x_1, \ldots, x_r$ such that $x_0 = x$, $x_r = y$, and $q_{x_0 x_1}, \ldots, q_{x_{r-1} x_r} > 0$. Note that all $\sigma(x, y)$ are finite since the Q-chain is irreducible.

- $\tau := \max_{x,y} \sigma(x, y)$.

- $\vartheta := \min\{q_{xy} \mid q_{xy} > 0\}$.

The main result of this section is the following theorem. In the proof we will assume that the proposal matrix Q has the property that all q_{xx} are strictly positive, this will facilitate the argument slightly.

Theorem 21.5 *Let the cooling schedule $\beta_1 \leq \beta_2 \leq \cdots$ be such that $\beta_k \to \infty$, and*

$$
\beta_k \leq \frac{1}{\tau\Delta} \log k.
$$

Then, if M is the set where $\mathcal{H}$ attains its minimum, simulated annealing converges to the uniform distribution on M. More precisely: if $(p_x)_{x \in S}$ is any initial distribution and if $p_x^{(k)}$ denotes the probability that the walk is in state x after k steps, then $p_x^{(k)}$ tends to zero (resp. to $1/\operatorname{card}(M)$) for $x \notin M$ (resp. for $x \in M$) with $k \to \infty$.

Proof. We want to apply the preceding proposition, with $P_k := P_{\beta_k} =$ the Metropolis matrix associated with β_k. We know that each P_k is irreducible and aperiodic with equilibrium given by $(\pi_k(x))_{x \in S} = (\exp(-\beta_k \mathcal{H}(x))/Z_{\beta_k})_{x \in S}$, and the theorem will be proved as soon as we have shown that

1. $\sum_k \|\pi^{(k)} - \pi^{(k+1)}\|_1 < \infty$;

2. for every k', $L_{P_{k'} P_{k'+1} \cdots P_{k'+k}} \to 0$ with $k \to \infty$;

3. the x-component of π_k tends to 0 resp. to $1/\mathrm{card}(M)$ for $x \notin M$ resp. for $x \in M$.

As is to be expected, the proof uses the concrete form of the equilibrium. We put $a(x,\beta) := \exp(-\beta\mathcal{H}(x))/Z_\beta$, our *claim* is that for every x there exists a β_x such that $\beta \mapsto a(x,\beta)$ is monotone (i.e., increasing or decreasing) for $\beta \geq \beta_x$.

First, we suppose that $x \in M$. Then

$$a(x,\beta) = \frac{1}{m + \sum_{y \notin M} \exp(-\beta(\mathcal{H}(y) - c))},$$

where $m := \mathrm{card}\,(M)$ and $c := \min \mathcal{H}$. Thus $a(x,\cdot)$ increases on all of $\mathbb{R}$, and the limit is $1/m$.

For $x \notin M$ the argument is a little bit subtler. Let, for $y \in S$, a_y be the number $\mathcal{H}(y) - c$. Then the inequality $a(x,\beta) \geq a(x,\beta + t)$ is equivalent with

$$
\begin{aligned}
f(t) \; :=\; & e^{ta_x}\Big(m + \sum_{y \notin M} e^{-(\beta+t)a_y}\Big) \\
\geq\; & m + \sum_{y \notin M} e^{-\beta a_y} \\
=:\; & b.
\end{aligned}
$$

Clearly $f(0) \geq b$ holds, we claim that the derivative of f is positive for $t \geq 0$ provided that β is sufficiently large. In fact, up to the factor e^{ta_x} this derivative is

$$a_x m + \sum_{y \notin M}(a_x - a_y)e^{-(\beta+t)a_y},$$

where $a_x m$ is strictly positive and the second summand can be made arbitrarily small uniformly in $t \geq 0$ for large β (note that all a_y under consideration are greater than zero). This proves the claim.

That "3." holds follows easily from the concrete form of the $a(x,\beta)$, the assertion "1." will be proved next. Fix any x and and choose k_x so large that $\beta_k \geq \beta_x$ for $k \geq k_x$. Then the sum $\sum_{k \geq k_x}|a(x,\beta_k) - a(x,\beta_{k+1})|$ is in fact a telescoping series since $a(x,\cdot)$ is monotone, and therefore it equals $|a(x,\beta_{k_x}) - \lim_k a(x,\beta_k)|$. In particular, the sum $\sum_k |a(x,\beta_k) - a(x,\beta_{k+1})|$ is finite, and this is essentially the statement "1.".

It remains to prove "2.". Fix any k' and consider the Lipschitz constants $l_k := L_{P_{k'+1}\cdots P_{k'+k}}$ for $k \geq 0$ (it will be more convenient to work with these numbers than with the $L_{P_{k'}\cdots P_{k'+k}}$). The sequence (l_k) is decreasing since all L_P are bounded by one, and therefore it suffices to prove that a subsequence tends to zero. The claim is that the subsequence $(l_\tau, l_{2\tau}, \ldots)$ has this property.

To prove this claim we first analyse l_τ. This number is the best possible Lipschitz constant of the map induced by the stochastic matrix $R := P_{\beta_{k'+1}} \cdots P_{\beta_{k'+\tau}}$. In R we find the transition probabilities r_{xy} to get from any x to any y in τ steps, where Metropolis samplers with inverse temperatures $\beta_{k'+1}, \beta_{k'+2}, \ldots, \beta_{k'+\tau}$ are used. The probability of a *single Metropolis step* from x to y under a general β is $\geq \vartheta e^{-\beta\Delta}$ if a transition is possible at all; this is due to the definition[5]. It follows that

---

[5] *Here* it is important that we have assumed that $q_{xx} > 0$, otherwise we would have difficulties to deal with the case $x = y$.

$$\begin{aligned}
r_{xy} &\geq \vartheta^\tau \exp(-\Delta(\beta_{k'+1} + \cdots + \beta_{k'+\tau})) \\
&\geq \vartheta^\tau \exp(-\tau \beta_{k'+\tau} \Delta) \\
&=: \delta,
\end{aligned}$$

i.e., R is a matrix the entries of which are strictly bounded from below by δ. Proposition 10.5 implies that $C_R \leq 1 - N\delta$, and this leads us to

$$L_{P_{k'+1} \cdots P_{k'+\tau}} \leq C_R \leq 1 - N\vartheta^\tau \exp(-\tau \beta_{k'+\tau} \Delta),$$

with $N := \mathrm{card}(S)$.

A similar argument provides estimates for the $L_{P_{k'+\tau+1} \cdots P_{k'+2\tau}}$, $L_{P_{k'+2\tau+1} \cdots P_{k'+3\tau}}, \cdots$, and putting these together we obtain

$$\begin{aligned}
l_{r\tau} &\leq L_{P_{k'+1} \cdots P_{k'+\tau}} \cdots L_{P_{k'+(r-1)\tau+1} \cdots P_{k'+r\tau}} \\
&\leq \prod_{j=1}^{r} \left(1 - N\vartheta^\tau \exp(-\tau \beta_{k'+j\tau} \Delta)\right).
\end{aligned}$$

To complete the proof it will suffice to show that the right hand side tends to zero with $r \to \infty$.

When does a sequence $\prod_{j=1}^{r}(1 - a_j)$ converge to zero, where the a_j lie in $[0, 1[$? Here it is useful to know that $\log(1 - a) \leq -a$ for $a < 1$, this fact implies that $\prod_{j=1}^{r}(1 - a_j) \to 0$ provided that $a_1 + a_2 + \cdots = \infty$. In the present case we have to check the series

$$\vartheta^\tau \left(\exp(-\tau \beta_{k'+\tau} \Delta) + \exp(-\tau \beta_{k'+2\tau} \Delta) + \exp(-\tau \beta_{k'+3\tau} \Delta) + \cdots\right),$$

now the assumption of logarithmic increase comes into play. Since we know that $\beta_k \leq (\log k)/\tau\Delta$ we can estimate a typical summand by

$$\exp(-\tau \beta_{k'+j\tau} \Delta) \geq \frac{1}{k' + j\tau}.$$

And since the harmonic series diverges we have in fact shown that the above products tend to zero with $r \to \infty$. Now the proof is complete. $\qquad\square$

Exercises

21.1: Prove that, under the assumptions of definition 21.1, the Metropolis chain coincides with the Q-chain iff the energy is constant.

21.2: In the definition of the Metropolis chain we have assumed that the proposal chain is symmetric. With this assumption it was possible to show that the Metropolis chain is reversible. Is the assumption really necessary? Is it sufficient to start with a reversible Q-chain to arrive at a reversible Metropolis chain?

21.3: Let $\mathcal{H}$ be an arbitrary energy function and (β_n) a sequence of positive numbers with $\beta_n \to 0$. Prove that the measures $\mathbb{P}_{\beta_n}$ converge to the uniform distribution.

21.4: Consider in proposition 21.3 the special case where the proposal matrix Q has in each row the uniform distribution. What can be said about the spectral gap of the Metropolis chain as a function of the number a from this proposition?

21.5: Here we consider a simple example of an inhomogeneous Markov chain. Let two stochastic $N \times N$-matrices Q_1 and Q_2 be given, we suppose that they have strictly positive entries. They give rise to a Markov chain on $\{1, \ldots, N\}$ if we prescribe that transitions in the k'th step are governed by Q_1 resp. Q_2 if k is even resp. odd. For example, if $(p_1, \ldots, p_N)$ denotes the initial distribution, then the probabilities to find the walk in the various states after the fifth step can be read off from

$$(p_1, \ldots, p_N) Q_1 Q_2 Q_1 Q_2 Q_1.$$

a) Does there exist a distribution $(\pi_1, \ldots, \pi_N)$ such that the probability to find the walk after k steps in state i tends to π_i with $k \to \infty$ for every i?

b) Prove that this is true provided that there is a probability $(\pi_1, \ldots, \pi_N)$ with

$$(\pi_1, \ldots, \pi_N) Q_1 = (\pi_1, \ldots, \pi_N) = (\pi_1, \ldots, \pi_N) Q_2.$$

21.6: Let $\alpha_1 < \alpha_2 < \cdots$ be any real sequence which tends to infinity. Prove that there are suitable integers $r_1, r_2, \ldots$ such that the sequence $\beta_1, \beta_2, \ldots$, defined by

$$\alpha_1, \ldots, \alpha_1, \alpha_2, \ldots, \alpha_2, \ldots$$

(with r_1 repetitions of α_1, r_2 repetitions of α_2, $\ldots$) satisfies the assumptions of theorem 21.5. This means that one can decrease the temperature arbitrarily provided that one stays for "sufficiently many" steps at the various levels.

21.7: In the proof of proposition 21.3 we have omitted two steps which should be proved now:

a) The Dirichlet form $\mathcal{E}(g, g)$ can be rewritten as $\sum_{x,y} (g(x) - g(y))^2 p_{xy} \pi_x / 2$.

b) Let a random variable $X : S \to \mathbb{R}$ on a finite probability space be given. The expectation of X is the unique number c such that the expectation of $(X - c)^2$ is minimal.

21.8: The assertion (i) of proposition 21.4 is a special case of the following general fact: whenever $x_1, \ldots$ is a sequence in a Banach space such that $\|x_1\| + \|x_2\| + \cdots < \infty$, then the series $x_1 + x_2 + \cdots$ converges. Prove this fact.

21.9: In the final step of the proof of theorem 21.5 we have demonstrated that $a_1 + a_2 + \cdots = \infty$ implies that $\prod_{j=1}^r (1 - a_j)$ converges to zero. Show that the converse also holds.

22   Notes and remarks

The material of *chapter 18* is mainly from [70], there one can find the tedious calculations which have only been sketched in our text. One point should be emphasized again: there are situations which are provable untractable if exact solutions are needed but which nevertheless can be solved up to arbitrary precision with the help of Markov chains in polynomial time.

Of course, in order to deal with such assertions rigorously it would be necessary to develop the basic definitions of algorithmic complexity. This is not our concern here; readers who want to learn more of this should first read [35] and then consult [70] again. A particularly important role play problems which are of type P resp. of type NP. The former are problems which are characterized by the property that the time to solve them is bounded by a polynomial in the number of bytes by which they are formulated, the latter are problems which – with positive probability – can be reduced to ones of type P. At present nobody knows a problem which is NP but not P. Can one prove nonexistence, is it possible to find an example? This is considered as one of the outstanding open questions of our time. Here the satisfiability problem has its place: it is typical in that it is NP and that *all* NP problems would be P once the problem of satisfiability could be proven to be of class P.

Chapter 19 mainly contains the basic standard definitions for random fields. Supplementary material can be found in chapter 7 of [20] and – together with some historical comments – in part II of [76]. What has been stated in theorem 19.4(ii) and (iii) seems to be new: usually the Markov chain defined by the local characteristics is only considered in the case of Gibbs fields.

Because of their importance in so many areas the Gibbs fields from *chapter 20* have attracted the attention of many mathematicians. For some references, see again [20] and [76], a more ambitious treatment can be found in [53]. Our presentation mainly follows [20], the proof of proposition 20.6 is inspired by that in [76]. Readers who want to learn more on the Ising model should consult [23], there one also finds an extensive bibliography.

As far as *chapter 21* is concerned I have profited much from discussions with P. Mathé. The Metropolis sampler has been presented in [57], since then it has played an important role whenever it is necessary to sample from a space where one only knows the relative probabilities π_i/π_j (for a survey see [67]). Standard references for simulated annealing are [9] and [52], further interesting material can be found in part II of [76] and in [2]. That simulated annealing is not only an appealing idea but can be stated as a mathematical theorem is due to S. and D. Geman ([36]). This is our theorem 21.5, the proof – which is based on Dobrushin's theorem (proposition 21.4) – follows [76].

In part III we have tried to present some typical examples where Markov chain methods are of importance. We close this book with a brief sketch of some others.

A classical application of random methods is *Monte-Carlo integration*. Let a bounded measurable function $f : D \to [0, \infty[$ be given, where D is a bounded measurable subset of $\mathbb{R}^n$ with Lebesgue measure one. Then the integral of f can be thought of as its expectation if f is considered as a random variable on D. On the other hand, by the law of large numbers, the expectation can be approximated by

$$\frac{f(x_1) + \cdots + f(x_r)}{r}, \tag{22.1}$$

if r is sufficiently large and the $x_1, \ldots, x_r$ are independent and uniformly distributed samples from D. (The complete truth is a little bit more complicated: for every $\varepsilon > 0$ and every $\delta > 0$ there is an r such that (22.1) is ε-close to the expectation with a probability at least $1 - \delta$.)

This so-called Monte-Carlo integration necessitates to find uniformly distributed samples in D. To this end D is replaced by a sufficiently fine finite grid, and on this grid points are sampled with prescribed probabilities using Markov chain methods. We refer the reader to [56] and the literature cited there.

Another field of research which should be mentioned here is *volume estimation*. More precisely, one is given a convex body K in a high-dimensional euclidean space $\mathbb{R}^n$, and the problem is to find reasonable approximations of the euclidean volume $V(K)$ of K. It is in general hopeless to evaluate $V(K)$ by analytical methods, here also Markov chain techniques have been applied successfully.

The idea is as follows. As a preparation the problem is reduced to the approximate evaluation of the quotient $V(L)/V(K)$ of two volumina, where K and L are convex bodies with $L \subset K$ such that $V(L)/V(K)$ lies between $1/2$ and 1. This ratio is evaluated by approximating K by a fine grid G. Then a Markov chain is run on G which has the uniform distribution as its equilibrium. It is stopped after "sufficiently many" steps at some state i, and one observes whether i lies in L or not. Then the proportion of successes after many such experiments serves as an approximation of $V(L)/V(K)$. It can be shown that in this way one can get good approximations with a high probability in a number of steps which is bounded by a polynomial in the input complexity (for details the reader is referred to [18]). On the other hand, deterministic methods give good results only after exponentially many steps, and therefore we have here – as in the case of the permanent – another example where stochastic approaches are provably superior to exact ones.

Finally another field of applications should be mentioned. Markov chain methods have proven to be extremely useful in *image analysis*. How can a computer recognize the essential features of a picture given by a grid of black-and-white pixels? How can a picture be reconstructed if some of the pixels are destroyed? Readers who have mastered the theory presented in this book should have the necessary prerequisites to understand the answers to these questions given in [76].

Bibliography

[1] D. ALDOUS. *Random walks on groups and rapidly mixing Markov chains.* Séminaire de probabilité XVII, Lecture Notes in Mathematics 986, Springer-Verlag, Berlin-Heidelberg-New York, 1983.

[2] D. ALDOUS. *On the Markov chain simulation method for uniform combinatorial distributions and simulated annealing.* Probability in the Engineering and Informational Sciences **1** (1987), 33–46.

[3] D. ALDOUS, P. DIACONIS. *Shuffling cards and stopping times.* American Math. Monthly **93** (1986), 333–348.

[4] D. ALDOUS, P. DIACONIS. *Strong uniform times and finite random walks.* Advances in Appl. Math. **8** (1987), 69–97.

[5] D. ALDOUS, J. A. FILL. *(Preliminary version of a book on finite Markov chains).* http://www.stat.berkeley.edu/users/aldous.

[6] D. ALDOUS, J. PROPP (EDS.). *Microsurveys in Discrete Probability.* DIMACS Series **41** (1998).

[7] N. ALON. *Eigenvalues and expanders.* Combinatorica **6** (1986), 83–96.

[8] N. ALON, V. MILMAN. λ_1, *isoperimetrical inequalities for graphs and superconcentrators.* Journal of Combinatorial Theory Series B **38** (1985), 73–88.

[9] R. AZENCOTT (ED.). *Simulated Annealing – Parallelization Techniques.* John Wiley & Sons, 1992.

[10] A. BARVINOK. *Polynomial time algorithms to approximate permanents.* Random Structures and Algorithms **14** (1999), 29–61.

[11] H. BAUER. *Probability Theory.* De Gruyter Studies in Mathematics, vol. 21, Walter de Gruyter Verlag, Berlin-New York, 1995.

[12] E. BEHRENDS. *Maß- und Integrationstheorie.* Springer Hochschultext, Springer-Verlag, Berlin-Heidelberg-New York, 1987.

[13] E. BEHRENDS. *Überall Zufall.* BI Wissenschaftsverlag, 1994.

[14] E. BEHRENDS. *Mathematik und Musik.* Mitteilungen der DMV **2/97** (1997), 44–48.

[15] E. BEHRENDS. *Faire Entscheidungen.* Elem. Math. **54** (1999), 107–117.

[16] P. BILLINGSLEY. *Probability and Measure.* 3rd edition, John Wiley & Sons, 1995.

[17] R. M. BLUMENTHAL, R. K. GETOOR. *Markov Processes and Potential Theory.* Academic Press, New York, London, 1968.

[18] B. BOLLOBÁS. *Volume estimates and rapid mixing.* In: Flavors of Geometry, Edited by S. Levy, Mathematical Sciences Research Institute Publications **31**, Cambridge University Press, (1997), 151–182.

[19] H. BOERNER. *Representations of Groups.* North Holland, 1970.

[20] P. BRÉMAUD. *Markov Chains, Gibbs Fields, Monte Carlo Simulation and Queues.* Springer-Verlag, Berlin-Heidelberg-New York, 1999.

[21] A. Z. BRODER. *How hard is it to marry at random? (On the approximation of the permanent).* Proceedings of the 18th ACM Symposium on Theory of Computing (1986), 50–58.

[22] K. L. CHUNG. *Lectures from Markov Processes to Brownian Motion.* Grundlehren der mathematischen Wissenschaften, vol. 249, Springer-Verlag, Berlin-Heidelberg-New York, 1982.

[23] B. CIPRA. *Introduction to the Ising model.* American Math. Monthly **94** (1987), 937–959.

[24] P. DIACONIS. *Group Representations in Probability and Statistics.* Institute of Mathematical Statistics Lecture Notes Series 11, Hayward, California, 1988.

[25] P. DIACONIS, L. SALOFF-COSTE. *What do we know about the Metropolis algorithm?.* Proc. 27. Symp. Theory of Computation (1995), 112–129.

[26] P. DIACONIS, M. SHAHSHAHANI. *Generating a random permutation with random transpositions.* Z. Wahrscheinlichkeitstheorie verw. Gebiete **57** (1981), 159–179.

[27] P. DIACONIS, M. SHAHSHAHANI. *On square roots of the uniform distribution on compact groups.* Proc. Amer. Math. Soc. **98** (1986), 341–348.

[28] J. DIEUDONNÉ. *Abrégé d'histoire des mathématiques.* Hermann, Paris, 1970.

[29] W. DOEBLIN. *Exposé de la théorie des chaînes simples constantes de Markov à un nombre fini d'états.* Rev. Math. de l'Union Interbalkanique **2** (1933), 77–105.

[30] W. FELLER. *An Introduction to Probability Theory and its Applications I.* John Wiley & Sons, New York 1968.

[31] W. FELLER. *An Introduction to Probability Theory and its Applications II.* John Wiley & Sons, New York 1971.

[32] S. FELSNER, L. WERNISCH. *Markov Chains for Linear Extensions, the Two-Dimensional Case.* In: Proc. of the eights annual ACM-SIAM symposium on discrete algorithms, New Orleans, 1997, 239–247.

[33] D. R. FOX. *Renewal Theory.* Methuen & Company Ltd., Science Paperback, 1970.

[34] W. FULTON, J. HARRIS. *Representation Theory.* Springer-Verlag, Berlin-Heidelberg-New York, Graduate Texts, 1991.

[35] M. R. GAREY-D. S.JOHNSON. *Computers and Intractability.* Freeman and Company, 1978.

[36] S. GEMAN, D. GEMAN. *Stochastic relaxation, Gibbs distributions, and the Bayesian restoration of images.* IEEE Trans. PAMI **6** (1984), 721–741.

[37] E. GINÉ, G. R. GRIMMETT, L. SALOFF-COSTE. *Lectures on Probability theory and Statistics.* Lecture Notes in Mathematics **1665**, Springer-Verlag, Berlin-Heidelberg-New York, 1997.

[38] S. GOLDSTEIN. *Maximal coupling.* Z. Wahrscheinlichkeitstheorie verw. Gebiete **46** (1979), 193–204.

[39] D. GRIFFEATH. *Coupling methods for Markov processes.* Thesis, Cornell University (1975).

[40] D. GRIFFEATH. *A maximal coupling for Markov chains.* Z. Wahrscheinlichkeitstheorie verw. Gebiete **31** (1975), 95–106.

[41] G. A. HUNT. *Some theorems concerning Brownian motion.* Trans. Amer. Math. Soc. **81** (1956), 294–319.

[42] M. IOSIFESCU. *Finite Markov Processes and Their Applications.* John Wiley & Sons, 1980.

[43] D. ISAACSON, R. MADSEN. *Markov Chains: Theory and Applications.* John Wiley & Sons, 1976.

[44] A. ISERLES. *Numerical Analysis of Differential Equations.* Cambridge University Press, 1996.

[45] E. ISING. *Beitrag zur Theorie des Ferromagnetismus.* Zeitschrift für Physik **31** (1925), 253–258.

[46] N. JACOBSON. *Basic Algebra I.* Freeman & Company, New York, 1974.

[47] G. JAMES, A. KERBER. *The Representation Theory of the Symmetric Group.* Addison Wesley, Reading (MA), 1981.

[48] S. KARLIN, H. M. TAYLOR. *A First Course in Stochastic Processes.* Academic Press, 1975.

[49] J. G. KEMENY, J. L. SNELL. *Finite Markov Chains.* Van Nostrand, New York, 1960.

[50] J. G. KEMENY, J. L. SNELL, S. KNAPP. *Denumerable Markov Chains (2. ed.* Springer-Verlag, Berlin-Heidelberg-New York, 1970.

[51] R. KURTH. *Axiomatics of Classical Statistical Mechanics.* Pergamon Press (1960).

[52] P.J. VAN LAARHOVEN, E.H. AARTS. *Simulated Annealing – Theory and Applications.* Kluwer Academic Publishers, Dordrecht, 1987.

[53] TH. M. LIGGETT. *Interacting Particle Systems.* Springer-Verlag, Berlin-Heidelberg-New York, 1985.

[54] T. LINDVALL. *Lectures on the Coupling Method.* John Wiley & Sons, 1992.

[55] L. LOVÁSZ, P. WINKLER. *Mixing times.* In: D. Aldous, J. Propp, Microsurveys in Discrete Probability, DIMACS series **41** (1998), 85–134.

[56] P. MATHÉ. *Numerical integration using Markov chains.* (To appear in: Markov Chain Methods and its Applications).

[57] N. METROPOLIS, A. ROSENBLUTH, M. ROSENBLUTH, A. TELLER, E. TELLER. *Equations of state calculations by fast computing machines.* J. Chem. Phys. **21** (1953), 1087–1092.

[58] H. MINC. *Permanents.* Addison-Wesley, Reading MA, 1978.

[59] S. B. MORRIS. *Magic Tricks, Card Shuffling and Dynamic Computer Memories.* The Math. Association of America (1998).

[60] J. R. NORRIS. *Markov Chains.* Cambridge Series in Statistical and Probabilistic Mathematics, Cambridge University Press, 1998.

[61] P. PHILIP. *Cardinality and structure of extreme infinite doubly stochastic matrices.* Archiv der Mathematik **71** (1998), 417–424.

[62] J. W. PITMAN. *On coupling of Markov chains.* Z. Wahrscheinlichkeitstheorie Verw. Gebiete **35** (1976), 315–322.

[63] J. PROPP, D. WILSON. *Exact sampling with coupled Markov chains and applications to statistical physics.* Random Structures and Algorithms **9** (1996), 223–252.

[64] J. PROPP, D. WILSON. *Coupling from the past: A user's guide.* In: D. Aldous, J. Propp, Microsurveys in Discrete Probability, DIMACS series **41** (1998), 181–192.

[65] J. S. ROSENTHAL. *On generalizing the cut-off phenomenon for random walks on groups.* Advances in Applied Mathematics **16** (1995), 306–320.

[66] D. RUELLE. *Thermodynamic Formalism.* Addison Wesley Publ. Comp. (1978).

[67] L. SALOFF-COSTE. *Lectures on finite Markov chains.* In: Lectures on Probability Theory and Statistics, Lecture Notes in Mathematics 1665, Springer-Verlag, Berlin-Heidelberg-New York, 1997.

[68] I. SCHNEIDER. *Die Entwicklung der Wahrscheinlichkeitstheorie von den Anfängen bis 1933.* Wissenschaftliche Buchgesellschaft, Darmstadt, 1988.

[69] E. SENETA. *Nonnegative Matrices and Markov Chains.* 2nd. edition, Springer-Verlag, Berlin-Heidelberg-New York, 1981.

[70] A. SINCLAIR. *Algorithms for Random Generating and Counting.* Birkhäuser, Boston, 1993.

[71] H. THORISSON. *On maximal and distributional coupling.* Ann. Probability **14** (1986), 873–876.

[72] G. TURNWALD. *Roots of Haar measure and topological hamiltonian groups.* in: Springer-Verlag, Lecture Notes in Mathematics **1379** (1989), 364–375.

[73] L. G. VALIANT. *The complexity of computing the permanent.* SIAM Journal on Computing **8** (1979), 410–421.

[74] U. VAZIRANI. *Rapidly mixing Markov chains.* Proc. Symp. Applied Math. **44** (1991), 99–121.

[75] D. B. WILSON. *Annoted bibliography of perfectly random sampling with Markov chains.* In: D. Aldous, J. Propp, Microsurveys in Discrete Probability, DIMACS series **41** (1998), 209–220.

[76] G. WINKLER. *Image Analysis, Random Fields and Dynamic Monte Carlo Methods.* Springer-Verlag, Berlin-Heidelberg-New York, 1995.

Index

Milnor's Textbook on Dynamics

Dynamics in One Complex Variable

Introductory Lectures

John Milnor

1999. viii, 257 pp.
Softc. DM 49,80
ISBN 3-528-03130-1

Chronological Table - Riemann Surfaces - Iterated Holomorphic Maps - Local Fixed Point Theory - Periodic Points: Global Theory - Stucture of the Fatou Set - Using the Fatou Set to study the Julia Set – Appendices

This text studies the dynamics of iterated holomorphic mappings from a Riemann surface to itself, concentrating on the classical case of rational maps of the Riemann sphere. It is based on introductory lectures given by the author at Stony Brook, NY, in the past ten years. The subject is large and rapidly growing. These notes are intended to introduce the reader to some key ideas in the field, and to form a basis for further study. The reader is assumed to be familiar with the rudiments of complex variable theory and of two-dimensional differential geometry, as well as some basic topics from topology. The exposition is clear and enriched by many beautiful illustrations.

vieweg

Abraham-Lincoln-Straße 46
65189 Wiesbaden
Fax 0180.57878-80
www.vieweg.de

November 1999
Änderungen vorbehalten.
Erhältlich beim Buchhandel oder beim Verlag.

Modern Algorithmic Techniques

Hypergeometric Summation

An Algorithmic Approach
to Hypergeometric
Summation and Special
Function Identities

Wolfram Koepf

1998. x, 230 pp.
(Advanced Lectures in Mathematics)
Softc. DM 69,00
ISBN 3-528-06950-3

The Gamma Function - Hypergeo-
metric Identities - Hypergeometric
Database - Holonomic Recurrence
Equations - Gosper`s Algorithm -
The Wilf-Zeilberger Method -
Zeilberger´s Algorithm - Extensions
of the Algorithms - Petkovšek`s

Algorithm - Differential Equations
for Sums - Differential Antideri-
vatives -Hyperexponential Antideri-
vatives - Holonomic Equations for
Integrals - Rodrigues Formulas and
Generating Functions

In this book modern algorithmic
techniques for summation, most of
which have been introduced within
the last decade, are developed and
carefully implemented in the computer algebra system Maple. The
algorithms of Gosper, Zeilberger
and Petkovšek on hypergeometric
summation and recurrence equations and their q-analogues are covered, and similar algorithms on differential equations are considered. An
equivalent theory of hyperexponen-
tial integration due to Almkvist and
Zeilberger completes the book. The
combination of all results considered gives work with orthogonal
polynomials and (hypergeometric
type) special functions a solid algorithmic foundation. Hence, many
examples from this very active field
are given.

Abraham-Lincoln-Straße 46
65189 Wiesbaden
Fax 0180.57878-80
www.vieweg.de

November 1999
Änderungen vorbehalten.
Erhältlich beim Buchhandel oder beim Verlag.

MIX
Papier aus verantwortungsvollen Quellen
Paper from responsible sources
FSC® C105338

If you have any concerns about our products,
you can contact us on
ProductSafety@springernature.com

In case Publisher is established outside the EU,
the EU authorized representative is:
Springer Nature Customer Service Center GmbH
Europaplatz 3, 69115 Heidelberg, Germany

Printed by Libri Plureos GmbH
in Hamburg, Germany